—/ / / 75
)

LONDON MATHEMATICAL SOCIETY STUDI

Managing editor: Professor E.B. Davies, Departm
King's College, Strand, London WC2R 2LS

1 Introduction to combinators and λ-calculus, J.R. HINDLEY & J.P. SELDIN
2 Building models by games, WILFRID HODGES
3 Local fields, J.W.S. CASSELS
4 An introduction to twistor theory, S.A. HUGGETT & K.P. TOD
5 Introduction to general relativity, L. HUGHSTON & K.P. TOD
6 Lectures on stochastic analysis: diffusion theory, DANIEL W. STROOCK
7 The theory of evolution and dynamical systems, J. HOFBAUER & K. SIGMUND
8 Summing and nuclear norms in Banach space theory, G.J.O. JAMESON
9 Automorphisms of surfaces after Nielsen and Thurston, A.CASSON & S. BLEILER
10 Non-standard analysis and its applications, N.CUTLAND (ed)
11 The geometry of spacetime, G. NABER
12 Undergraduate algebraic geometry, MILES REID

London Mathematical Society Student Texts. 7

The Theory of Evolution and Dynamical Systems

Mathematical Aspects of Selection

JOSEF HOFBAUER
Institute of Mathematics, University of Vienna

KARL SIGMUND
IIASA, Laxenburg, Austria

The right of the University of Cambridge to print and sell all manner of books was granted by Henry VIII in 1534. The University has printed and published continuously since 1584.

CAMBRIDGE UNIVERSITY PRESS
Cambridge
New York New Rochelle Melbourne Sydney

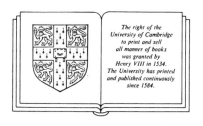

Published by the Press Syndicate of the University of Cambridge
The Pitt Building, Trumpington Street, Cambridge CB2 1RP
32 East 57th Street, New York, NY 10022, USA
10 Stamford Road, Oakleigh, Melbourne 3166, Australia

Originally published in German as *Evolutionstheorie und dynamische System: Mathematische Aspekte der Selektion* by Paul Parey Verlag

First published in English 1988

Printed in Great Britain at the University Press, Cambridge

British Library cataloguing in publication data:
Hofbauer, Josef, *1956-*
The Theory of evolution and dynamical systems: mathematical aspects of selection.
1. Organisms. Evolution. Natural selection. Dynamical systems
I. Title II. Sigmund, Karl, *1945-*
II. Series IV. Evolutionsthreorie and dynamische system. *English*
575.01'62

Library of Congress cataloguing in publication data:
Hofbauer, Josef. 1956-
[Evolutionstheorie und dynamische System. English]
The theory of evolution and dynamical systems: mathematical aspects of selection / Josef Hofbauer, Karl Sigmund.
p. cm. -- (London Mathematical Society Student Texts : 7)
Bibliography: p.
Includes index.
ISBN 0 521 35288 6. ISBN 0 521 35283 8 (pbk.)
1. Evolution--Mathematics. 2. Differential dynamical systems.
I. Sigmund, Karl. 1945- . II. Title. III. Series.
QH3/1.H63 1288
575.01'62--dc19

ISBN 0 521 35288 6 hardcover
ISBN 0 521 35838 8 paperback

Table of Contents

Part 5: Compete, predate or cooperate: the struggle for permanence

Part 6: Back to the gene pool: gradients and cycles

Part 7: On sex and games: strategic and genetic evolution

PREFACE

Biomathematics is a fairly young science, but one forerunner dates back to the middle ages, when Leonardo da Pisa, better known as Fibonacci, computed the reproductive success of rabbits. He assumed that a couple of rabbits produces, from its second month onwards, a new couple of rabbits every month. If a_n is the number of couples in the n-th month, this yields the recurrence relation

$$a_n = a_{n-1} + a_{n-2} \quad .$$

Indeed, a_n is the sum of a_{n-1} (the number of couples from last month) and a_{n-2} (the number of newborn couples, equal to the number of couples at least two months old). If one starts from one pair ($a_0 = a_1 = 1$), one obtains the famous Fibonacci sequence

$$1,1,2,3,5,8,13,\ldots$$

This first example of biological modelling displayed already a strong tendency to neglect factors which could mar the beauty of the analysis: neither the mortality of rabbits nor fluctuations in the number of offspring are taken into account. Biomathematics has developed a good deal since, but there is no denying that a certain highhandedness towards the complications of "real life" persists. Idealizations (and oversimplifications) are not much more blatant than in mathematical physics, but they are much less readily accepted. On the other hand, it has been established at least since the "golden age" of biomathematics in the twenties and thirties that crucial aspects of theoretical biology can only be captured by mathematical modelling; and just as important as the mathematical applications in biology are the biological motivations to mathematics. We have tried to do justice to both points of view in this book. It should be

(a) an introduction to the theory of dynamical systems (and in particular the qualitative theory of differential equations), based entirely on examples from biology; and

(b) a survey of recent developments on four branches of the theory of evolution, namely population genetics, mathematical ecology, prebiotic evolution of macromolecules, and game theoretic modelling of animal behaviour.

The choice of material is, needless to say, highly subjective. The survey does not attempt to be complete, and the introduction is not meant to replace a textbook on differential equations. Nevertheless, here is a field where the path from undergraduate mathematics to current research is particularly short; and we hope to point out a few interesting sights on the way.

The book is organized in such a way that the mathematical level grows steadily. The same models keep reappearing and are analysed with increasingly sophisticated tools. The first half of the book contains the biological background: its mathematics is fairly elementary. In the second half the mathematics is still motivated biologically, but it assumes centre stage now. The different parts are less tightly ordered, and many results are given in the form of exercises.

The book is divided into chapters and sections. The theorems and exercises are referred to by their section. If there are several of them in one section, they are numbered consecutively. Some of the exercises are quite hard, at least for one of the authors.

The text grew out of lectures which we delivered at the Institute of Mathematics at the University of Vienna. It is a pleasure to thank our audience, and in particular G. Kirlinger, M. Koth, E. Amann, I. Bomze, R. Bürger and V. Losert for many helpful comments and useful criticisms. We are also indebted to E. Akin, G. Butler, M. Eigen, J. Franks, H. Freedman, K.P. Hadeler, P. Hammerstein, V. Hutson, A. Kurzhanski, J. Maynard Smith, C. Robinson, R. Selten, P. Waltman and E.C. Zeeman for valuable discussions and unpublished material.

Financial help from the Forschungsförderungsfonds, projects P 5994 and J 0092, is gratefully acknowledged. We also wish to thank Ms. E. Herbst and E. Gruber for preparing the manuscript. Our largest debt is to our colleague P. Schuster, whose ideas form an essential part of this book and who is responsible for stirring up our interest in biomathematics.

PART I: SELECTION DYNAMICS AND POPULATION GENETICS: A DISCRETE INTRODUCTION

1. The Biological Background

1.1. Somatic cells

The cells of higher organisms may present very different aspects, depending on the type of tissue. Apart from a few exceptions, however, they all contain a nucleus where the genetic program is stored in the form of *chromosomes.* The number of chromosomes varies with the species (there are 46 of them in our cells), but they always occur in homologous pairs. If a cell divides, each chromosome duplicates, and the two daughter cells each receive a complete set of chromosomes (see Figure 1.1).

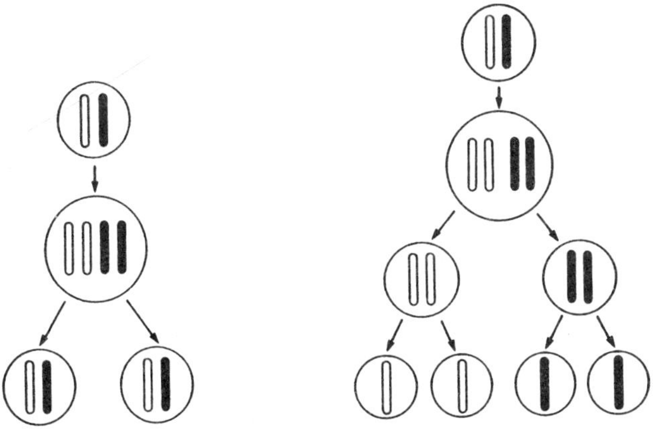

Figures 1.1 and 1.2

The mechanisms of cell division are extremely complex and not yet understood in full detail. They guarantee that daughter cells contain the same genetic program as their predecessors. Each one of our cells, from the bone of our toes to the grey matter of our brain, contains copies of the chromosomes of the unicellular organism that we have been at conception, that is, copies of our complete genome.

Let us consider now some inherited trait expressed in the *phenotype* (i.e. the set of manifested attributes) of an organism, for instance eye colour or blood group. In the simplest case, it is determined by the joint action of two *genes* sitting at corresponding positions on a pair of homologous chromosomes: this site is called the chromosomal *locus* of the trait. In many cases, there are several types of genes which may occupy a locus, the so-called *alleles* $A_1,\ldots,A_n$. The *genotype* is determined by that pair which actually occurs. In the case A_iA_i, that is when the same allele appears twice, the genotype is *homozygous*; in the case A_iA_j $(i\neq j)$ it is *heterozygous*. In the latter case, one allele may suppress the effect of the other, in the sense that A_iA_j manifests itself as A_iA_i. The allele A_i, then, is called *dominant* and A_j *recessive*. It may also happen that the heterozygous genotypes lead to an expression which is different from the homozygous ones. The genotypes A_jA_i and A_iA_j cannot be distinguished however.

1.2. Germ cells

The somatic cells described so far are *diploid*, which means that their chromosomes occur in pairs. Our organism also contains *haploid* cells having only half the number of chromosomes, one from each pair. These are the germ cells or *gametes* (sperms and eggs). Such cells issue from somatic cells by the process of reduction or *meiosis* (see Figure 1.2), which splits the chromosomal pairs.

At mating, two gametes fuse, thereby restoring the full number of chromosomes. The newly formed diploid cell, called a *zygote*, is the starting point of a new organism, which inherits from each parent half of its genome.

At meiosis, different pairs of chromosomes split up independently of each other.

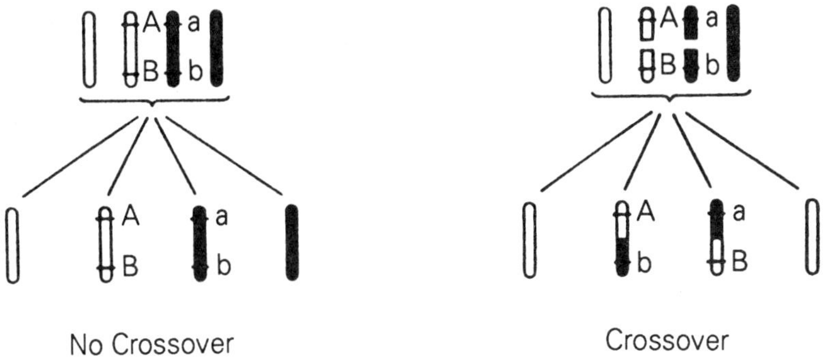

Figure 1.3

1.3. Recombination

The last sentence is only valid as a first approximation. In the *tetrad stage* between the duplication of chromosomes and the first meiotic division, there can be a *recombination* - or "*crossing over*" - of genetic material between corresponding parts of the homologous chromosomes (see Figure 1.3). The frequency of recombination between two chromosomal loci is a measure of their distance. Statistical investigations allow therefore the establishment of "chromosome charts" relating the sequential arrangement of the genes.

The previous description has also to be modified in another way. Not all pairs of chromosomes are homologous. There is one exception, the pair of sex chromosomes. Thus for mammals, the X-chromosome is longer than the Y-chromosome. Females have two X-chromosomes, males one X- and one Y-chromosome. Half of the male sperm cells contain the X, half the Y-chromosome, while all female egg cells carry X. Fusion leads with equal probability to XX (a daughter) or XY (a son).

Genes situated on the sex chromosomes are called *sex-linked*; the other genes are called *autosomal*.

1.4. Mendel's experiments

The discovery of the laws of inheritance preceded that of the corresponding cellular processes. The monk Gregor Mendel set up a (now classical) series of experiments in the garden of his monastery and deduced from them the basic laws of genetics.

In one such experiment, he bred two pure lines of peas, one of them having red, the other one yellow seeds, and subsequently crossed them. The daughter generation F_1 had uniformly yellow seeds. If he crossed peas from F_1, he obtained peas with yellow and peas with red seeds in the ratio 3 to 1.

Mendel recognized that two alleles (say r and y) belong to the locus determining the colour of the seeds, y being dominant. The genotypes yy and ry lead to the phenotype "yellow seeds", the genotype rr to "red seeds".

The two uniform parental lines were homozygous, their genotypes rr and yy respectively. The gametes of the first line all contained the gene r, those of the second the gene y. Fusion led to the heterozygous daughter generation F_1, with genotype ry and phenotype "yellow". Half of the gametes of this generation contained the gene r, the other half the gene y. Random fusion of these germ cells leads with equal probability to the gene pairs (r,r), (r,y), (y,r) and (y,y). The phenotype is therefore with probability $\frac{1}{4}$ "red", with probability $\frac{3}{4}$ "yellow".

In another experiment Mendel investigated the inheritance of two traits of his peas: colour of the seeds (red or yellow) and form of the seeds (smooth or wrinkled). One of his pure lines had smooth yellow seeds, the other wrinkled red seeds. Again, the crossing of the two lines led to a uniform daughter generation: all seeds were yellow and smooth. But crosses of the peas of this daughter generation led to all possible phenotypes, namely red and wrinkled, red and smooth, yellow and wrinkled and yellow and smooth, in the proportions 1:3:3:9.

The interpretation of this result is immediate. The gene loci for colour and form of the seeds are on different chromosomal pairs, and segregate independently. We already know r and y, the two alleles for the colour of the seeds. Let us denote those for the form of the seeds by s (smooth) and w (wrinkled), s being dominant.

The genotypes of the two lines of the parental generation are homozygous, yy/ss and rr/ww, the corresponding gametes being y/s and r/w, respectively. The first daughter generation is heterozygous on both loci, its genotype being yr/sw. Their gametes, in turn, are with equal probability r/s, r/w, y/s and y/w. Random fusion leads to 16 possible (and equiprobable) gene combinations: nine of them yield the genotype "yellow and smooth", three "yellow and wrinkled", three "red and smooth" and one "red and wrinkled".

	r/s	r/w	y/s	y/w
r/s	rr/ss	rr/sw	ry/ss	ry/sw
r/w	rr/sw	rr/ww	ry/sw	ry/ww
y/s	ry/ss	ry/sw	yy/ss	yy/sw
y/w	ry/sw	ry/ww	yy/sw	yy/ww

1.5. Selection

While Mendel's laws remained shrouded in obscurity for many years, the theory of Darwin gained instant fame.

Darwin proposed explaining biological evolution (which was already fairly well recognized) as the effect of *natural selection*. Some of the first pages of his "Origin of the Species" deal with the methods and results of pigeon breeders, who had managed to create, out of the ordinary rock pigeon, an astonishing variety of races of domestic pigeons. The technique was *artificial selection*. A breeder aiming, say, for pigeons having uncommonly many feathers in the tail, would choose in each generation those pigeons having most tail feathers, and use them for further breeding. This method led from the rock pigeon with its modest number of twelve or fourteen feathers to the fantail pigeon boasting forty feathers and more.

Natural selection acts in much the same way. There is no conscious aim behind it, but the method is that of the breeder. Selection acts upon the variability of traits. Those which are best adapted to the environment lead to more offspring. If the traits are inherited, their frequency increases. A well studied case in point is the famous sickle cell gene, which is a recessive allele in the human genome. It kills homozygotes, but confers to heterozygotes a higher resistance to malaria. Within the black population in America, this allele is consequently rarer than in Africa, where the threat of malaria is considerably higher.

1.6. Mutations

In the early years of this century, Darwinism and Mendelism were at loggerheads. In particular, experiments seemed to show that breeding could only select already existing traits, but not develop them any further.

For example, if one classified beans according to weight, one obtained a Gaussian curve for the distribution. Selecting the largest beans and crossing them, one obtained an offspring generation whose Gaussian distribution had a higher mean. Selecting the heaviest beans again did not lead to further increase, however. Artificial breeding had reached its limit.

The interpretation was clear. Even beans with identical genetic make up are of variable size. A mixture of pure lines in the parental generation can be unscrambled by selection. But breeding does not lead any further. The remaining variability in weight is due to other factors, does not affect the genome at all and cannot, therefore, be passed on through inheritance: no new species can emerge in this way.

It was soon recognized, however, that variability stems from two different sources. Variations caused by environmental factors, so-called *modifications*, cannot be transmitted by inheritance. But variations of the genome itself, so-called *mutations*, do get copied and transmitted to the next generation. Such mutations are caused by "chance". They occur only rarely, but they provide the variability upon which natural selection acts.

1.7. Notes

Darwin's epoch-making book (1859) has seen several re-editions. Many biologists feel now that the earliest editions were more correct and "Darwinian". The work of Mendel (1866) remained neglected for a long time. Some classics of the synthesis of Darwinism and genetics are Haldane (1932), Dobzhanski (1937), Simpson (1953), Mayr (1963), Ford (1964), Lewontin (1974) and Maynard Smith (1975).

2. The Hardy-Weinberg Law

2.1. Gene and genotype frequencies

If two alleles A_1 and A_2 may occupy a given locus, then three genotypes are possible, namely A_1A_1, A_1A_2 and A_2A_2. Let us denote by N_{11}, N_{12} and N_{22} the numbers of individuals of each genotype and by $N=N_{11}+N_{12}+N_{22}$ the total population size. The *genotype frequencies* are

$$x = \frac{N_{11}}{N} \quad y = \frac{N_{12}}{N} \quad z = \frac{N_{22}}{N} \quad . \tag{2.1}$$

Each individual carries (at a given locus) two genes, so that the total number of genes is $2N$. Since A_1A_1-individuals contribute two genes A_1 to the gene pool and A_1A_2-individuals only one, the number of genes A_1 is $2N_{11}+N_{12}$ and that of genes A_2 is $2N_{22}+N_{12}$. The *gene frequencies* are

$$p = \frac{2N_{11}+N_{12}}{2N} = x+\frac{y}{2} \quad \text{and} \quad 1-p = \frac{2N_{22}+N_{12}}{2N} = z+\frac{y}{2} \quad . \tag{2.2}$$

The point $(p,1-p)$ lies (in the coordinate plane) on the segment between (0,1) and (1,0). Similarly, the point (x,y,z) lies in $\mathbf{R}^3$ on the triangle with vertices (1,0,0), (0,1,0) and (0,0,1) (see Fig. 2.1).

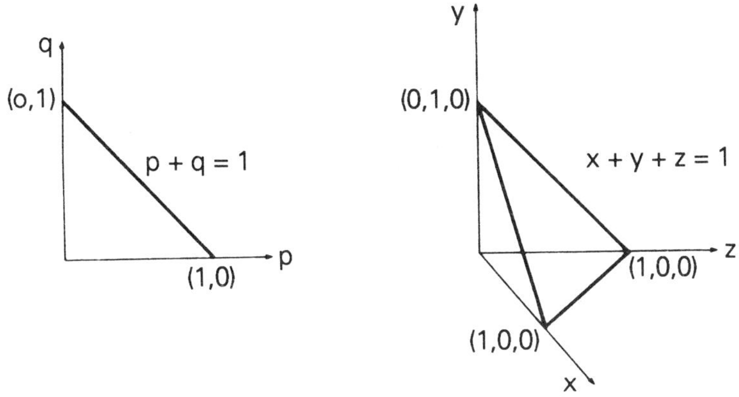

Figure 2.1

The gene frequencies refer to the gene pool of the population, the genotype frequencies describe how the gene pool is organized into genotypes. The quantities p and $1-p$ are uniquely determined by x,y and z, but not conversely. As we shall presently see, however, the gene frequencies are enough to specify the genotype frequencies if the population mates at random.

But first, a remark on notation. For the genotype of a given individual, it does not matter which one of his two genes stems from the father and which one from the mother. Nevertheless, we shall use the convention of describing the *gene pair* of an individual by writing first the gene from the father, and then the gene from the mother. We shall distinguish therefore the gene pairs (A_1,A_2) and (A_2,A_1), although both correspond to the same genotype A_1A_2.

2.2. The Hardy-Weinberg law

If the gene frequency of the allele A_1 in the parental generation is p, then p is the probability, for a given offspring, that the gene inherited from the father is A_1, i.e. that the first component of its gene pair is A_1. Similarly, p is the probability that A_1 occurs as the second component of the gene pair. The probabilities of the gene pairs (A_1,A_1), (A_1,A_2), (A_2,A_1) and (A_2,A_2) are therefore p^2, $p(1-p)$, $(1-p)p$ and $(1-p)^2$. The frequencies of the genotypes A_1A_1, A_1A_2 and A_2A_2 are p^2, $2p(1-p)$ and $(1-p)^2$. The gene frequency of A_1 is

$$x + \frac{y}{2} = p^2 + p(1-p) = p$$

just as in the parental generation.

The *Hardy-Weinberg law* summarizes this as follows:

1) *gene frequencies remain unchanged from generation to generation*;

2) *from the first daughter generation onwards the genotype frequencies are*

$$x = p^2 \qquad y = 2p(1-p) \qquad z = (1-p)^2 \quad . \tag{2.3}$$

Let us discuss the derivation of this law.

In order to compute the probability of the gene pair (A_1,A_1), we multiplied the probability that the paternal gene was A_1 with the probability that the maternal gene was A_1. We assumed, therefore, that the genes were inherited independently of each other. This means *random union* of the gametes. For sea urchins, which release their gametes directly into water, or for wind pollinated plants, this assumption seems reasonable enough. For many other types of organisms, it is the individuals, rather than the gametes, which meet randomly. It is easy to see, however, that this leads to the same result as random fusion of the gametes (see exercise 2). Of course, there are situations where random union is precluded, for example through *mating systems* based on phenotypic traits (like tall people preferring tall partners) or genetic relations (like brother marrying sister). Nevertheless, random mating is often a valid assumption.

A few more premises were used implicitly in the derivation. For instance, we equated "frequency" with "probability". This is admissible in the limiting case of very large populations. If the number of offspring is small, however, the *sampling error* may significantly affect the composition of the gene pool. We furthermore excluded the effect of selection, that is the possibility that some genotypes (or some pairings of genotypes) produce more offspring than others. We assumed non-overlapping generations - for otherwise, we would have had to take into account the age structure of the population. On top of all this, the genes were supposed to be autosomal: for sex linked genes, the evolution is slightly more complicated, as we shall see in **2.4.** All these conditions restrict, of course, the applicability of the Hardy-Weinberg law. Nevertheless, it is the starting point of every treatment of population genetics: the next step being to drop one assumption or another and see what happens then.

Exercise 1: Show that (x,y,z) (with $x+y+z=1$) satisfies the Hardy-Weinberg relation (2.3) if and only if $y^2=4xz$. Draw the "Hardy-Weinberg"-parabola given by this relation.

Exercise 2: Derive the Hardy-Weinberg law for random mating. (Hint: write down all possible mating types: $A_1A_1 \times A_1A_1$, $A_1A_1 \times A_1A_2$ etc.; write down the probability of each mating: x^2, $2xy$, etc.; write down the genotypes of the offspring of each mating type, and sum them up to obtain (2.3)).

2.3. The Hardy-Weinberg law for more than two alleles

Let us now extend the previous considerations to the case of n alleles. We shall assume that the probabilities of the alleles $A_1,\dots,A_n$ are given by $p_1,\dots,p_n$ $(p_1+\dots+p_n=1)$, and those of the gene pairs (A_i,A_j) by p_{ij} $(1 \le i,j \le n)$. A randomly chosen gene will sit with probability $\frac{1}{2}$ at a "first place" (according to our convention, it has been transmitted by the father) and with probability $\frac{1}{2}$ at a "second place". In the former case, it is of type A_i if the gene pair is of the form (A_i,A_j) (for arbitrary j), which happens with probability p_{ij}. In the latter case, it is of type A_i if the given pair is of the form (A_j,A_i) (for arbitrary j), which happens with probability p_{ji}. Thus

$$p_i = \frac{1}{2}\sum_j p_{ij} + \frac{1}{2}\sum_j p_{ji} \; . \tag{2.4}$$

Let us denote by p_i' resp. p_{ij}' the corresponding probabilities in the next generation. Assuming random mating of gametes we obtain

$$p_{ij}' = p_i p_j \tag{2.5}$$

(the paternal gene is with probability p_i of type A_i, the maternal one with probability p_j of type A_j). Consequently

$$p_i' = \frac{1}{2}\Big(\sum_j p_{ij}' + \sum_j p_{ji}'\Big) = \frac{1}{2}\cdot 2\cdot \sum_j p_i p_j = p_i$$

The Hardy-Weinberg law therefore says:

a) *the gene probabilities remain unchanged from generation to generation;*

b) *from the first daughter generation onward, the probabilities of the homozygous genotypes A_jA_j are given by p_j^2, those of the heterozygous genotypes A_iA_j $(i \neq j)$ by $2p_ip_j$.*

The "dynamics" of the evolution are rather simple. The state in the gene pool, given by $(p_1,\dots,p_n)$, does not change at all. The state in the genotype space, given by the p_{ij}, attains within one generation the equilibrium values given in (b) and remains constant thereafter.

2.4. The case of sex-linked genes

Let us consider the case of two alleles A_1 and A_2 sitting on the X-chromosome. Females have two X-chromosomes. We shall denote the frequencies of the (female) genotypes A_1A_1, A_1A_2 and A_2A_2 in the n-th generation by $x(n)$, $y(n)$ and $z(n)$ respectively, and the frequencies of the alleles A_1 and A_2 in the female gene pool by $p^f(n)$ and $1-p^f(n)$. Males have only one X-chromosome, which they inherit from the mother. Their gene frequencies in the n-th generation will be denoted by $p^m(n)$ and $1-p^m(n)$.

Since a male offspring inherits his gene from the mother, we have

$$p^m(n+1) = p^f(n) \; . \tag{2.6}$$

Female offspring receive one gene from the father and one from the mother. Assuming random union again, we obtain

$$\begin{aligned} x(n+1) &= p^f(n)p^m(n) \\ y(n+1) &= p^f(n)(1-p^m(n))+(1-p^f(n))p^m(n) \\ z(n+1) &= (1-p^f(n))(1-p^m(n)) \; . \end{aligned} \tag{2.7}$$

From $p^f(n+1) = x(n+1) + \frac{1}{2}y(n+1)$ it follows that

$$p^f(n+1) = \frac{1}{2}(p^f(n)+p^m(n)) .$$

It is obviously enough to consider the sequence of gene frequencies in the male gene pool. Now since

$$p^m(n+2) = p^f(n+1) = \frac{1}{2}(p^f(n)+p^m(n)) = \frac{1}{2}(p^m(n+1)+p^m(n)) , \quad (2.8)$$

the sequence $u_n = p^m(n)$ satisfies the recurrence relation

$$u_{n+2} = \frac{1}{2}u_{n+1}+\frac{1}{2}u_n . \quad (2.9)$$

Such a relation allows the stepwise computation of the elements u_n, starting from given values of u_0 and u_1. But it is also possible to compute the u_n directly. Quite generally, a linear recurrence relation

$$u_{n+2} = au_{n+1}+bu_n$$

can be written in vector notation as

$$\begin{pmatrix} u_{n+2} \\ u_{n+1} \end{pmatrix} = A \begin{pmatrix} u_{n+1} \\ u_n \end{pmatrix} \quad \text{with} \quad A = \begin{pmatrix} a & b \\ 1 & 0 \end{pmatrix} .$$

The matrix A has two eigenvalues λ_1 and λ_2, solutions of the characteristic equation

$$\det\,(A-\lambda I) = \lambda^2-a\lambda-b = 0 \quad . \quad (2.10)$$

If $\lambda_1 \neq \lambda_2$, the matrix A is diagonalizable. There exists an invertible 2×2-matrix S such that

$$A = SDS^{-1} \quad \text{with} \quad D = \begin{pmatrix} \lambda_1 & 0 \\ 0 & \lambda_2 \end{pmatrix}$$

and hence

$$A^n = SD^nS^{-1} \quad \text{with} \quad D^n = \begin{pmatrix} \lambda_1^n & 0 \\ 0 & \lambda_2^n \end{pmatrix} . \quad (2.11)$$

Taking $\mathbf{x} = \begin{pmatrix} u_1 \\ u_0 \end{pmatrix}$, we obtain u_n as second component of $A^n\mathbf{x}$. It follows from (2.11) that u_n is a linear combination of λ_1^n and λ_2^n, i.e.

$$u_n = c_1\lambda_1^n+c_2\lambda_2^n \quad (2.12)$$

with constants c_1 and c_2 which are determined by the initial values u_1 and u_0.

In our case, $a=b=\frac{1}{2}$, $\lambda_1=1$, $\lambda_2=-\frac{1}{2}$, $u_0=p^m(0)$, and $u_1=p^m(1)=p^f(0)$. Hence

$$p^m(n) = c_1+(-\frac{1}{2})^n c_2 \tag{2.13}$$

with

$$c_1 = \frac{1}{3}\left[p^m(0)+2p^f(0)\right] \quad \text{and} \quad c_2 = \frac{2}{3}\left[p^m(0)-p^f(0)\right].$$

According to (2.13) the sequence $p^m(n)$ converges to the limit c_1. Denoting this value by p, we obtain the further relations $p^f(n)\to p$, $x(n)\to p^2$, $y(n)\to 2p(1-p)$ and $z(n)\to(1-p)^2$. Hence a Hardy-Weinberg-equilibrium is attained, but as a limiting value only rather than already after one generation.

Exercise: Deal with the Fibonacci sequence in a similar way.

2.5. Notes

The law of genetic equilibrium was discovered independently - and almost simultaneously - by the German physician Weinberg (1908) and the British mathematician Hardy (1908). First class texts on population genetics are due to Crow and Kimura (1970), Roughgarden (1979) and Ewens (1979).

3. Selection and the Fundamental Theorem

3.1. The selection model

Let us now investigate the effect of selection upon a population. As a first step, we shall study a classical model involving selection through differential viabilities of the genotypes.

Consider a single chromosomal locus with n alleles $A_1,\ldots,A_n$. Let $p_1,\ldots,p_n$ denote the gene frequencies at the mating stage in the parental generation. The assumption of random mating leads to p_ip_j for the probability that a zygote carries the gene pair (A_i,A_j). Let w_{ij} be the probability that an (A_i,A_j)-individual survives to adult age. The *selective values* w_{ij} satisfy $w_{ij}\geq 0$ and $w_{ij}=w_{ji}$, since the gene pairs (A_i,A_j) and (A_j,A_i) belong to the same genotype. The selection matrix

$$W = \begin{bmatrix} w_{11} & \cdots & w_{1n} \\ \cdot & & \cdot \\ \cdot & & \cdot \\ \cdot & & \cdot \\ w_{n1} & \cdots & w_{nn} \end{bmatrix}$$

is therefore symmetric.

If N is the number of zygotes in the new generation, then $p_i p_j N$ of them carry the gene pair (A_i, A_j), of which $w_{ij} p_i p_j N$ survive to adulthood. The total number of individuals reaching the mating stage is $\sum_{r,s=1}^{n} w_{rs} p_r p_s N$. We shall assume that this number is distinct from 0. Denoting by p'_{ij} the frequency of the gene pair (A_i, A_j) in the adult stage of the new generation, and by p'_i the frequency of the allele A_i, we obtain

$$p'_{ij} = \frac{w_{ij} p_i p_j N}{\sum_{r,s=1}^{n} w_{rs} p_r p_s N}.$$

For p'_{ji}, we obtain the same expression, since $w_{ij} = w_{ji}$. Clearly

$$p'_i = \frac{1}{2} \sum_j p'_{ij} + \frac{1}{2} \sum_j p'_{ji}.$$

This leads to the relation

$$p'_i = p_i \frac{\sum_j w_{ij} p_j}{\sum_{r,s} w_{rs} p_r p_s} \qquad i = 1, \ldots, n \tag{3.1}$$

which describes the evolution of the gene frequencies from one generation to the next.

Let us recall the assumptions: (i) separate generations; (ii) very large population size; (iii) random union of gametes; (iv) selection acting solely upon one chromosomal locus, through different survival probabilities; no other effects like mutation, migration etc.

3.2. The selection equation as a dynamical system

The state of the gene pool of the parental generation is given by the vector $\mathbf{p} = (p_1, \ldots, p_n)$ of gene frequencies. The state of the next generation is given by $\mathbf{p}' = (p'_1, \ldots, p'_n)$. Both $\mathbf{p}$ and $\mathbf{p}'$ have nonnegative components summing up to one, and belong therefore to the *simplex*

$$S_n = \{\mathbf{x} = (x_1, \ldots, x_n) \in \mathbf{R}^n : \sum x_i = 1,\ x_i \geq 0 \text{ for } i = 1, \ldots, n\}$$

which is a convex $(n-1)$-dimensional subset of $\mathbf{R}^n$ whose extremal points are the unit vectors

$$\mathbf{e}_i = (0,0,\ldots,\underset{\substack{\uparrow \\ i\text{-th place}}}{1},0,\ldots,0)\ .$$

Equation (3.1) describes the action of selection from one generation to the next. The map sending $\mathbf{p}$ to $\mathbf{p}'$ defines a dynamical system on the space S_n.

Quite generally, a *(discrete) dynamical system* consists of a map T from a space X into itself. This map $\mathbf{x} \to T\mathbf{x}$ can be iterated. The sequence $\mathbf{x}$, $T\mathbf{x}$, $T(T\mathbf{x}) = T^2\mathbf{x}, \ldots$, $T^k\mathbf{x}, \ldots$ is called the *orbit* of $\mathbf{x}$. Usually, $\mathbf{x}$ is interpreted as the initial state of a system and k as the number of time intervals elapsed. We shall only consider dynamical systems where X is a subset of $\mathbf{R}^n$ and T is differentiable. It is often convenient to write the map $\mathbf{x} \to T\mathbf{x}$ in the form $\mathbf{x}' = T\mathbf{x}$ or $\mathbf{x}' - \mathbf{x} = T\mathbf{x} - \mathbf{x}$, which displays the increment $\mathbf{x}' - \mathbf{x}$ explicitly. Equations of this type are also called *recurrence* or *difference equations.*

In our example (3.1), the state space is S_n and the map T is given by $T\mathbf{p} = \mathbf{p}'$. After k generations, the gene pool is in the state $T^k\mathbf{p}$. In principle, one could compute $T^k\mathbf{p}$ recursively, for $k = 1,2,\ldots$. In practice, it would be better to let a computer do the job. It can churn tirelessly through the calculations, print out tons of paper and display the evolution of the orbit on a screen.

The basic question is, of course: What happens in the long run? And here, we may expect some helpful hints from the computer, but not the answer, obviously. Computing cannot tackle questions like: will the dynamics settle down to some steady state, or oscillate periodically, or keep up some "chaotic motion"? Will it be stable with respect to minute variations in the initial values of the parameters, or will it respond with an altogether different behaviour? The finiteness in the precision and the time horizon of numerical computations is not suited to a study of asymptotic regimes. *Qualitative analysis* is what one needs. We shall get acquainted with some of its basic methods by studying models provided by evolutionary biology.

One of the principal features of any dynamics is what remains static, i.e. the *fixed points* (or *rest points*, or *equilibria*, or *stationary states*). These are the points $\mathbf{x}$ such that $T\mathbf{x} = \mathbf{x}$.

We shall presently see that the set of fixed points of the selection model (3.1) is easily described, and that every orbit converges to it.

3.3. The Fundamental Theorem of natural selection

The Fundamental Theorem of natural selection states that under selection, the average fitness increases from generation to generation. Before stating it more precisely, we will introduce some notation. We shall write $\mathbf{x}\cdot\mathbf{y}$ for the Euclidean inner product of the two vectors $\mathbf{x} = (x_1,\ldots,x_n)$ and $\mathbf{y} = (y_1,\ldots,y_n)$ in $\mathbf{R}^n$:

$$\mathbf{x}\cdot\mathbf{y} = \sum_{i=1}^{n} x_i y_i \ . \tag{3.2}$$

We may then write the denominator in (3.1) as inner product

$$\mathbf{p}\cdot W\mathbf{p} = \sum p_i (W\mathbf{p})_i \ , \tag{3.3}$$

where

$$(W\mathbf{p})_i = \sum_{j=1}^{n} w_{ij} p_j \tag{3.4}$$

is the i-th component of the vector $W\mathbf{p}$. The expression $\mathbf{p}\cdot W\mathbf{p}$, which we shall also denote by $\bar{w}(\mathbf{p})$, can be viewed as the *average fitness* or, more precisely, as the average selective value of the population. Indeed, w_{ij} is the selective value of the gene pair (A_i,A_j) and $p_i p_j$ is its frequency. The Fundamental Theorem then says:

Theorem. *For the dynamical system* $\mathbf{p}\rightarrow T\mathbf{p} = \mathbf{p}'$ *given by* (3.1), *i.e. by*

$$p_i' = p_i \frac{(W\mathbf{p})_i}{\mathbf{p}\cdot W\mathbf{p}} \ , \tag{3.5}$$

the average fitness $\bar{w}(\mathbf{p}) = \mathbf{p}\cdot W\mathbf{p}$ *increases along every orbit in the sense that*

$$\bar{w}(\mathbf{p}') \geq \bar{w}(\mathbf{p}) \tag{3.6}$$

with equality if and only if $\mathbf{p}$ *is an equilibrium point.*

We defer the proof to the next section, and consider a few corollaries first.

The point $\mathbf{p}$ is an equilibrium if and only if $p_i' = p_i$, i.e.

$$p_i = p_i \frac{(W\mathbf{p})_i}{\bar{w}(\mathbf{p})} \qquad i = 1,\ldots,n \ . \tag{3.7}$$

This means that for every i, one has $p_i = 0$ or $(W\mathbf{p})_i = \bar{w}(\mathbf{p})$.

The *boundary* of the simplex S_n consists of the points $\mathbf{x}\in S_n$ for which $x_i = 0$ for some i, i.e. of the states where some alleles are missing. The *interior* of S_n is the set of points $\mathbf{x}\in S_n$ with $x_i>0$ for all i. We shall denote the boundary by bdS_n and the interior by intS_n. For any $\mathbf{p}\in S_n$, we shall denote by supp ($\mathbf{p}$) (the *support* of $\mathbf{p}$) the set of all indices i with $p_i>0$. Thus

$$\text{int } S_n = \{\mathbf{p}\in S_n\text{: supp }(\mathbf{p}) = \{1,\ldots,n\}\} .$$

According to (3.7), a point $\mathbf{p}\in$int S_n, i.e. with $p_i>0$ for all i, is an equilibrium if and only if

$$(W\mathbf{p})_1 = (W\mathbf{p})_2 = \cdots = (W\mathbf{p})_n \tag{3.8}$$

and furthermore

$$p_1+\cdots+p_n = 1 . \tag{3.9}$$

(The common value of the expressions in (3.8) must be $\mathbf{p}\cdot W\mathbf{p} = \bar{w}(\mathbf{p})$.) This is a set of n linear equations in the n variables p_1 to p_n. Depending on W, there may be zero, or one, or infinitely many solutions in int S_n. In the last case (which is degenerate, since it occurs only if the w_{ij}'s satisfy certain linear relations) the fixed points form a linear manifold in the interior of S_n.

Exercise 1: Characterize the degenerate fitness matrices W.

If J is a nontrivial subset of the set $\{1,\ldots,n\}$ of indices, then

$$S_n(J) = \{\mathbf{x}\in S_n\text{:}x_i = 0 \text{ for all } i\notin J\} \tag{3.10}$$

is a lower dimensional subsimplex which constitutes a face of the boundary of S_n. Each such face $S_n(J)$ is invariant: if $\mathbf{p}$ lies in $S_n(J)$, then so does $\mathbf{p}'$. (The "missing" alleles A_i, $i\notin J$, cannot be introduced by selection alone). In fact the restriction of the dynamical system $\mathbf{p}\to\mathbf{p}'$ on $S_n(J)$ is again of the form (3.1), but with fewer indices: all $i\notin J$ are dropped. A point $\mathbf{p}$ in $S_n(J)$ with $p_i>0$ for all $i\in J$ is an equilibrium if and only if all $(W\mathbf{p})_i$, for $i\in J$, are equal. This leads to a system of linear equations similar to (3.8) and (3.9), but of lower dimension. Considering each of the faces $S_n(J)$, one can find all equilibrium points in S_n. In general, there is at most one fixed point in the interior of each face, just as for int S_n. Of course, each corner of S_n - i.e., each unit vector $\mathbf{e}_i$ - is an equilibrium.

We shall now see that every orbit $T^k\mathbf{p}$ converges to the set of equilibria, or more precisely that every accumulation point of the orbit is a fixed point. Indeed, the sequence $\bar{w}(T^k\mathbf{p})$ increases monotonically by (3.6) and hence converges to some limit L. Let $\mathbf{y}$ be an accumulation point of $T^k\mathbf{p}$: there exists, then, a subsequence $k_j\to+\infty$ such that

$T^{k_j}\mathbf{p}\to\mathbf{y}$. The continuity of T implies that $T^{k_j+1}\mathbf{p} = T(T^{k_j}\mathbf{p})$ converges to $T\mathbf{y}$. But

$$\bar{w}(\mathbf{y}) = \bar{w}(\lim_{j\to+\infty} T^{k_j}\mathbf{p}) = L$$

and

$$\bar{w}(T\mathbf{y}) = \bar{w}(\lim T^{k_j+1}\mathbf{p}) = L$$

so that $\bar{w}(\mathbf{y}) = \bar{w}(T\mathbf{y})$. Thus $\mathbf{y}$ is an equilibrium point.

If the fixed points are isolated (which is the typical case, as we have seen), then every orbit converges to a single equilibrium. Indeed, it cannot jump to and fro between two isolated equilibria, since being near a fixed point implies being approximately stationary.

Exercise 2: Prove this in a precise way.

3.4. Proof of the Fundamental Theorem

Let us start by recalling a few inequalities. A function f defined on some interval I is said to be *strictly convex* if for any two distinct points x_1 and x_2 in I and any two positive numbers p_1,p_2 with $p_1+p_2 = 1$, one has

$$f(p_1x_1+p_2x_2)<p_1f(x_1)+p_2f(x_2) \ . \tag{3.11}$$

The geometric interpretation is that the segment joining any two points of the graph of f lies above this graph. By induction, one can easily show that (3.11) implies the *Jensen inequality*:

$$f(\sum p_ix_i) \leq \sum p_if(x_i) \tag{3.12}$$

for all $x_1,\dots,x_n\in I$ and every $\mathbf{p} = (p_1,\dots,p_n)\in\text{int } S_n$, with equality if and only if all x_i coincide. Recall that f is strictly convex if $\frac{d^2f}{dx^2}> 0$ on I. Applying (3.12) to the function $f(x) = x^\alpha$ with $\alpha>1$, which is strictly convex on $[0,+\infty)$, we obtain

$$(\sum p_ix_i)^\alpha \leq \sum p_ix_i^\alpha \tag{3.13}$$

for $x_1,\dots,x_n\geq0$ and $(p_1,\dots,p_n)\in S_n$, with equality iff there is some value c such that $x_j = c$ for all j with $p_j>0$.

We shall also use the *inequality of the arithmetic and the geometric mean*

$$\sqrt{ab} \leq \frac{a+b}{2} \tag{3.14}$$

which is valid for $a,b \geq 0$.

Exercise 1: Prove (3.12), (3.14) and the *inequality of Cauchy-Schwarz-Bunyakowski*

$$(\sum_{i=1}^{n} a_i b_i)^2 \leq (\sum_{i=1}^{n} a_i^2)(\sum_{i=1}^{n} b_i^2) \quad .$$

Let us now prove (3.6). Since we assume $\mathbf{p}\cdot W\mathbf{p}>0$, we have to show that

$$(\mathbf{p}\cdot W\mathbf{p})^2(\mathbf{p}'\cdot W\mathbf{p}') \geq (\mathbf{p}\cdot W\mathbf{p})^3 \; . \tag{3.15}$$

Clearly

$$(\mathbf{p}\cdot W\mathbf{p})^2(\mathbf{p}'\cdot W\mathbf{p}') = (\mathbf{p}\cdot W\mathbf{p})^2 \sum_{i,k} p_i' \, w_{ik} p_k' \; .$$

On replacing p_i' and p_k' by the expression in (3.1) we obtain

$$\sum_{i,k} p_i (W\mathbf{p})_i w_{ik} p_k (W\mathbf{p})_k = \sum_{i,j,k} p_i w_{ij} p_j w_{ik} p_k (W\mathbf{p})_k = s(1) \; .$$

Exchange of the indices j and k yields

$$\sum_{i,j,k} p_i w_{ik} p_k w_{ij} p_j (W\mathbf{p})_j = s(2) \; .$$

The expressions $s(1)$ and $s(2)$ are equal, and hence coincide with their arithmetic mean

$$\sum_{i,j,k} p_i w_{ij} p_j w_{ik} p_k (\frac{1}{2}[(W\mathbf{p})_j + (W\mathbf{p})_k]) \; .$$

By (3.14), this is not smaller than

$$\sum_{i,j,k} p_i w_{ij} p_j w_{ik} p_k (W\mathbf{p})_j^{\frac{1}{2}} (W\mathbf{p})_k^{\frac{1}{2}}$$

$$= \sum_i p_i \sum_j w_{ij} p_j (W\mathbf{p})_j^{\frac{1}{2}} \sum_k w_{ik} p_k (W\mathbf{p})_k^{\frac{1}{2}}$$

$$= \sum_i p_i [\sum_j w_{ij} p_j (W\mathbf{p})_j^{\frac{1}{2}}]^2$$

By (3.13), with $\alpha = 2$, this last expression is not smaller than

$$\left[\sum_i p_i \sum_j w_{ij} p_j (W\mathbf{p})_j^{1/2}\right]^2 = \left[\sum_j p_j (W\mathbf{p})_j^{1/2} \sum_i p_i w_{ij}\right]^2 \; .$$

Since $w_{ij} = w_{ji}$, we obtain thereby

$$\left[\sum p_j (W\mathbf{p})_j^{1/2} (W\mathbf{p})_j\right]^2 = \left[\sum p_j (W\mathbf{p})_j^{3/2}\right]^2 .$$

Using (3.13) again, with $\alpha = \frac{3}{2}$, we see that this expression is not smaller than

$$\left[\sum p_j (W\mathbf{p})_j\right]^{\frac{3}{2}\cdot 2} = (\mathbf{p}\cdot W\mathbf{p})^3$$

which proves (3.15).

If $\bar{w}(\mathbf{p}) = \bar{w}(\mathbf{p}')$, the last estimate must be an equality, i.e. there must be a value c such that $(W\mathbf{p})_j = c$ for all j with $p_j > 0$. This means, as we have seen in the last section, that $\mathbf{p}$ is an equilibrium.

Exercise 2: Try to give an alternative proof for the Fundamental Theorem:

1) Show that $(\frac{\mathbf{x}\cdot A\mathbf{x}}{\mathbf{x}\cdot\mathbf{x}})^m \leq \frac{\mathbf{x}\cdot A^m\mathbf{x}}{\mathbf{x}\cdot\mathbf{x}}$ for symmetric $n\times n$ matrices A with nonnegative elements $a_{ij} \geq 0$ and $x_i \geq 0$. (Use induction on n.)
2) Set $m = 3$ and choose $a_{ij} = p_i^{1/2} w_{ij} p_j^{1/2}$ to obtain the Fundamental Theorem.

Exercise 3: Show that the following convexity inequalities

$$\prod_{i=1}^{n} x_i^{p_i} \leq \sum_{i=1}^{n} p_i x_i$$

and

$$\left(\sum_{i=1}^{n} x_i\right)^2 \leq n\sum_{i=1}^{n} x_i^2$$

hold for $x_i \geq 0$ and $\mathbf{p} \in S_n$.

3.5. The case of two alleles

We shall write p and $1-p$ instead of p_1 and p_2. Equation (3.5) defines a dynamical system on the interval $[0,1]$, namely

$$p' = F(p) = \frac{a_1}{a_1+a_2} \tag{3.16}$$

with

$$a_1 = p(w_{11}p+w_{12}(1-p)) \text{ and } a_2 = (1-p)(w_{12}p+w_{22}(1-p)) .$$

The average fitness is

$$\begin{aligned}\bar{w}(p) &= \mathbf{p}\cdot W\mathbf{p} = a_1+a_2 \\ &= p^2[(w_{11}-w_{12})+(w_{22}-w_{12})]-2p(w_{22}-w_{12})+w_{22}\,. \end{aligned} \tag{3.17}$$

As we can check directly, the increment within one generation is given by

$$p'-p = \frac{p(1-p)}{2\bar{w}(p)}\frac{d}{dp}\,\bar{w}(p)\,. \tag{3.18}$$

The fixed points are therefore the two endpoints of $[0,1]$ and the critical points of $\bar{w}(p)$ in the interior $(0,1)$. There are several possibilities.

(1) In the (degenerate) case $w_{12} = \dfrac{w_{11}+w_{22}}{2}$, the function $\bar{w}(p)$ is linear. If $w_{11} = w_{12} = w_{22}$, selection does not operate, $\bar{w}(p)$ is constant, and all points are equilibria. Otherwise, the slope of $\bar{w}(p)$ is nonzero, and p converges to 0 or to 1. Thus, the homozygote with the highest fitness gets established.

(2) In the generic case $w_{12} \neq \dfrac{w_{11}+w_{22}}{2}$, $\bar{w}(p)$ is a parabola whose extremum is at the point

$$\bar{p} = \frac{w_{22}-w_{12}}{(w_{11}-w_{12})+(w_{22}-w_{12})}\,. \tag{3.19}$$

If

$$(w_{11}-w_{12})(w_{22}-w_{12}) \leq 0 \tag{3.20}$$

i.e. w_{12} lies between w_{11} and w_{22}, then $\bar{p}\notin(0,1)$. All orbits in the interior of the state space converge to that end point where $\bar{w}$ is highest. In this case, which includes dominance of one allele ($w_{12} = w_{11}$ or $w_{12} = w_{22}$), the homozygote with highest fitness becomes established in the population.

If (3.20) does not hold, however, then $\bar{p}$ is an equilibrium in $(0,1)$. Two cases are possible.

(a) In the case of ***heterozygote advantage***, $w_{12} > w_{11}$ and $w_{12} > w_{22}$, the average fitness $\bar{w}(p)$ has its maximum at $\bar{p}$;

(b) otherwise, $\bar{w}(p)$ has its minimum at $\bar{p}$.

As we shall presently see, an orbit cannot jump from one side of $\bar{p}$ to the other side. This implies that an orbit starting in $(0,\bar{p})$ or $(\bar{p},1)$ remains in that interval and converges to the endpoint with highest fitness. In case (a), this leads to the ***polymorphic*** state $\bar{p}$ (both alleles are present). In case (b), one or the other allele vanishes from the gene pool, depending on the initial state (see Fig. 3.1).

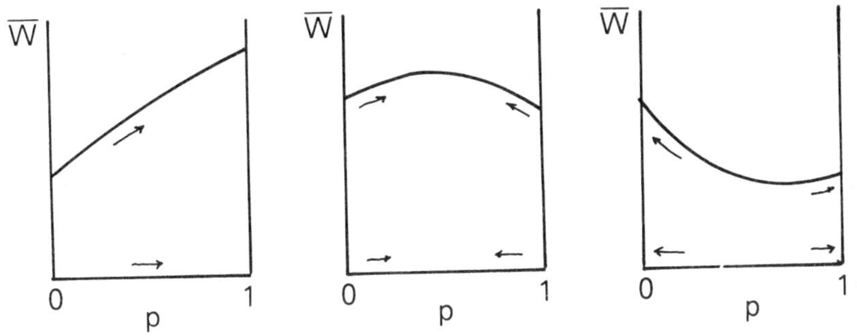

Figure 3.1

That $p<\bar{p}$ implies $p'<\bar{p}$ follows from the monotonicity of F, which in turn is a consequence of

$$\frac{d}{dp}F(p) = (a_1+a_2)^{-2}[a_2\frac{d}{dp}a_1-a_1\frac{d}{dp}a_2] > 0\,.$$

Indeed, a straightforward computation shows that

$$a_2\frac{d}{dp}a_1-a_1\frac{d}{dp}a_2 = w_{11}w_{12}p^2 + 2w_{11}w_{22}p(1-p) + w_{22}w_{12}(1-p)^2$$

is positive in [0,1] both in case (a) and case (b). Let us remark that the heterozygote genotype survives only in the case of heterozygote advantage.

Exercise 1: Prove these results without referring to the Fundamental Theorem. (Hint: Calculations are simplified, if $x = p/(1-p)$ is used as variable.)

Exercise 2: In case (a) compute the maximal value of $\bar{w} = \bar{w}(\bar{p})$. The difference $w_{12}-\bar{w}(\bar{p})$ can be interpreted as a *segregational load*, that is the loss in fitness for the heterozygote caused by Mendelian segregation. (Assume fitness parameters $w_{11}{:}w_{12}{:}w_{22} = 1-s_1{:}1{:}1-s_2$).

3.6. Notes

R.A. Fisher, the founding father of modern statistics and population genetics, presented a simple version of the Fundamental Theorem in his book "The genetical theory of natural selection" (1930). The general result for n alleles was proved by Mulholland and Smith (1959), Scheuer and Mandel (1959) and Kingman (1961), see also Baum and Eagon (1967). The proof by Kingman is presented in **3.4**, and the proof by Mulholland and Smith in Exercise 3.4.2. Recently, Lyubich et al. (1980) and Losert and Akin (1983) showed that gene frequencies always converge to a single equilibrium point. More results on the selection equation can be found in Karlin (1984). For a discussion of the continuous time selection equation (generations blending into each other), we refer to Chapter 23.

4. Mutation and Recombination

4.1. Recurrent mutations

The copying of genes does not occur with absolute precision. An allele may mutate to another one. Recurrent mutations, although very rare, help to maintain a supply of genetic variation; even if selection pressure at a given time acts against an allele, it will be maintained by mutation and thus kept available for changed circumstances.

Let us assume first that the two alleles A_1 and A_2 are selectively neutral. We shall denote by p and $1-p$ the frequencies of A_1 and A_2, by μ the rate of mutation from A_1 and A_2 and by ν the rate of mutation from A_2 to A_1. The frequency of A_1 in the next generation will then be

$$p' = G(p) = p-\mu p+\nu(1-p) = (1-\mu-\nu)p+\nu \,. \tag{4.1}$$

This is a dynamical system in the interval $[0,1]$. Its only equilibrium is

$$\bar{p} = \frac{\nu}{\mu+\nu} \,. \tag{4.2}$$

Since (4.1) may be written as

$$p' - \bar{p} = (1-\mu-\nu)(p - \bar{p}) \tag{4.3}$$

we see that all orbits converge at the rate $(1-\mu-\nu)$ towards the equilibrium $\bar{p}$:

$$T^n p - \bar{p} = (1-\mu-\nu)^n (p - \bar{p}) \to 0 \,.$$

It should be noted, however, that this model is of dubious value. The mutation rates are extremely small (of the order of 10^{-6}), so that convergence is ridiculously slow. Genetic drift will be much more effective and thus blur the deterministic behaviour predicted by our dynamical system. However, an investigation of stochastic fluctuations is outside the scope of this book.

4.2. The mutation selection equation

Let us now assume that selection acts betwen the zygote and the adult stage and that during the subsequent meiosis, mutations may occur. In the model with two alleles A_1 and A_2, the frequency p of A_1 is changed first into $F(p)$ and then into $G(F(p))$, with F given by (3.16) and G by (4.1). Thus the frequency p' of A_1 after one generation is given by

$$p' = H(p) = G(F(p)) = (1-\mu-\nu)F(p)+\nu\ . \tag{4.4}$$

Since the mutation rates μ and ν are very small, we have obviously $1-\mu-\nu>0$. F is monotonic, as we have seen in **3.5**. Therefore, so is H. The map $p\to p'$ of $[0,1]$ into itself is invertible. Setting

$$\Phi(p) = p^{2\nu}(1-p)^{2\mu}[\bar{w}(p)]^{1-\mu-\nu} \tag{4.5}$$

where $\bar{w}$ is the average fitness as in (3.17), one can check by a straightforward computation that for $p\in(0,1)$

$$p'-p = \frac{p(1-p)}{2\Phi(p)}\frac{d}{dp}\Phi(p) \tag{4.6}$$

which is similar to (3.18). The equilibria of the dynamical system $p\to H(p)$ in the interior of the state space $[0,1]$ are therefore the critical points of Φ. The orbits are monotonic: the function Φ is increasing along them.

Exercise 1: Prove this! (Consider each monotonicity interval of Φ separately.)

In the absence of mutation (i.e. for $\mu = \nu = 0$), this is nothing but the Fundamental Theorem. In the absence of selection (i.e. if the fitness is constant), Φ reduces to the function $p^{2\nu}(1-p)^{2\mu}$, whose maximum is at $\frac{\nu}{\mu+\nu}$. The interplay between selection and mutation leads to a compromise between two maximum principles.

In particular, recurrent mutations from A_2 to A_1 will sustain the allele A_1 even if it is threatened by elimination through selection pressure. Let us, for simplicity, assume that the mutation rate μ from A_1 to A_2 is zero. If A_1 is dominant, then the fitness of the genotypes A_1A_1 and A_1A_2 will be some fraction $1-s$ of the fitness of A_2A_2 (with

$s \in (0,1)$). The unique stable equilibrium $\bar{p}$ in $(0,1)$ is the critical point given (approximately) by $\frac{\nu}{s}$. If A_1 is recessive, however, then A_1A_2 has the same fitness as A_2A_2, and $\bar{p}$ is approximately $\sqrt{\nu/s}$.

Exercise 2: Check this. Can there exist a second equilibrium in $(0,1)$? What happens for $\mu > 0$?

Typical values for s and ν would be 10^{-2} and 10^{-6}, respectively. Then $\bar{p}$ is in the dominant case 10^{-4} and in the recessive case 10^{-2}. The unfavourable allele A_1 can subsist in the gene pool at a much higher frequency if it is recessive, because selection cannot discern it in heterozygous individuals.

4.3. The selection recombination equation

Recombination is, like mutation, a source of genetic variability, but it acts on another level. It does not produce new genes, but new gene combinations. It also works, in general, at an altogether different speed. The recombination rate between different chromosomes can be as high as 1/2.

In order to study recombination, we have to leave one locus genetics. There is no hope, of course, of grasping the full complexity of a real genetic system with its tens of thousands of interacting parts in analytical detail. But even the absurdly oversimplified case of two loci with two alleles each provides some insight into the intricacies of gene shuffling.

Let us assume that the alleles A_1 and A_2 belong to the first genetic locus, and the alleles B_1 and B_2 to the second one. The possible gametes, then, are A_1B_1, A_2B_1, A_1B_2 and A_2B_2. We shall denote them by G_1 to G_4, and their frequencies at the zygote stage (that is, immediately after fusion) with x_1 to x_4. The state of the gamete pool, then, is described by the point $\mathbf{x} = (x_1,x_2,x_3,x_4)$ in the simplex S_4.

Each individual is obtained by fusion of a G_i-sperm with G_j-egg, and will be denoted by (G_i,G_j). The probability of its survival from zygote to adult stage will be denoted by w_{ij}. We have, of course, $w_{ij} = w_{ji}$, and shall assume furthermore that $w_{23} = w_{14}$, since (G_2,G_3) and (G_1,G_4) are both on the first locus of genotype A_1A_2 and on the second locus of genotype B_1B_2. The fitness, then, is given by the following table:

	A_1A_1	A_1A_2	A_2A_2
B_1B_1	w_{11}	$w_{12}=w_{21}$	w_{22}
B_1B_2	$w_{13}=w_{31}$	$w_{14}=w_{41}=w_{23}=w_{32}$	$w_{24}=w_{42}$
B_2B_2	w_{33}	$w_{34}=w_{43}$	w_{44}

(4.7)

We shall assume random union again. A gamete G_i will fuse with probability x_j with a gamete G_j, and then belongs to an individual with fitness w_{ij}. We may define

$$w_i = \sum_j w_{ij}x_j = (W\mathbf{x})_i \tag{4.8}$$

as fitness of the gamete G_i. The average fitness in the gamete pool of the population is

$$\bar{w} = \sum_i w_i x_i = \sum_{ij} w_{ij}x_i x_j \,. \tag{4.9}$$

At meiosis, a cross over between the two loci may occur with probability r ($r = \frac{1}{2}$ if the loci are independent, for example on non-homologous chromosomes). If the individual is homozygous on at least one locus, a crossover does not change anything. It is only for the "double heterozygotes" (G_2,G_3), (G_3,G_2), (G_1,G_4) and (G_4,G_1) that recombination makes a difference. Let us consider for instance type (G_2,G_3), whose frequency in the adult stage is $w_{23}x_2x_3\bar{w}^{-1}$. A fraction $\frac{r}{2}$ of its gametes are of types G_1 and G_4 each, and a fraction $\frac{1-r}{2}$ of type G_2 and G_3. Similar considerations hold for the other gamete pairs. The gametic frequencies in the zygote stage of the next generation are given by

$$\begin{aligned} x_1' &= \frac{1}{\bar{w}}(x_1w_1 - rbD) \\ x_2' &= \frac{1}{\bar{w}}(x_2w_2 + rbD) \\ x_3' &= \frac{1}{\bar{w}}(x_3w_3 + rbD) \\ x_4' &= \frac{1}{\bar{w}}(x_4w_4 - rbD) \end{aligned} \tag{4.10}$$

where b is the selection coefficient of the double heterozygotes ($b = w_{23} = w_{32} = w_{14} = w_{41}$) and $D = x_1x_4 - x_2x_3$. The case $r = 0$ leads to the selection equation (3.1) for the four "alleles" G_1 to G_4.

4.4. Linkage

The *linkage disequilibrium coefficient* $D = x_1x_4 - x_2x_3$ is a measure for the statistical dependence between the two gene loci. Indeed, since the frequencies of the alleles A_1, A_2, B_1 and B_2 are given by x_1+x_3, x_2+x_4, x_1+x_2 and x_3+x_4, and since

$$x_1x_4 - x_2x_3 = x_1 - (x_1+x_3)(x_1+x_2) \tag{4.11}$$

it follows that

$$D = \text{Prob}(A_1B_1) - (\text{Prob } A_1)(\text{Prob } B_1)\ . \tag{4.12}$$

Hence, $D = 0$ if and only if

$$\text{Prob }(A_iB_j) = (\text{Prob } A_i)(\text{Prob } B_j) \tag{4.13}$$

for $i,j = 1,2$. In this case, the population is said to be in *linkage equilibrium*. The coefficient D vanishes for the states $\mathbf{x}$ belonging to the *Wright-manifold*

$$\mathbf{W} = \{\mathbf{x} \in S_4 : x_1x_4 = x_2x_3\}\ . \tag{4.14}$$

Exercise 1: Check that $\mathbf{W}$ looks like in Fig. 4.1.

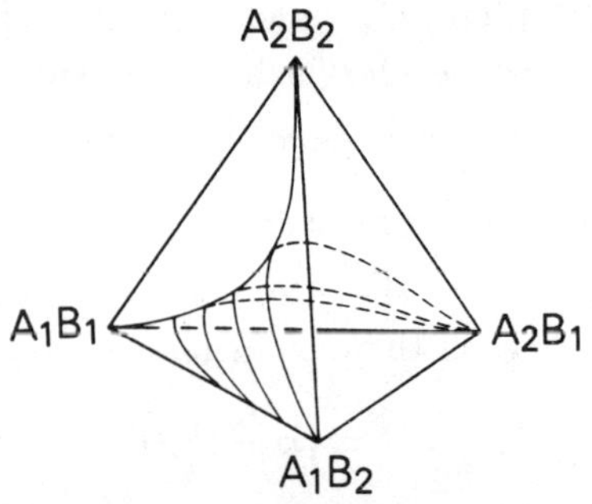

Figure 4.1

If all w_{ij} are equal, that is, if no selection operates, then

$$x_i' = x_i - rD \qquad (i = 1,4)$$

$$x_i' = x_i + rD \qquad (i = 2,3)$$

and the linkage disequilibrium coefficient in the next generation is given by

$$D' = x_1' x_4' - x_2' x_3' = (1-r)D . \tag{4.15}$$

Since $D = 0$ implies $D' = 0$, the Wright-manifold **W** is invariant: A population in linkage equilibrium remains in linkage equilibrium forever. If $r>0$, the linkage disequilibrium coefficient converges to zero. Recombination, thus, drives the population into linkage equilibrium. The situation resembles that of Chapter 2, where random mating produces genotypes by independent fusion of the genes. The Hardy-Weinberg equilibrium is reached in one generation. Recombination leads to gametes constituted of independently assorted genes, but the manifold of linkage equilibria is reached only in the long run. The states on **W** are uniquely determined by the allelic frequencies of A_1 and B_1.

If the population is subject to selection, however, then **W** is not invariant in general, and the state need not converge to linkage equilibrium.

Exercise 2: Show this by an example.

4.5. Fitness

If the fitness depends on more than one locus, the Fundamental Theorem need not be valid. It is easy, indeed, to find examples where the average fitness does not increase. For the two-locus two-alleles model, it is sufficient to choose parameter values w_{ij} so that $\bar{w}$ attains an uniquely determined maximum at a point $\mathbf{x}$ in int S_n which does not lie on the Wright manifold.

Exercise 1: Find such a fitness matrix.

If the recombination rate were zero, $\mathbf{x}$ would have to be a fixed point of the resulting selection equation, since the average fitness can no longer increase. This would imply $w_1=w_2=w_3=w_4=\bar{w}$ (cf. (4.10)). As it is, $w_i=\bar{w}$ must still be valid for $r>0$, since neither $\bar{w}$ nor the w_i's depend on r. From (4.10) it follows that

$$x_1' = x_1 - \frac{rbD}{\bar{w}} , \tag{4.16}$$

which implies $\mathbf{x} \neq \mathbf{x}'$ since $D \neq 0$. The function attains its maximum at $\mathbf{x}$, and hence

$$\bar{w}(\mathbf{x}') < \bar{w}(\mathbf{x}) . \tag{4.17}$$

This means that the average fitness can drop from one generation to the next.

In some special cases, the average fitness does increase. This happens, for instance, if the fitness depends additively upon the contributions of the loci:

	A_1A_1	A_1A_2	A_2A_2
B_1B_1	a_1+b_1	a_2+b_1	a_3+b_1
B_1B_2	a_1+b_2	a_2+b_2	a_3+b_2
B_2B_2	a_1+b_3	a_2+b_3	a_3+b_3

(4.18)

In this case the average fitness is given by

$$\begin{aligned}\bar{w}(\mathbf{x}) &= \sum_{i,j} w_{ij}x_ix_j \\ &= a_1(x_1+x_3)^2 + 2a_2(x_1+x_3)(x_2+x_4) + a_3(x_2+x_4)^2 \\ &\quad + b_1(x_1+x_2)^2 + 2b_2(x_1+x_2)(x_3+x_4) + b_3(x_3+x_4)^2 \ .\end{aligned} \tag{4.19}$$

$\bar{w}(\mathbf{x}')$ depends only on the gene frequencies $x_1' + x_3'$ etc. ..., which are obtained via (4.10): the value of r does not enter the computation. For $r = 0$ the function $\bar{w}$ increases, as we know from the Fundamental Theorem applied to the 4×4 matrix W: hence $\bar{w}$ increases for $r > 0$ too.

Exercise 2: Assume $a_2>a_1,a_3$ and $b_2>b_1,b_3$ in the additive fitness scheme (4.18) (overdominance at both loci). Show that there is a unique stationary solution $\hat{\mathbf{x}}$ in the interior of S_4. Prove that all orbits in S_4 converge to $\hat{\mathbf{x}}$.
Exercise 3: Is the Wright manifold $\mathbf{W}$ invariant for additive fitness?
Exercise 4: Consider a multiplicative fitness scheme analogous to (4.18). Show that D does not change sign along an orbit. In particular, $\mathbf{W}$ is invariant. Prove that on $\mathbf{W}$ the mean fitness $\bar{w}$ increases.

4.6. Notes

The book by Roughgarden (1979) offers an excellent introduction to models of mutation and recombination. There is a large literature on two-locus two-alleles systems: we only refer to Karlin (1975), Nagylaki (1977), Levin (1978), Hastings (1981b) and Akin (1982). A particularly exciting aspect is the theory of modifier genes: a survey by Karlin and McGregor (1974) deals with many of its aspects. For a global analysis of dominance modifiers, we refer to Bürger (1983a,b). We shall return to some of these questions for continuous time models in Chapter 25.

PART II: GROWTH RATES AND ECOLOGICAL MODELS: AN ABC ON ODE'S

5. The Ecology of Populations

5.1. The three levels of ecology

Ecology deals with the relationships between organisms and their environment, including in particular the interactions between different species. Usually, one distinguishes three levels of ecology. *Physiological ecology* investigates the exchanges between individuals and their surroundings, for example the influence of factors like temperature or food abundance upon the metabolism and behaviour of an organism. The *ecology of populations* is concerned with the growth and interaction of entire populations, for instance the patterns of variation, in time and space, of the densities of certain insect and bird species living together. The *study of ecosystems*, finally, deals with whole communities and their environment, for example deserts, lakes, coral reefs or tropical forests. This includes, at the highest level of integration, the study of the biosphere in its entirety.

5.2. Ecological interdependence

In part, no doubt, because of the environmental quandaries caused by human interference, there is a growing awareness of the complexities of ecological networks. However, precise statements about the range and intensity of such dependencies are difficult to obtain. It is a rather forbidding task, for example, to keep track of the overall exchanges of energy and matter between a square mile of spruce forest and its environment. The comprehensive investigation of the biogeochemical cycles on Earth is yet in its infancy.

It is generally thought that, with ecosystems as with organisms, selfregulating mechanisms lead to some kind of steady flow equilibrium. According to a widely held belief, the stability of such systems increases with their complexity (tropical rain forests offering a case in point), but to date, system analysis has not provided a convincing reason for this. In fact, the arguments against this belief seem to predominate. The details of the couplings between the components of an ecosystem are still mostly unknown, so that one is often reduced to rough "black box diagrams" listing inputs and outputs only. Much of the arsenal of control theory is of no use, because the functional form of regulatory responses is so hard to determine. Often, one has to make do with the sign of the coupling term only - positive, negative or zero.

A simple hierarchical ordering within ecosystems is provided by the different trophic levels. At the beginning of every trophic chain are the primary producers, which for the major part use photosynthesis to transform the energy of light. The next link in the food chain consists of herbivores, which in their turn are eaten by predators. At every step - from pine needle to mite to ladybird to spider to wasp, for example - a large part of the consumed energy gets lost. The densities of populations decrease, as a general rule, with every step. Of great importance in the food web are the decomposers, since ecosystems would be choked by their own dead organic matter if it were not recycled by scavengers. The fallen leaves of an oak forest, for example, get opened by springtails and perforated by earth worms, whose excrements are processed by fungi and microorganisms, until the resulting humus, in its turn, furnishes the essential substrates for the living trees.

5.3. Adaptation and evolution

No population can grow forever at a constant rate. This fact, first forcefully articulated by Malthus, represented one of the starting points of Darwin's thinking. An excess productivity can be sustained for a few generations at most; uncontrolled growth quickly reaches the limits of natural resources. Some individuals are better adapted than others to the resulting selection pressure. Their offspring will be more numerous. The driving force of selection is therefore competition, both within and between species.

The profusion of plant and animal species - far more than one million - indicates that the "struggle for life" can be successfully tackled by very different strategies. In unpredictable environments, populations with high fecundity and facilities for dispersal are favoured. In stable, well established habitats, a less numerous but more intensely cared for offspring promises more success. The better the adaptation of species living together, the less their resources will overlap. Competition leads to a gradual alteration of traits. With populations evolving in reproductive isolation, this can lead to *allopatric speciation*: a famous example is that of the Galapagos finches, issued from a common continental stock, but evolved differently on different islands; in particular, their staple diets and, accordingly, the shape of their bills diverged remarkably. In contrast to this, ***sympatric speciation*** may occur between populations with overlapping ranges through character displacement: geographical proximity, in such cases, is compensated by a more pronounced ecological separation of the niches. The mutual influence of closely interacting species leads to their ***coevolution***: one species reacts to changes in the other one, and vice versa, a phenomenon that may lead, for example between predators and preys, to veritable "arms races" lasting millions of years.

Ecosystems evolve in an autogenous way. It is usually assumed that a succession of species leads from pioneer states through intermediate phases towards a stable ecological climax, but such scenarios with static final outcome get increasingly challenged. The "Red Queen" hypothesis, for example, asserts that every species is subject to a deterioration of its environment caused by the adaptive improvement of the other species, and so has "to keep running just in order to stay in place". Apart from that, of course, allogenic effects like geological or meteorological changes keep most ecosystems on the move.

5.4. Competitors, symbionts and parasites

It is easier to marvel at the web of intricate dependencies of an ecosystem with its thousands of components than to model it. Even the analysis of the interaction between two species can be quite complicated, involving the effects of exterior and interior parameters, like seasonal variations, age structure and the like. As a first approximation, however, one may distinguish (apart from the case of zero interaction) three basic situations.

(a) *Competition.* Two species are rivals in the exploitation of a common resource. The more there is of one species, the worse for the other one. Because of the importance of competition as a limiting factor in evolution, such situations have attracted considerable attention.

(b) *Symbiosis.* This is the reverse situation: both species benefit from each other. The more there is of one species, the better for the other one. As an example, we may mention lichen, which is a coalition between algae and fungi, or the strange partnerships between hermit crabs and sea anemones. Such *mutualistic* relationships have received comparatively little attention in theoretical ecology, but their importance is of the highest order. In particular, there are good reasons to think that the living cells of the type occurring in higher organisms are the outcome of a symbiosis between more primitive organisms.

(c) *Host-parasite relationship.* The situation, here, is asymmetrical. The parasites benefit from the host but they do it no good. The cuckoo is a parasite of its foster species, the tapeworm a parasite of man, and man is a parasite of cod fish and a lot besides. Quite generally, *predators* can be viewed as parasites of their *prey.*

5.5. Notes

The books by May (1973), Maynard Smith (1974a) and Roughgarden (1979) are excellent introductions to mathematical modelling in ecology. The "Red Queen" hypothesis has been put forward by van Valen (1977). Margulis (1981) develops an interesting theory on the role of symbiosis in cell evolution.

6. The Logistic Equation

6.1. Exponential growth

Let R be the rate of growth of a population with discrete generations. This means that

$$x' = Rx \tag{6.1}$$

where x is the density in one generation and x' in the next. If R remains constant, then the density after n generations will be $R^n x$, which for $R>1$ means explosive growth to infinity.

Of course such a *population explosion* - a kind of chain reaction - will get checked sooner or later. Before dealing with limits to growth, however, let us consider unrestricted multiplication for populations whose generations are not discrete, but blend continuously into each other. If $x(t)$ is the population number at time t, then

$$\frac{x(t+\Delta t)-x(t)}{\Delta t}$$

is the average speed of growth in the time interval $[t,t+\Delta t]$. The function $x(t)$ is integer valued and hence not differentiable, of course. Still, if the density is very large, then the jumps caused by individual births and deaths will look negligibly small on a graph of $x(t)$. So let us postulate the existence of a time derivative

$$\frac{dx(t)}{dt} = \lim_{\Delta t \to 0} \frac{x(t+\Delta t)-x(t)}{\Delta t} \tag{6.2}$$

which we shall usually denote by $\dot{x}(t)$. The quantity $\frac{\dot{x}}{x}$ may be viewed as the growth rate of the population or average contribution of one individual to the population growth. We note that

$$\frac{\dot{x}}{x} = \frac{1}{x}\frac{dx}{dt} = \frac{d}{dt}(\log x) = (\log x)^{\cdot} .$$

If the rate of growth is constant, i.e. if

$$\dot{x} = rx \tag{6.3}$$

then

$$(\log x)^{\cdot} = r$$

and hence by integration

$$\log x(t) - \log x(0) = rt \tag{6.4}$$

or

$$x(t) = x(0)e^{rt} . \tag{6.5}$$

For the continuous model, therefore, we have exponential growth again.

6.2. Logistic growth

There are various ways in which exponential growth will eventually be checked. A larger population means fewer resources, and this implies a smaller rate of growth. We shall consider the simplest case, where the rate decreases linearly as a function of x. It is then of the form $r(1-\frac{x}{K})$, with positive constants r and K. This yields the *logistic equation*

$$\dot{x} = rx(1-\frac{x}{K}) \quad . \tag{6.6}$$

The factor rx, which represents unhampered growth, is reduced by the term $\frac{r}{K}x^2$, which corresponds to competition within the population. It can be viewed as a kind of "social friction" proportional to the number of encounters between individuals.

The behaviour of (6.6) is easy to analyse. The increase $\dot{x}$ is zero if $x = 0$ (obviously) and if $x = K$. In these two cases, the density does not change. For $0<x<K$, it increases, and for $x>K$ it decreases. The solution of (6.6) is given by

$$x(t) = \frac{Kx(0)e^{rt}}{K+x(0)(e^{rt}-1)} \tag{6.7}$$

as can be verified directly.

Exercise: Show that (6.7) can be obtained by the method of *separation of variables.* (Write (6.6) as

$$\frac{dt}{dx} = \frac{1}{r}\left(\frac{1}{x}+\frac{1}{K-x}\right)$$

integrate to obtain $t(x)$ and invert this function.)

The shape of the solution is sketched in Fig. 6.1. The behaviour in the region between 0 and K is called *logistic growth:* for very small values of x, it is almost exponential, then slackens down and levels off asymptotically to the constant K.

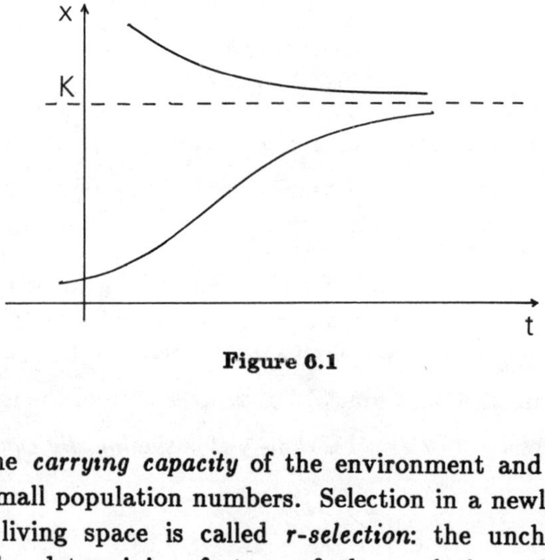

Figure 6.1

K is the *carrying capacity* of the environment and r the rate of growth for small population numbers. Selection in a newly opened, still unexploited living space is called *r-selection:* the unchecked rate of growth is the determining feature of the evolution. Selection in a saturated habitat is called *K-selection:* here, fine adjustments to the limiting factors are what counts.

6.3. The recurrence relation $x' = Rx(1-x)$

Let us return to populations with non-overlapping generations and denote by y and y' the population numbers in one generation and the next. Obviously

$$\frac{y'-y}{y}$$

can be viewed as rate of increase. If we assume, as with the logistic equation, that the rate decreases linearly in y, we obtain

$$y' = Ry(1-\frac{y}{K}) \ . \tag{6.8}$$

Equation (6.8) has the unrealistic feature that if the population number y exceeds the number K, then $y' < 0$. Thus it cannot be used for values y larger than K. On the other hand if the parameter R is larger than 4, and y is near $\frac{K}{2}$, then y' is larger than K and the model becomes nonsensical again. Hence we have to restrict R to values between 0 and 4. It may well be asked whether it is worthwhile to pursue the study of an equation with such glaring defects. It is highly to be recommended, in fact, but less for biological than for mathematical reasons. (6.8) has been called a "mathematical morality play" because it exemplifies some types of dynamic behaviour so well.

In order to simplify the discussion, we set $x = \frac{y}{K}$ and $x' = \frac{y'}{K}$. This leads to the difference equation $x' = F(x)$ with

$$F(x) = Rx(1-x) \ . \tag{6.9}$$

This is surely the simplest kind of a nonlinear recurrence relation: however, as we shall presently see, its dynamics can be of amazing complexity, quite in contrast to its continuous counterpart (6.6). We shall not attempt to give a complete discussion: larger volumes than this one would be needed for such a task. But we shall use it to introduce two central concepts: *bifurcation* and *chaotic behaviour*.

6.4. Stable and unstable equilibria

For values of R between 0 and 4, the map

$$x \to Rx(1-x)$$

defines a dynamical system on the interval $[0,1]$. An obvious equilibrium is the point 0. If $R \leq 1$, one has $x' < x$ for all $x \in (0,1]$. The orbit of x, then, decreases monotonically to 0. From now on we shall only consider $R > 1$.

The graph of F is a parabola intersecting the x-axis at the points 0 and 1 and achieving its maximum $\frac{R}{4}$ at $\frac{1}{2}$. It intersects the diagonal $y = x$ in a unique point **P** in the interior of the unit square. The abscissa p of **P** satisfies $p' = p$ and corresponds therefore to a further equilibrium $p = \frac{R-1}{R}$ (see Fig. 6.2).

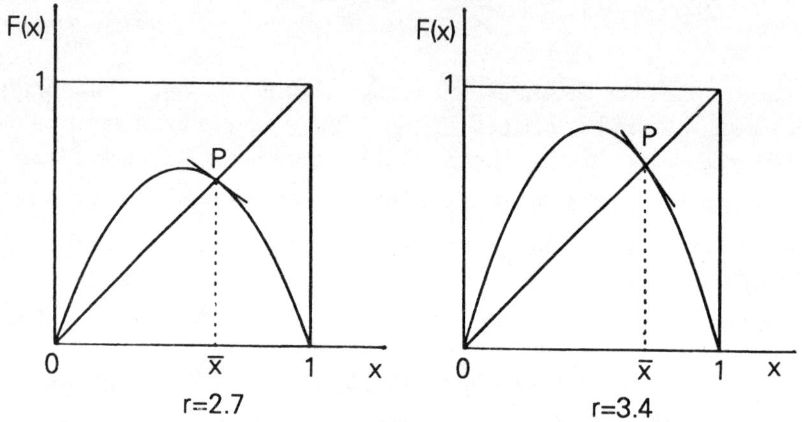

Figure 6.2

What about the stability of p? From the mean value theorem, we know that

$$F(x)-p = F(x)-F(p) = (x-p)\frac{d}{dx}F(c)$$

holds for some suitable c between x and p. If $|\frac{d}{dx}F(p)|<1$ then $|\frac{d}{dx}F(c)|<1$ whenever x (and hence c) is sufficiently close to p. Then

$$|F(x)-p|<|x-p|$$

i.e. x' is closer than x to the fixed point p. This implies that the orbit of x converges to p if x is in some suitable neighbourhood of p. The equilibrium p is *asymptotically stable* in this sense. If $|\frac{d}{dx}F(p)|>1$, however, then

$$|F(x)-p|>|x-p|$$

i.e. the orbit of x is driven away from the equilibrium. In such a case p is *unstable.*

Now $F(x) = Rx(1-x)$, hence $\frac{d}{dx}F(p) = 2-R$, i.e. p is asymptotically stable for $1 < R < 3$, but not for $3 < R < 4$. What happens then?

Exercise 1: Show that for $1 < R \leq 3$, p is even a global attractor, i.e. all orbits in (0,1) converge to p. For $R \leq 2$ this convergence is ultimately monotone, for $2 < R \leq 3$ the orbits oscillate ultimately around p.

Exercise 2: Study the difference equation $x' = F(x)$ with $F(x) = xe^{h(1-x)}$.

6.5. Bifurcations

In two generations, x is transformed into $F(F(x)) = F^{(2)}(x)$. In our case, $F^{(2)}(x)$ is a polynomial of degree 4, with a local minimum at 1/2 and two local maxima, symmetrically to the left and to the right of 1/2. (See Fig. 6.3.) Again, the diagonal and the graph of $F^{(2)}$ intersect at P. Indeed, $p = F(p) = F^{(2)}(p)$. The derivative of $F^{(2)}$ at p is given by the chain rule

$$\frac{d}{dx}F^{(2)}(p) = \frac{dF}{dx}(p)\cdot\frac{dF}{dx}(p) = (2-R)^2$$

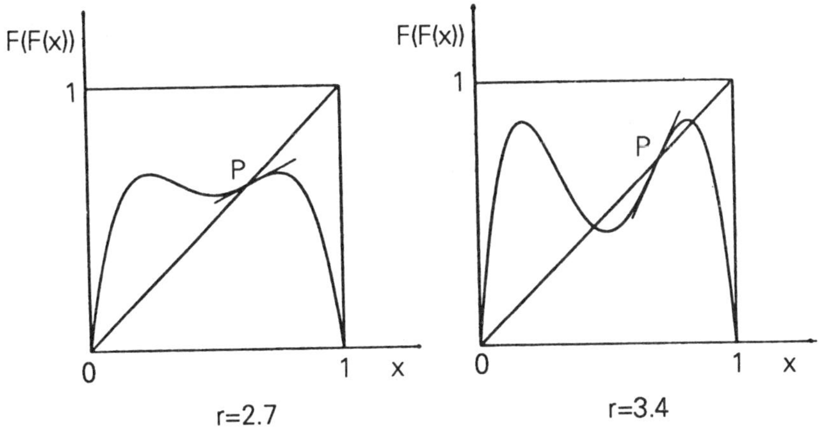

Figure 6.3

If $1<R<3$, i.e. if p is asymptotically stable, P is the only intersection of the diagonal with the graph of $F^{(2)}$; but for $3<R<4$ one obtains to the left and to the right an additional intersection - since the slope of the tangent at P is larger than 1. These two intersections correspond to points p_1 and p_2 with period 2: indeed, since $F^{(2)}(p_1) = p_1$ and $F^{(2)}(p_2) = p_2$, one must have $F(p_1) = p_2$ and $F(p_2) = p_1$.

Quite generally, a point $\mathbf{x}$ is said to be a *periodic point* for the dynamical system $T{:}X \to X$ if there exists a $k>1$ such that $T^k\mathbf{x} = \mathbf{x}$ (but $T^j\mathbf{x} \neq \mathbf{x}$ for $j = 1,..,k-1$). The integer k is called the *period* of $\mathbf{x}$.

The occurrence of periodic oscillations in our model of population growth is not too surprising. In contrast to the continuous adjustment of the rate of growth in the logistic equation (6.6), the regulation in the discrete model (6.8) operates with a delay of one generation. This delay can lead to "overshooting", i.e. to jumping from one side of the equilibrium to the other one, and to oscillations which refuse to settle down. Similar effects plague many steering mechanisms and control devices.

The parameter value $R = 3$ is a "bifurcation point": for slightly smaller values of R, the fixed point p is asymptotically stable and no periodic oscillations take place; for slightly larger values of R, p is no longer stable and periodic points appear. Thus the behaviour of the dynamical system (6.9) changes drastically if the parameter R (under some exterior influence, say) crosses the level 3.

One can show that if R is not much larger than 3, the points p_1 and p_2 are asymptotically stable as fixed points of $x \to F^{(2)}(x)$. For larger values of R, they become unstable. More bifurcations occur: the number of periodic points grows. The periods, at first, are all of the form 2^n. For $R = 3{,}5700$ there are already infinitely many periodic points. For $R = 3{,}6786$ the first odd periods appear, from $R = 3{,}82...$ onwards, all periods occur. The structure of the orbits becomes extremely complicated, and the system behaves "chaotically" in a way we shall briefly describe by an example.

Exercise: Show a similar phenomenon for the difference equation from exercise 6.4.2.

6.6. Chaotic motion

Let us consider (6.9) with $R = 4$. The maximum of the parabola reaches 1. The intervals $I_0 = [0,\frac{1}{2}]$ and $I_1 = [\frac{1}{2},1]$ are both mapped by F in a bijective manner onto the entire interval $[0,1]$. I_0 can be decomposed into two compact subintervals $I_{00} = [0,q]$ and $I_{01} = [q,\frac{1}{2}]$ which are mapped into I_0 and I_1, respectively (here $q = \frac{1}{2}(1-\frac{1}{\sqrt{2}})$). Similarly, I_1 can be decomposed into $I_{10} = [1-q,1]$ and $I_{11} = [\frac{1}{2},1-q]$: the first interval is mapped into I_0, the second one into I_1 (see Fig. 6.4).

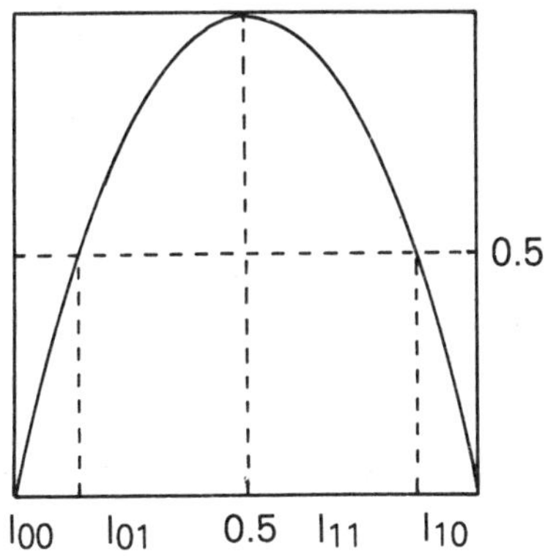

Figure 6.4

To recapitulate: $[0,1]$ is decomposed into two compact intervals of rank 1, namely I_0 and I_1, each of which is mapped by F in a bijective way onto $[0,1]$. I_0 and I_1, in turn, are decomposed into compact intervals of rank 2, namely I_{00}, I_{01}, I_{10} and I_{11}, each of which is mapped by $F^{(2)}$ in a bijective way onto $[0,1]$. Each of these intervals, in turn, can be decomposed into two compact subintervals of rank 3; one of them is mapped by $F^{(2)}$ onto I_0, the other one onto I_1.

We can repeat this procedure inductively, and obtain 2^n compact intervals of rank n. Each of them gets mapped by $F^{(n)}$ in a bijective way onto $[0,1]$, and hence may be decomposed into one subinterval mapped onto I_0 and one mapped onto I_1. These are the subintervals of rank $n+1$.

The notation for these intervals is suggestive of the *binary expansion*. If $x \in I_{0100}$, for example, then $x \in I_0$, $F(x) \in I_1$, $F^{(2)}(x) \in I_0$, $F^{(3)}(x) \in I_0$.

Hence, if we are given any finite, or even infinite, sequence of zeros and ones - obtained, for example, by repeatedly tossing a coin - we can find a point which visits I_0 and I_1 in the corresponding sequence. On the other hand, even if two points are extremely close, they will lie eventually in different subintervals of rank n, and hence have quite uncorrelated fates in store. Since we know the initial value x only up to some error, we cannot predict its orbit too far into the future. (In fact, even if we knew it precisely, our computer would introduce some round-off error.) This "randomness" of (6.9) is not restricted to $R = 4$. It holds for many other parameter values as well.

What is the lesson, then, of this "morality play"? Essentially that "computable" is not "predictable", and that deterministic motion may be undistinguishable from a random one in the long run. Even an innocent looking recurrence relation may display bifurcations and chaotic motion.

Another lesson is that the corresponding model for continuous generations may be quite tame, compared with its discrete counterpart. This is, to be sure, not always so, and continuous dynamics can offer all kinds of exciting behaviour. Still, if one looks for regularity, it often pays to use as model a differential rather than a difference equation.

Exercise: What is the maximal length of a binary interval of rank n?

6.7. Notes

The logistic differential equation was first investigated by Verhulst (1845). It plays a central role in population dynamics, but also in chemical kinetics and many other fields. For a discussion of r- and K-selection we refer to MacArthur and Wilson (1967). The difference equation (6.9) has become a favourite with mathematicians: we refer to Sharkovski (1964), Smale and Williams (1976), Guckenheimer (1977), Block et al. (1980) and Feigenbaum (1978), and recommend Devaney (1986) for an introduction to chaos. More biologically oriented treatments, as well as alternative models for one-population growth, can be found in May (1976) and May and Oster (1976).

7. Lotka-Volterra Equations for Predator-Prey Systems

7.1. A predator-prey equation

In the years after the First World War, the amount of predatory fishes in the Adriatic was found to be considerably higher than in the years before. The hostilities between Austria and Italy had disrupted fishery to a great extent, of course, but why was this more favorable to predators than to their prey? When this question was posed to the famous mathematician Volterra, he did what he had to do, denoted by x the density of the prey fishes, by y that of the predators, and came up with a differential equation. Volterra assumed that the rate of growth of the prey population, in the absence of predators, is given by some constant a, but decreases linearly as a function of the density y of predators. This leads to

$$\frac{\dot{x}}{x} = a - by \qquad (a,b>0) \ .$$

In the absence of prey, the predatory fishes would have to die, which means a negative rate of growth; but this rate picks up with the density x of prey fishes, hence

$$\frac{\dot{y}}{y} = -c + dx \qquad (c,d>0) \ .$$

Together, this yields

$$\begin{aligned} \dot{x} &= x(a-by) \\ \dot{y} &= y(-c+dx) \ . \end{aligned} \tag{7.1}$$

This is a differential equation. What can one do with it?

7.2. Solutions of differential equations

We shall write

$$\dot{\mathbf{x}} = \mathbf{f}(t,\mathbf{x}) \tag{7.2}$$

for the *ordinary differential equation* (or ODE)

$$\dot{x}_i = f_i(t,x_1,\ldots,x_n) \qquad i = 1,..,n \ . \tag{7.3}$$

Here, the functions f_i are defined on some open subset of $\mathbf{R}^{n+1}$ and continuously differentiable in all variables. A *solution* is a map

$$t \to \mathbf{x}(t) = (x_1(t),\ldots,x_n(t)) \tag{7.4}$$

from some interval I into $\mathbf{R}^n$ such that the components $x_i(t)$ are differentiable and satisfy (for all $t \in I$):

$$\dot{x}_i(t) = f_i(t,x_1(t),\ldots,x_n(t)) \qquad i = 1,\ldots,n \ .$$

We picture this in the following way: at every time t there is attached to every point $\mathbf{x}$ in $\mathbf{R}^n$ (or in some open subset of $\mathbf{R}^n$, the *domain of definition* of the differential equation) an n-dimensional vector $\mathbf{f}(t,\mathbf{x})$ whose components are the $f_i(t,x_1,\ldots,x_n)$. This vector may be viewed as "velocity of the wind" at the point $\mathbf{x}$ and at time t. A path $t \to \mathbf{x}(t)$ is the description of the motion of a "particle" in n-space: both position $\mathbf{x}(t) = (x_1(t),\ldots,x_n(t))$ and velocity $\dot{\mathbf{x}}(t) = (\dot{x}_1(t),\ldots,\dot{x}_n(t))$ at time t are

n-dimensional vectors. A solution of the differential equation (7.2) is a path whose velocity $\dot{\mathbf{x}}(t)$ coincides at every instant with the "velocity of the wind" $\mathbf{f}(t,\mathbf{x}(t))$ at the point $\mathbf{x}(t)$.

An *initial condition* is the specification of the position of the particle at some given time. One can find in every textbook on differential equations the *theorem on the existence and uniqueness* of solutions, which states that *for every initial condition, there exists a unique solution to the differential equation* (7.2). Our initial condition will always be that the position at time 0 is at some point $\mathbf{x}$: the corresponding solution will usually be denoted by $\mathbf{x}(t)$. With this convention, then, $\mathbf{x}(0) = \mathbf{x}$. For fixed t, $\mathbf{x}(t)$ depends continuously (and even differentiably) on the initial point $\mathbf{x}$.

Let us note that a solution need not be defined for all times. There certainly exists a maximal open interval (a,b) such that the solution $\mathbf{x}(t)$ is defined for all $t \in (a,b)$, but (a,b) is not necessarily $(-\infty,+\infty)$. A case in point is the equation $\dot{x} = 1+x^2$ in $\mathbf{R}$. The function $x(t) = \tan t$ is a solution satisfying the initial condition $x(0) = 0$, but it is only defined (as a solution) for $t \in (-\frac{\pi}{2},\frac{\pi}{2})$. In a way, the solution $x(t)$ picks up speed so quickly (i.e. travels so rapidly through the real line $\mathbf{R}$), that it explodes at time $\frac{\pi}{2}$ to infinity.

Of particular interest are the *time-independent* differential equations

$$\dot{\mathbf{x}} = \mathbf{f}(\mathbf{x}) \quad \text{or} \quad \dot{x}_i = f_i(x_1,\ldots,x_n) \quad . \tag{7.5}$$

The right-hand side, here, does not depend on t, i.e. the "velocity of the wind" at $\mathbf{x}$ does not change with time. If we let two "particles" start at point $\mathbf{x}$, one of them T time units after the other one, then both will travel through the same trajectory, one of them always T time units late. In other words, $\mathbf{x}(T) = \mathbf{y}$ implies $\mathbf{x}(T+t) = \mathbf{y}(t)$ for all t for which the solutions exist.

In an important case the solution is defined for all times t: if (a,b) is the maximal open interval for which the solution $\mathbf{x}(t)$ exists, and if there is some compact set K in the domain of definition of $\mathbf{f}$ which $\mathbf{x}(t)$ does not leave, then $(a,b) = (-\infty,+\infty)$.

If there exists a set X in the domain of definition of $\mathbf{f}$ such that for all $\mathbf{x} \in X$ and all $t \in \mathbf{R}$, the solution $\mathbf{x}(t)$ is defined and lies in X, then the differential equation (7.5) determines a *continuous dynamical system* on X. To every $\mathbf{x} \in X$ corresponds the *orbit* $\{\mathbf{x}(t): t \in \mathbf{R}\}$. (In some cases one considers "semi"-dynamical systems defined only for $t \in \mathbf{R}^+$).

Three types of solutions $\mathbf{x}(t)$ can occur:
(A) if $\mathbf{x}(t) \equiv \mathbf{x}$ for all $t \in \mathbf{R}$, i.e. if $\mathbf{x}(t)$ is a constant, then $\mathbf{x}$ is called an *equilibrium*. These points are characterized by $\mathbf{f}(\mathbf{x}) = \mathbf{0}$. They are also called rest points, or fixed points, or steady or stationary states. If one starts at such a point, one remains there forever.
(B) If $\mathbf{x}(T) = \mathbf{x}$ for some $T>0$, but $\mathbf{x}(t) \neq \mathbf{x}$ for all $t \in (0,T)$, then $\mathbf{x}$ is called a *periodic point* and T is called the *period*. All other points on the orbit are periodic with period T. The motion describes an endless periodic oscillation. Topologically, i.e. up to a continuous transformation, the orbit looks like a circle and the solution travels round and round.
(C) If $t \to \mathbf{x}(t)$ is injective, then the orbit never intersects itself. The orbit may be bent, knotted and twisted, but topologically it looks like a line.

7.3. Analysis of the Lotka-Volterra predator-prey equation

Let us now return to the differential equation (7.1). We may write down three solutions immediately:
(i) $x(t) = y(t) = 0$
(ii) $x(t) = 0 \quad y(t) = y(0)e^{-ct}$ (for any $y(0)>0$);
(iii) $y(t) = 0 \quad x(t) = x(0)e^{at}$ (for any $x(0)>0$).
This means that if the density of predator or prey is zero at some given time, then it is always zero. In the absence of prey, predators will become extinct ($y(t)$ converges to 0, for $t \to +\infty$). In the absence of predators, the prey population explodes ($x(t) \to +\infty$). This last feature is somewhat silly, and will be amended in subsequent models.

To the three solutions (i), (ii) and (iii) correspond three orbits: (i) the origin (0,0), which is a rest point; (ii) the positive y-axis; (iii) the positive x-axis. Together, these three orbits form the boundary of the positive orthant

$$\mathbf{R}^2_+ = \{(x,y) \in \mathbf{R}^2 : x \geq 0, y \geq 0\} . \tag{7.6}$$

Since population densities have to be nonnegative, we shall only consider the restriction of (7.1) to $\mathbf{R}^2_+$. This set is *invariant* in the sense that any solution which starts in it remains there for all (positive and negative) time for which it is defined. Indeed, as we have seen, the boundary bd $\mathbf{R}^2_+$ is invariant. Since no orbit can cross another, the interior

$$\text{int } \mathbf{R}^2_+ = \{(x,y) \in \mathbf{R}^2 : x>0, y>0\}$$

is also invariant.

There is a unique equilibrium in int $\mathbf{R}^2_+$. Indeed, such an equilibrium $\mathbf{F} = (\bar{x},\bar{y})$ must satisfy $\bar{x}(a-b\bar{y}) = 0$ and $\bar{y}(-c+d\bar{x}) = 0$. Since $\bar{x}>0$ and $\bar{y}>0$, this implies

$$\bar{x} = \frac{c}{d} \quad \bar{y} = \frac{a}{b} . \tag{7.7}$$

The signs of $\dot{x}$ and $\dot{y}$ depend on whether y is larger or smaller than $\bar{y}$, and x larger or smaller than $\bar{x}$. Thus int $\mathbf{R}^2_+$ is divided into four regions I, II, III, IV (see Fig. 7.1). As we shall presently see, $\mathbf{F}$ is surrounded by periodic orbits which travel from I to II, from II to III, etc. ... in a counter-clockwise rotation.

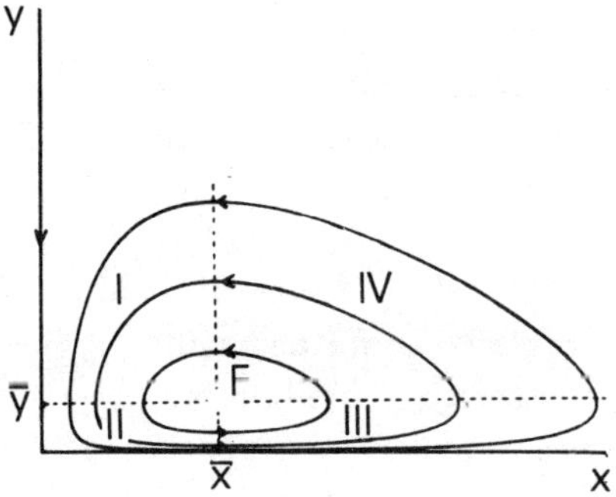

Figure 7.1

Indeed, if we multiply the equation $\frac{\dot{x}}{x} = a - by$ by $c-dx$, and the equation $\frac{\dot{y}}{y} = -c+dx$ by $a-by$, and add, then we obtain

$$(\frac{c}{x}-d)\dot{x} + (\frac{a}{y}-b)\dot{y} = 0$$

or

$$\frac{d}{dt}[c\log x - dx + a\log y - by] = 0 . \tag{7.8}$$

We shall write this in a slightly different way. With

$$H(x) = \bar{x}\log x - x \quad G(y) = \bar{y}\log y - y \tag{7.9}$$

and

$$V(x,y) = dH(x)+bG(y) , \tag{7.10}$$

(7.8) turns into

$$\frac{d}{dt}V(x(t),y(t)) = 0 \tag{7.11}$$

or

$$V(x(t),y(t)) = const \quad . \tag{7.12}$$

The function V, which is defined in int $\mathbf{R}^2_+$, remains constant along the orbits of (7.1): it is a so-called *constant of motion.*

Since $H(x)$ satisfies

$$\frac{dH}{dx} = \frac{\bar{x}}{x}-1 \quad \frac{d^2H}{dx^2} = -\frac{\bar{x}}{x^2} < 0$$

it attains its maximum at $x = \bar{x}$; $G(y)$ takes its maximum at $y = \bar{y}$. Thus $V(x,y)$ has its unique maximum at the equilibrium $\mathbf{F} = (\bar{x},\bar{y})$: it decreases along every half-line issuing from $\mathbf{F}$. The constant level sets $\{(x,y)\in\text{int } \mathbf{R}^2_+ : V(x,y) = const\}$ are closed curves around $\mathbf{F}$. (We may interpret $V(x,y)$ as the "height" at the point (x,y): then $\mathbf{F}$ is the unique summit of this landscape). The solutions have to remain on the constant level sets, and thus return to their starting point. The orbits, therefore, are periodic.

7.4. Volterra's principle

The densities of predator and prey will oscillate periodically, with both the amplitude and frequency of the oscillations depending on the initial conditions. The *time averages* of the densities, however, will remain constant, and in fact equal to the corresponding equilibrium values: that is,

$$\frac{1}{T}\int_0^T x(t)\,dt = \bar{x}, \quad \frac{1}{T}\int_0^T y(t)\,dt = \bar{y} , \tag{7.13}$$

where T is the period of the solution. Indeed, from

$$\frac{d}{dt}(\log x) = \frac{\dot{x}}{x} = a-by$$

it follows by integration that

$$\int_0^T \frac{d}{dt} \log x(t)\,dt = \int_0^T (a - by(t))\,dt$$

i.e.

$$\log x(T) - \log x(0) = aT - b\int_0^T y(t)\,dt\,. \tag{7.14}$$

Since $x(T) = x(0)$, this implies that

$$\frac{1}{T}\int_0^T y(t)\,dt = \frac{a}{b} = \bar{y}\,,$$

and an analogous result holds for the x-averages.

If the two populations in model (7.1) are not in equilibrium, the values $\bar{x}$ and $\bar{y}$ are nevertheless important as time-averages.

We are ready now for Volterra's explanation of the increase of predatory fishes during the war. Fishing reduces the rate of increase of the prey (instead of a, we now have some smaller value $a-k$) and it augments the rate of decrease of the predators (instead of c, we get, some larger value $c+m$). However, the interaction constants b and d do not change. The average density of predators is now $\frac{a-k}{b}$ and hence smaller, that of the prey $\frac{c+m}{d}$ and hence larger than in the unperturbed state. A stoppage of fishing leads to an increase of predators and a decrease of prey.

Volterra's principle remains valid for much more realistic equations than (7.1). It shows (among other things) that insecticides are of dubious value. Almost all of these chemicals are unspecific, and hence not only deleterious to the insect "pest" (plant feeders like aphids, caterpillars or weevils), but also to their natural enemies like ladybirds or avian predators. The result is quite often an increase of the pest and a decrease of the birds.

7.5. The predator prey equation with intraspecific competition

We have seen that the differential equation (7.1) displays the rather unrealistic feature that in the absence of predators, the prey population is subject to exponential growth: $\dot{x} = ax$. This is easily repaired by taking competition within the prey species into account and assuming logistic growth: $\dot{x} = x(a - ex)$. If we wish we may also allow for competition within the predators. (This is less crucial, however, as their population does not explode anyway.)

In lieu of (7.1), we obtain

$$\begin{aligned} \dot{x} &= x(a-ex-by) \\ \dot{y} &= y(-c+dx-fy) \end{aligned} \tag{7.15}$$

with $e>0$ and $f\geq 0$. Again, $\mathbf{R}^2_+$ is invariant. Its boundary consists of five orbits: the two equilibria $\mathbf{0} = (0,0)$ and $\mathbf{P} = (\frac{a}{e},0)$, the two intervals $(0,\frac{a}{c})$ and $(\frac{a}{c},+\infty)$ of the x-axis and the positive y-axis.

In order to get some rough feeling about what happens in int $\mathbf{R}^2_+$, i.e. in the presence of both populations, we shall have a look at the *isoclines*. The x-isocline is the set where $\dot{x} = 0$, i.e. where the vector field is vertical: in int $\mathbf{R}^2_+$, this is the set where

$$ex+by = a\ . \tag{7.16}$$

Similarly, the y-isocline, where the vector field is horizontal, is the set where

$$dx-fy = c\ . \tag{7.17}$$

Depending on the parameters, these lines may or may not intersect in int $\mathbf{R}^2_+$. If they don't, they divide int $\mathbf{R}^2_+$ into three regions I, II, III (see Fig. 7.2). In I, since $\dot{x}<0$, the vector field points to the left: so every orbit from I enters II. In region II, we still have $\dot{x}<0$, and furthermore $\dot{y}<0$: the vector field points to the left, and downwards. An orbit can either stay forever in II (then it will converge towards $\mathbf{P}$), or enter III. In region III, the direction is to the right, and downwards. It is *forward invariant*: no solution can leave it. Again, every orbit has to converge towards $\mathbf{P}$. The predators will therefore vanish; the prey density converges towards the limit $\frac{a}{e}$, which corresponds to the carrying capacity of the logistic equation $\dot{x} = x(a-ex)$ which, in the absence of predators, governs their growth.

If the isoclines intersect at some point $\mathbf{F} = (\bar{x},\bar{y})$ in int $\mathbf{R}^2_+$, this point is an equilibrium. Its coordinates solve the linear equation (7.16) - (7.17). In this case, int $\mathbf{R}^2_+$ is divided into four regions I to IV (see Fig. 7.3). The signs of $\dot{x}$ and $\dot{y}$ suggest that the orbits move counterclockwise around $\mathbf{F}$. But is this rotational motion periodic, again, or will it tend to the fixed point $\mathbf{F}$, or spiral away from it? The isoclines don't tell us enough to decide.

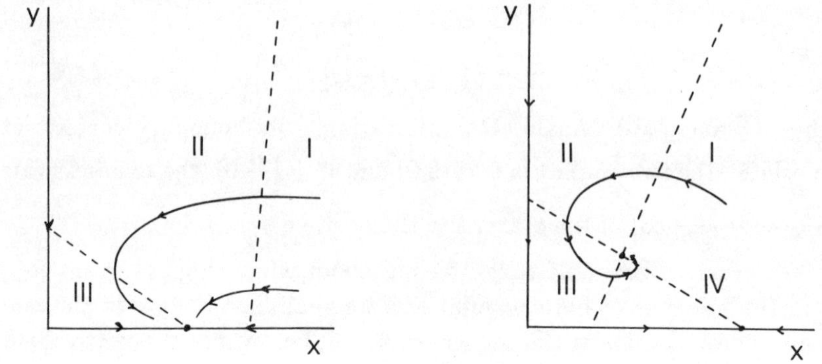

Figures 7.2 and 7.3

7.6. On ω-limits and Ljapunov functions

We know that the solutions of (7.15) exist, but we don't know how to compute them. That's the trouble with most differential equations. As in the case of discrete dynamical systems, we are left with two complementary courses: to calculate approximate orbits or to analyze the qualitative behaviour. We shall pursue the latter line, since it is the final outcome of the dynamics which interests us most.

The asymptotic features of a solution are encapsulated in its ω-limit. Let $\dot{\mathbf{x}} = \mathbf{f}(\mathbf{x})$ be a time independent ODE in some region of $\mathbf{R}^n$ and let $\mathbf{x}(t)$ be a solution defined for all $t \geq 0$ and satisfying the initial condition $\mathbf{x}(0) = \mathbf{x}$. The *$\omega$-limit* of $\mathbf{x}$ is the set of all accumulation points of $\mathbf{x}(t)$, for $t \to +\infty$:

$$\omega(\mathbf{x}) = \{\mathbf{y} \in \mathbf{R}^n : \mathbf{x}(t_k) \to \mathbf{y} \text{ for some sequence } t_k \to +\infty\} \quad . \tag{7.18}$$

The points in the ω-limit have the property that all of their neighbourhoods keep getting visited by the solution $\mathbf{x}(t)$, even after an arbitrarily long time (α-limits are defined in the same way, but for $t_k \to -\infty$).

The ω-limit of a point $\mathbf{x}$ may be empty. This is the case for the equation $\dot{x} = 1$ on $\mathbf{R}^1$, for instance. The solution $x(t) = x + t$ travels along the line at constant speed and never returns. But if the orbit, or at least the positive semiorbit $\{\mathbf{x}(t) : t \geq 0\}$, remains in some compact set K, then every sequence $\mathbf{x}(t_k)$ must admit accumulation points, and $\omega(\mathbf{x})$ can't be empty. Any point $\mathbf{z}$ on the orbit of $\mathbf{x}$ has the same ω-limit: indeed, in that case $\mathbf{z} = \mathbf{x}(T)$ for some time T, and hence $\mathbf{z}(t-T) = \mathbf{x}(t)$

for all t. If $\mathbf{x}(t_k)$ converges to $\mathbf{y}$, then so does $\mathbf{z}(t_k - T)$.

It is easy to see, furthermore, that $\omega(\mathbf{x})$ is *closed*, being a set of accumulation points: we can represent $\omega(\mathbf{x})$ as an intersection of closed sets:

$$\omega(\mathbf{x}) = \bigcap_{t \geq 0} \overline{\{\mathbf{x}(s): s \geq t\}} \, .$$

The set $\omega(\mathbf{x})$, finally, is *invariant*. Indeed let $\mathbf{y}$ be in $\omega(\mathbf{x})$, and t arbitrary. Since $\mathbf{x}(t_k) \to \mathbf{y}$ for some sequence t_k, and since solutions are continuous functions of their initial data, one has $\mathbf{x}(t_k + t) \to \mathbf{y}(t)$. Hence $\mathbf{y}(t)$ belongs to $\omega(\mathbf{x})$.

Equilibria and periodic orbits constitute their own ω-limit. If $\omega(\mathbf{x})$ is compact, it is also connected, in the sense that any pair $\mathbf{y}, \mathbf{z} \in \omega(\mathbf{x})$ can be joined by a continuous path in $\omega(\mathbf{x})$.

One can get a handle on ω-limits even if one fails to know the solution. This is the gist of *Ljapunov's theorem*:

Theorem. *Let $\dot{\mathbf{x}} = \mathbf{f}(\mathbf{x})$ be a time independent ODE defined on some subset G of $\mathbf{R}^n$. Let $V: G \to \mathbf{R}$ be continuously differentiable. If for some solution $t \to \mathbf{x}(t)$, the derivative $\dot{V}$ of the map $t \to V(\mathbf{x}(t))$ satisfies the inequality $\dot{V} \geq 0$ (or $\dot{V} \leq 0$), then $\omega(\mathbf{x}) \cap G$ (and $\alpha(\mathbf{x}) \cap G$) is contained in the set $\{\mathbf{x} \in G: \dot{V}(\mathbf{x}) = 0\}$.*

Proof: If $\mathbf{y} \in \omega(\mathbf{x}) \cap G$, there is a sequence $t_k \to +\infty$ with $\mathbf{x}(t_k) \to \mathbf{y}$. Since $\dot{V} \geq 0$ along the orbit of $\mathbf{x}$, one has $\dot{V}(\mathbf{y}) \geq 0$ by continuity. Suppose that $\dot{V}(\mathbf{y}) = 0$ does not hold. Then $\dot{V}(\mathbf{y}) > 0$. Since the value of V can never decrease along an orbit, this implies

$$V(\mathbf{y}(t)) > V(\mathbf{y}) \tag{7.19}$$

for $t > 0$. The function $V(\mathbf{x}(t))$ is also monotonically increasing. Since V is continuous, $V(\mathbf{x}(t_k))$ converges to $V(\mathbf{y})$, and hence

$$V(\mathbf{x}(t)) \leq V(\mathbf{y}) \tag{7.20}$$

for every $t \in \mathbf{R}$. From $\mathbf{x}(t_k) \to \mathbf{y}$ it follows that $\mathbf{x}(t_k + t) \to \mathbf{y}(t)$ and hence

$$V(\mathbf{x}(t_k + t)) \to V(\mathbf{y}(t)) \, , \tag{7.21}$$

so that by (7.19)

$$V(\mathbf{x}(t_k + t)) > V(\mathbf{y})$$

for k sufficiently large. However, this contradicts (7.20).

This theorem does not tell how to find such a *Ljapunov function* V. There is no general recipe, in fact.

7.7. Coexistence of predators and prey

Let us now return to (7.15) and consider the case of intersecting isoclines. There exists, then, an equilibrium $\mathbf{F} = (\bar{x},\bar{y})$ in int $\mathbf{R}^2_+$. At the corresponding densities, predators and prey can coexist. But is such an equilibrium stable?

Let us try the function V defined in (7.10), to wit

$$V(x,y) = dH(x)+bG(y)$$

with

$$H(x) = \bar{x} \log x - x \quad \text{and} \quad G(y) = \bar{y} \log y - y \ .$$

The derivative of the function $t \to V(x(t),y(t))$ is

$$\begin{aligned}\dot{V}(x,y) &= \frac{\partial V}{\partial x}\dot{x}+\frac{\partial V}{\partial y}\dot{y} \\ &= d(\frac{\bar{x}}{x}-1)x(a-by-ex)+b(\frac{\bar{y}}{y}-1)y(-c+dx-fy) \ .\end{aligned}$$

Since $\bar{x}$ and $\bar{y}$ are the solutions of (7.16) and (7.17), we may replace a and c by $e\bar{x}+b\bar{y}$ and $d\bar{x}-f\bar{y}$, respectively. This yields

$$\begin{aligned}\dot{V}(x,y) &= d(\bar{x}-x)(b\bar{y}+e\bar{x}-by-ex)+b(\bar{y}-y)(-d\bar{x}+f\bar{y}+dx-fy) \\ &= de(\bar{x}-x)^2+bf(\bar{y}-y)^2 \geq 0 \ . \qquad (7.22)\end{aligned}$$

We may, therefore, apply the theorem of Ljapunov. The ω-limit of every orbit in int $\mathbf{R}^2_+$ is contained in the set $\{(x,y): \dot{V}(x,y) = 0\}$. For $f>0$, this consists only of the point $\mathbf{F}$. For $f = 0$, it is the set $K = \{(x,y)\in\mathbf{R}^2: y>0, x = \bar{x}\}$. But the ω-limit must be an invariant subset of K, and hence as before reduces to $\mathbf{F}$. Every solution in int $\mathbf{R}^2_+$ converges to this equilibrium.

Let us interpret $V(x,y)$ again as the height of a landscape at the point (x,y). While the solution paths of (7.1) follow the contour lines and remain at the same altitude, those of (7.15) ascend. A *strict Ljapunov function* – one for which $\dot{V}(\mathbf{x})>0$ whenever $\mathbf{x}$ is not an equilibrium – describes a steady uphill movement. Systems admitting such a function are said to be *gradient-like*.

An equilibrium $\mathbf{z}$ of an ODE $\dot{\mathbf{x}} = \mathbf{f}(\mathbf{x})$ is said to be *stable* if, for any neighbourhood U of $\mathbf{z}$, there exists a neighbourhood W of $\mathbf{z}$ such that any orbit through W remains forever in U (i.e. $\mathbf{x} \in W$ implies $x(t) \in U$ for all $t \geq 0$). It is said to be *asymptotically stable* if, in addition, such orbits converge to $\mathbf{z}$ (i.e.$\mathbf{x}(t) \to \mathbf{z}$ for all $\mathbf{x} \in W$). The set of points $\mathbf{x}$ with $\mathbf{x}(t) \to \mathbf{z}$, as $t \to +\infty$, is called the *basin of attraction* of $\mathbf{z}$. It is an open invariant set. If it is the whole state space – or at least its interior – then $\mathbf{z}$ is said to be *globally stable*.

The equilibrium $\mathbf{F}$ is stable for (7.1), and asymptotically stable – in fact, even globally stable – for (7.15). As the neighbourhood W, we can choose the interior of a contour line of V contained in U.

Note that asymptotic stability means much more than stability alone. A small perturbation away from the equilibrium state $\mathbf{F}$ will be promptly offset by the dynamics of (7.15), since the solution will tend back towards equilibrium: it is like the action of a spring pulling the state back. In contrast to this, the response of the dynamics of (7.1) to a perturbation away from the equilibrium is flaccid and ineffective: the state will remain on a periodic orbit and not return to equilibrium. In fact, a sequence of perturbations may send the state from periodic orbit to periodic orbit, and ultimately to bd $\mathbf{R}^2_+$, corresponding to the extinction of one population.

There is another sense in which (7.15) is more stable than (7.1). A small change in the vector field (7.15) would slightly shift the position of the equilibrium $\mathbf{F}$, but not essentially alter the behaviour of the orbits, which will still spiral towards $\mathbf{F}$. In contrast to this, the behaviour of (7.1) is radically altered by the introduction of the competition term $-ex^2$ in the equation governing the growth of the prey, no matter how small it is. Instead of undamped periodic oscillations of (7.1), one would get damped oscillations and convergence to a steady state. The dynamical system (7.1) is not *structurally stable*. In fact, its main features – existence of a constant of motion, periodicity of all orbits, an equilibrium which is stable but not asymptotically stable – are highly degenerate.

7.8. Notes

In 1924, the biologist D'Ancona introduced Volterra to ecological problems. Equation (7.1) had been studied earlier by Lotka (1920) in the context of chemical kinetics, but Volterra's discussion went considerably further. We refer to Volterra (1931), as well as to Scudo and Ziegler (1978) for a re-edition of classical papers from the "golden age" of mathematical ecology. More realistic models for predator-prey interactions are discussed in Holling (1973). The behaviour of such interactions will be investigated in Chapter 18. An excellent textbook on differential equations is by Hirsch and Smale (1978); and expositions of more advanced aspects of the theory of dynamical systems can be found in Guckenheimer and Holmes (1983) and Arnold (1983).

8. The Lotka-Volterra equation for two competing species.

8.1. Linear differential equations

Most of the differential equations in this book are nonlinear. Still, the all-important question of the stability of equilibria can in many cases be reduced to an associated linear equation. So we shall briefly describe some of the main features of *linear systems*: for a fuller treatment we refer to the textbooks.

Let A be a real $n \times n$-matrix and

$$\dot{\mathbf{x}} = A\mathbf{x} \tag{8.1}$$

the corresponding linear differential equation in $\mathbf{R}^n$. The solution $\mathbf{x}(t)$ can be written as $e^{At}\mathbf{x}(0)$ where the matrix e^{At} is given by

$$e^{At} = I + A\frac{t}{1!} + A^2\frac{t^2}{2!} + \cdots \tag{8.2}$$

The eigenvalues of A can be real or complex: in the latter case, they occur in conjugate pairs. The components $x_i(t)$ of the solutions of (8.1) are linear combinations (with constant coefficients) of the following functions:

(i) $e^{\lambda t}$, whenever λ is a real eigenvalue of A;

(ii) $e^{at}\cos bt$ and $e^{at}\sin bt$, i.e. the real and the imaginary part of $e^{\mu t}$, whenever $\mu = a + ib$ is a complex eigenvalue of A;

(iii) $t^j e^{\lambda t}$, $t^j e^{at}\cos bt$ or $t^j e^{at}\sin bt$, with $0 \le j < m$, if the eigenvalue λ or μ occurs with multiplicity m.

We note that the complex eigenvalues μ introduce an oscillatory part into the solutions. These oscillations will be damped iff (i.e. if and only if) $a < 0$.

The origin $\mathbf{0}$ is obviously an equilibrium of (8.1). It is called

(a) a *sink* if the real parts of the eigenvalues, i.e. the λ's and the a's, are all negative. In this case, $e^{\lambda t} \to 0$ and $e^{at} \to 0$ for $t \to +\infty$, and hence $\{\mathbf{0}\}$ is the ω-limit of every orbit;

(b) a *source* if the real parts of the eigenvalues are all positive. In this case, $\{\mathbf{0}\}$ is the α-limit of every orbit;
(c) a *saddle* if some eigenvalues are in the left half and some in the right half of the complex plane C, but none on the imaginary axis. The orbits whose ω-limit is $\{\mathbf{0}\}$ form a linear submanifold of $\mathbf{R}^n$ called the *stable manifold*; those whose α-limit is $\{\mathbf{0}\}$ form the *unstable manifold*; and these subspaces span $\mathbf{R}^n$.

The equilibrium $\mathbf{0}$ is called *hyperbolic* if it is a source, a saddle or a sink, i.e. if no eigenvalue of A has real part 0. Eigenvalues on the imaginary axis correspond to "degenerate" solutions: an eigenvalue 0 corresponds to a linear manifold of rest points, and a pair of purely imaginary eigenvalues $\pm ib$ to a linear manifold of periodic orbits with period $2\pi/b$. If all eigenvalues are on the imaginary axis, $\mathbf{0}$ is called a *centre*. If $\mathbf{0}$ is not hyperbolic, then an arbitrarily small perturbation of the coefficients of A can change $\mathbf{0}$ into a source, a sink or a saddle and so lead to an altogether different behaviour. On the other hand, the hyperbolic case is *structurally stable*: if the perturbation of the coefficients is sufficiently small, the orbits will exhibit an essentially unchanged behaviour.

Similar results hold for discrete linear dynamics. A linear map $\mathbf{x} \to A\mathbf{x}$ from $\mathbf{R}^n$ to $\mathbf{R}^n$ has the origin $\mathbf{0}$ as fixed point. This origin is asymptotically stable iff all eigenvalues of A have absolute value less than 1. (The necessity is obvious by considering an eigenvector $\mathbf{x}$ to the eigenvalue λ. The sufficiency follows from the representation of A in triangular form.) If one of the eigenvalues has absolute value larger than 1, then there exists an orbit going exponentially fast to infinity, and $\mathbf{0}$ is unstable.

Exercise 1: If $\mathbf{0}$ is a sink for (8.1), i.e. if all eigenvalues of A have negative real part, then A is said to be a *stable* matrix. Show that A is stable iff there exists a positive definite matrix Q such that $QA + A^tQ$ is negative definite. Geometrically this means that the quadratic form $V(\mathbf{x}) = \mathbf{x} \cdot Q\mathbf{x}$ is a Ljapunov function for (8.1).

Exercise 2: Find a similar characterization for maps.

Exercise 3: Show that a 2×2 matrix

$$A = \begin{pmatrix} a & b \\ c & d \end{pmatrix}$$

is stable iff trace $A = a + d < 0$ and $\det A = ad - bc > 0$. $\mathbf{0}$ is a saddle for (8.1) iff $\det A < 0$.

8.2. Linearization

Let us now consider the local behaviour of the solutions of

$$\dot{\mathbf{x}} = \mathbf{f}(\mathbf{x}) \tag{8.3}$$

near a point $\mathbf{z}$ in $\mathbf{R}^n$.

If $\mathbf{z}$ is not an equilibrium, there exists a neighbourhood U of $\mathbf{z}$ where the orbits can be "straightened out" by a continuous transformation, so that they turn into parallel lines. This *flow box theorem* is plausible enough, and follows from the fact that the first term of the Taylor expansion of $\mathbf{f}$ at $\mathbf{z}$ is the constant $\mathbf{f}(\mathbf{z}) \neq \mathbf{0}$.

If $\mathbf{z}$ is a rest point, however, then the local behaviour is less easy to sketch. The constant term $\mathbf{f}(\mathbf{z})$ vanishes now. The next term – the linear one – is given by the *Jacobian matrix* $D_{\mathbf{z}}\mathbf{f} = A$ of first order partial derivatives:

$$A = \begin{bmatrix} \frac{\partial f_1}{\partial x_1}(\mathbf{z}) & \cdots & \frac{\partial f_1}{\partial x_n}(\mathbf{z}) \\ \cdot & & \cdot \\ \cdot & & \cdot \\ \cdot & & \cdot \\ \frac{\partial f_n}{\partial x_1}(\mathbf{z}) & \cdots & \frac{\partial f_n}{\partial x_n}(\mathbf{z}) \end{bmatrix}$$

(Instead of $\frac{\partial f_i}{\partial x_j}$ one sometimes writes $\frac{\partial \dot{x}_i}{\partial x_j}$). The linear equation

$$\dot{\mathbf{y}} = A\mathbf{y} \tag{8.4}$$

can be solved explicitly. What has it to do with (8.3)? A lot, as long as $\mathbf{z}$ is hyperbolic in the sense that all eigenvalues of A have nonvanishing real part. This is the content of the *Theorem of Hartman and Grobman:*

Theorem. *For any hyperbolic equilibrium* $\mathbf{z}$ *of* (8.3), *there exists a neighbourhood* U *and a homeomorphism* $\mathbf{h}$ *from* U *to some neighbourhood* V *of the origin* $\mathbf{0}$ *such that* $\mathbf{y} = \mathbf{h}(\mathbf{x})$ *implies* $\mathbf{y}(t) = \mathbf{h}(\mathbf{x}(t))$ *for all* $t \in \mathbf{R}$ *with* $\mathbf{x}(t) \in U$. (Here, $\mathbf{x}(t)$ is the solution of (8.3) with $\mathbf{x}(0) = \mathbf{x}$ and $\mathbf{y}(t)$ the solution of the linear equation (8.4) with $\mathbf{y}(0) = \mathbf{y}$. A homeomorphism is a continuous map with a continuous inverse.)

Locally, then, the orbits of (8.3) near $\mathbf{z}$ look like those of (8.4) near $\mathbf{0}$. We may speak again of sinks, sources, saddles, stable and unstable manifolds, just as in the linear case. In particular, if $\mathbf{z}$ is a sink, then it is asymptotically stable. If an eigenvalue of A has strictly positive real part, then $\mathbf{z}$ is unstable.

We emphasize that the theorem of Hartman and Grobman says nothing about the nonhyperbolic case and in particular about centres (where all eigenvalues have real part 0). The local behaviour, there, depends on higher order terms of the Taylor expansion of $\mathbf{f}$.

Exercise 1: Linearize (7.15) at the interior fixed point $\mathbf{F}$.

Again, similar results hold for discrete dynamical systems. If $\mathbf{z}$ is a hyperbolic fixed point of $\mathbf{x}\to\mathbf{f}(\mathbf{x})$, i.e. if no eigenvalue of $D_{\mathbf{z}}\mathbf{f}$ has absolute value 1 or 0, then the local behaviour of $\mathbf{f}$ is completely reflected by its linearization $\mathbf{x}\to D_{\mathbf{z}}\mathbf{f}(\mathbf{x})$. There exists a homeomorphism g as above such that $g(\mathbf{f}(\mathbf{x})) = D_{\mathbf{z}}\mathbf{f}(g(\mathbf{x}))$ for all $\mathbf{x}$. In particular, if all eigenvalues of $D_{\mathbf{z}}\mathbf{f}$ are in the interior of the unit circle, then $\mathbf{z}$ is an asymptotically stable fixed point.

It is instructive to compare these stability criteria for differential and difference equations.

To a differential equation (8.3) in $\mathbf{R}^n$ corresponds in a natural way the difference equation

$$\mathbf{x}' = \mathbf{x}+h\mathbf{f}(\mathbf{x}) \tag{8.5}$$

whose increment $\mathbf{x}'-\mathbf{x}$ points into the same direction as the vector field (8.3), with h as "step length" (often $h=1$). In numerical analysis this is called the *Euler scheme*. If h is small then the orbits of (8.5) will stay close to those of (8.3), for some finite time at least.

Suppose now that $\mathbf{z}$ is a fixed point of (8.3), and hence of (8.5), and $A=D_{\mathbf{z}}\mathbf{f}$. Let us compare the local behaviour near $\mathbf{z}$ in both equations by considering their linearizations

$$\dot{\mathbf{y}} = A\mathbf{y} \tag{8.6}$$

and

$$\mathbf{y}' = \mathbf{y}+hA\mathbf{y} = (I+hA)\mathbf{y} \quad . \tag{8.7}$$

$\mathbf{0}$ is stable for (8.6) iff all eigenvalues of A have negative real part. If λ is an eigenvalue of A then $1+h\lambda$ is an eigenvalue of $I+hA$. Thus for stability of $\mathbf{0}$ in (8.7) we need

$$|1+h\lambda|<1$$

i.e. that λ lies in the interior of the circle with centre $-\dfrac{1}{h}$ and radius $\dfrac{1}{h}$. Thus the stability condition for the difference equation is more restrictive than for the differential equation. If $\mathbf{0}$ is stable for (8.6) it will be stable for (8.7) only if h is small enough. If h gets too large then **0** loses stability for discrete time. For example, if there is a real eigenvalue $\lambda<0$ then for $h>-\frac{2}{\lambda}$ overshooting will occur, which is often accompanied by period doubling for the nonlinear equation (8.5), as we observed in **6.5**.

On the other hand, if **0** is a centre for (8.6) then it will be unstable for (8.7) for *every* choice of h, although it could be asymptotically stable for (8.3).

Exercise 2: If $\mathbf{z}$ is a hyperbolic fixed point then for h small enough, the local behaviours of (8.3) and (8.5) are similar.

8.3. A competition equation

Let us return to ecology, and model the interaction of two *competing species.* If x and y denote their densities, then the rates of growth $\dot{x}/x$ and $\dot{y}/y$ will be decreasing functions of both x and y, since competition will act both within and between the species. The most simple-minded assumption would be that this decrease is linear. This leads to

$$\begin{aligned}\dot{x} &= x(a-bx-cy)\\ \dot{y} &= y(d-ex-fy)\end{aligned} \tag{8.8}$$

with positive constants a to f. Again, since the boundary of $\mathbf{R}^2_+$ is invariant, so is $\mathbf{R}^2_+$ itself. In fact, if one population is absent, the other obeys the familiar logistic growth law.

The x- and y-isoclines are given by

$$a-bx-cy = 0$$

$$d-ex-fy = 0$$

in int $\mathbf{R}^2_+$. These are straight lines with negative slopes.

Exercise 1: (a) If the isoclines coincide, then xy^{-k} (with $k=\frac{a}{d}$) is a constant of motion; this is a degenerate case.

(b) If the isoclines don't intersect in int $\mathbf{R}^2_+$, one of the species (which one?) tends to extinction. (See Fig. 8.1.) One species, in this sense, is *dominant.*

There remains the case of a unique intersection $\mathbf{F}=(\bar{x},\bar{y})$ of the isoclines in int $\mathbf{R}^2_+$, when

$$\bar{x}=\frac{af-cd}{bf-ce} \qquad \bar{y}=\frac{bd-ae}{bf-ce} \quad . \tag{8.9}$$

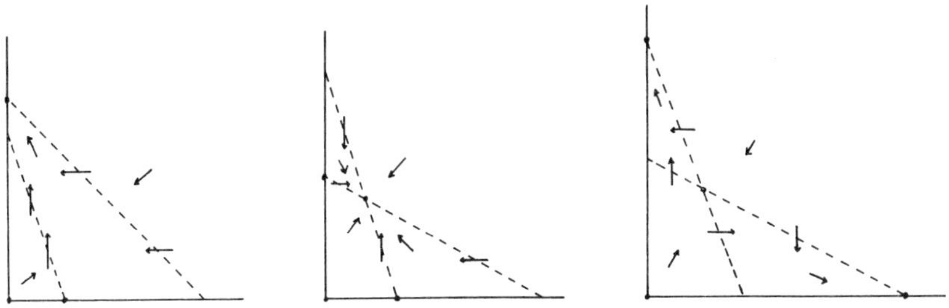

Figures 8.1, 8.2 and 8.3

The Jacobian of (8.8) at **F** is

$$A = \begin{bmatrix} -b\overline{x} & -c\overline{x} \\ -e\overline{y} & -f\overline{y} \end{bmatrix} . \tag{8.10}$$

We have to distinguish two situations:
(a) If $bf > ce$, the denominator in (8.6) is positive. This implies $af - cd > 0$, $bd - ae > 0$ and hence

$$\frac{b}{e} > \frac{a}{d} > \frac{c}{f} . \tag{8.11}$$

From the signs of $\dot{x}$ and $\dot{y}$ in the regions I, II, III, IV (see Fig. 8.2) we infer that every orbit in int $\mathbf{R}^2_+$ converges to **F**. This agrees with the fact that the eigenvalues of (8.10) are negative and **F**, consequently, is a sink. This is the case of *stable coexistence.*
(b) otherwise

$$\frac{c}{f} > \frac{a}{d} > \frac{b}{e} . \tag{8.12}$$

As seen from Fig. 8.3, all orbits in region I converge to the y-axis and all those of region III to the x-axis. Since $\det A = \overline{x}\overline{y}(bf - ce) < 0$, **F** is a saddle. Its stable manifold consists of two orbits converging to **F**. One of them must lie in region II, the other one in region IV. Together, they divide int $\mathbf{R}^2_+$ into two basins of attraction. All orbits from one basin converge to $\mathbf{F}_2 = (0,\frac{d}{f})$, all those from the other one to $\mathbf{F}_1 = (\frac{a}{e},0)$. This means that – depending on the initial condition – one or the other species gets eliminated. This is the so-called *bistable case.*

Exercise 2: Bring (8.8) into the form

$$\dot{x}_1 = x_1 r_1 \left(1 - \frac{x_1}{K_1} - \alpha_{12}\frac{x_2}{K_2}\right)$$

$$\dot{x}_2 = x_2 r_2 \left(1 - \alpha_{21}\frac{x_1}{K_1} - \frac{x_2}{K_2}\right)$$

and discuss it in terms of the carrying capacities K_1 and K_2 and the interaction coefficients α_{12} and α_{21}.

Exercise 3: Show that in the case (a) of stable coexistence, the interior equilibrium $\mathbf{F}$ lies above the line connecting the two one-species equilibria $\mathbf{F}_1$ and $\mathbf{F}_2$, and in the bistable case (b) below.

Exercise 4: Show that the quadratic form

$$Q(x,y) = be(x-\bar{x})^2 + 2ce(x-\bar{x})(y-\bar{y}) + cf(y-\bar{y})^2$$

is a Ljapunov function for the competition equation (8.8).

Exercise 5: Show that if two competing populations depend on one resource, one of them will vanish. (Assume that the resource R is given by $\bar{R} - c_1 x - c_2 y$ ($\bar{R}, c_1, c_2 > 0$) and $\frac{\dot{x}}{x} = b_1 R - \alpha_1$, $\frac{\dot{y}}{y} = b_2 R - \alpha_2$.) Show that coexistence is possible if the two species compete for two distinct resources.

Exercise 6: Set up and discuss a model for two symbiotic species. Show that it has unbounded solutions if det $A<0$ and a globally stable rest point if det $A > 0$.

8.4. Notes

For a proof of the theorem on linearization, we refer to Hartman (1964). The basic papers of Volterra on competition have been translated and published in Scudo and Ziegler (1978). More general models of competition between two species will be discussed in Chapter 18.

9. Lotka-Volterra Equations for More Than Two Populations

9.1. The general Lotka-Volterra equation

The general Lotka-Volterra equation for n populations is of the form

$$\dot{x}_i = x_i(r_i + \sum_{j=1}^{n} a_{ij}x_j) \quad i=1,..,n \; . \tag{9.1}$$

The x_i denote the densities; the r_i are intrinsic growth (or decay) rates, and the a_{ij} describe the effect of the j-th upon the i-th population, which is positive if it enhances and negative if it inhibits the growth. All sorts of interactions can be modelled in this way, as long as one is prepared to assume that the influence of every species upon the growth rates is linear (more general models will be considered later). The matrix $A=(a_{ij})$ is called the *interaction matrix.*

The state space is, of course, the positive orthant

$$\mathbf{R}^n_+ = \{\mathbf{x}=(x_1,\ldots,x_n)\in\mathbf{R}^n: x_i\geq 0 \text{ for } i=1,\ldots,n\} \; .$$

The boundary points of $\mathbf{R}^n_+$ lie on the coordinate planes $x_i=0$, which correspond to the states where species i is absent. These "faces" are invariant, since $x_i(t)\equiv 0$ is the unique solution of the i-th equation of (9.1) satisfying $x_i(0)=0$. In such a model, a missing species cannot "immigrate". Thus the boundary bd $\mathbf{R}^n_+$, and consequently $\mathbf{R}^n_+$ itself, are invariant under (9.1). So is the interior int $\mathbf{R}^n_+$, which means that if $x_i(0)>0$ then $x_i(t)>0$ for all t. The density $x_i(t)$ may approach 0, however, which means extinction.

The ecological equations of the last chapter were examples of (9.1) with $n=2$. We shall see in Chapter 18 that all possible two dimensional cases can be classified. In higher dimensions, many open questions remain. In particular, numerical simulation shows that even the case of 3 populations may lead to some kind of *chaotic motion* (see Fig. 9.1): the asymptotic behaviour of the solutions consists of highly irregular oscillations and depends in a very sensitive way upon the initial conditions. The long term outcome, in such a case, is unpredictable.

One is far from an understanding of this type of *chaos*, which in its erratic nature resembles that of the discrete system $x\rightarrow 4x(1-x)$ described in **6.6**. In spite of the fact that the dynamics is deterministic, the orbits look like the results of some random motion. Actually the trajectory of a ball in a roulette wheel is also completely specified by the initial condition - the throw - but cannot be controlled, since the smallest

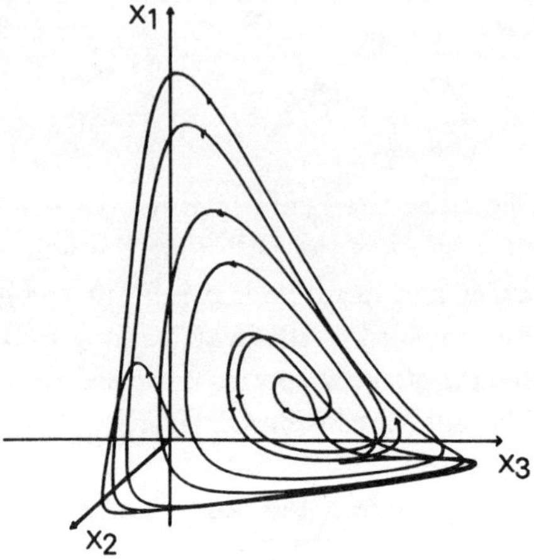

Figure 9.1

deviation leads to a different final outcome. One has to fall back upon probabilistic techniques and statements about mean values. These, however, are sufficiently precise to guarantee casino owners a comfortable income.

In this chapter, we shall describe a few general results about (9.1) and then turn to some special cases of biological interest.

9.2. Interior equilibria

The rest points of (9.1) in int $\mathbf{R}^n_+$ are the solutions of the linear equations

$$r_i + \sum_{j=1}^{n} a_{ij} x_j = 0 \quad i = 1, \ldots, n \tag{9.2}$$

whose components are positive. (The equilibria on the boundary faces of $\mathbf{R}^n_+$ can be found in a similar way: one has only to note that the restriction of (9.1) to any such face is again of Lotka-Volterra type.)

Theorem 1. Int $\mathbf{R}^n_+$ *contains* α- *or* ω- *limit points if and only if* (9.1) *admits an interior equilibrium.*

One direction of this proposition is trivial. A rest point coincides with its own α- and ω- limit. It is the converse which is of interest, since it is (in principle) not hard to check if (9.2) admits positive solutions. If it does not, then every orbit has to converge to the boundary, or to infinity. In particular, if int $\mathbf{R}^n_+$ contains a periodic orbit, it must also contain a rest point.

In order to prove the converse, let $L:\mathbf{x}\rightarrow\mathbf{y}$ be defined by

$$y_i = r_i + \sum_{j=1}^{n} a_{ij}x_j \quad i = 1,\ldots,n \ .$$

If (9.1) admits no interior equilibrium, the set $K = L(\text{int } \mathbf{R}^n_+)$ is disjoint from $\mathbf{0}$. A well known theorem from convex analysis implies that there exists a hyperplane H through $\mathbf{0}$ which is disjoint from the convex set K. Thus there exists a vector $\mathbf{c} = (c_1,\ldots,c_n) \neq \mathbf{0}$ which is orthogonal to H ($\mathbf{c}\cdot\mathbf{x} = 0$ for all $\mathbf{x}\in H$) such that $\mathbf{c}\cdot\mathbf{y}$ is positive for all $\mathbf{y}\in K$. Setting

$$V(\mathbf{x}) = \sum c_i \log x_i \ , \tag{9.3}$$

we see that V is defined on int $\mathbf{R}^n_+$. If $\mathbf{x}(t)$ is a solution of (9.1) in int $\mathbf{R}^n_+$, then the time derivative of $t\rightarrow V(\mathbf{x}(t))$ satisfies

$$\dot{V} = \sum c_i \frac{\dot{x}_i}{x_i} = \sum c_i y_i = \mathbf{c}\cdot\mathbf{y} > 0 \ . \tag{9.4}$$

Thus V is increasing along each orbit. But then no point $\mathbf{z}\in$int $\mathbf{R}^n_+$ may belong to its ω-limit: indeed, by Ljapunov's theorem from **7.6**, the derivative $\dot{V}$ would have to vanish there. This contradiction completes the proof. It also shows:

Corollary: *If* (9.1) *admits no interior equilibrium, then it is gradient-like in* int $\mathbf{R}^n_+$.

In general (9.2) will admit one solution in int $\mathbf{R}^n_+$, or none at all. It is only in the "degenerate" case $\det A = 0$ that (9.2) can have more than one solution: these will form a continuum of rest points.

Exercise 1: Construct an invariant of motion in the case of a continuum of fixed points. (Hint: Try (9.3) for suitable $\mathbf{c}$).

If there exists a unique interior equilibrium $\mathbf{p}$, and if the solution $\mathbf{x}(t)$ converges neither to the boundary nor to infinity, then its time average converges to $\mathbf{p}$.

Theorem 2. *If there exist positive constants a and A such that $a < x_i(t) < A$ for all i and all $t > 0$, and* **p** *is the unique rest point in* int $\mathbf{R}^n_+$, *then*

$$\lim_{T\to\infty} \frac{1}{T}\int_0^T x_i(t)\,dt = p_i \quad i=1,..,n\,. \tag{9.5}$$

Proof: Let us write (9.1) in the form

$$(\log x_i)^{\cdot} = r_i + \sum_j a_{ij}x_j \tag{9.6}$$

and integrate it from 0 to T. After division by T, we obtain

$$\frac{\log x_i(T) - \log x_i(0)}{T} = r_i + \sum a_{ij}z_j(T) \tag{9.7}$$

where

$$z_j(T) = \frac{1}{T}\int_0^T x_j(t)\,dt\,. \tag{9.8}$$

One has obviously $a < z_j(T) < A$ for all j and all $T > 0$.

Now consider any sequence T_k converging to $+\infty$. The bounded sequence $z_j(T_k)$ admits a convergent subsequence. By "diagonalization" we obtain a subsequence – which we are going to denote by T_k again – such that $z_j(T_k)$ converges for every j towards some limit which we shall denote by $\bar{z}_j$. The sequences $\log x_i(T_k) - \log x_i(0)$ are also bounded. Passage to the limit in (9.7) thus leads to

$$0 = r_i + \sum a_{ij}\bar{z}_j$$

Thus the point $\bar{\mathbf{z}}_j = (\bar{z}_1, \dots \bar{z}_n)$ is an equilibrium. Since $\bar{z}_j \geq a > 0$, it belongs to int $\mathbf{R}^n_+$. Hence it coincides with **p**. This implies (9.5).

Exercise 2: Give another proof of Theorem 1, using a time-average argument. (This will work at least in the generic case.)

Exercise 3: Show that a similar averaging principle holds for the difference equation $\mathbf{x} \to \mathbf{x}'$ with

$$x_i' = x_i \exp(r_i + \sum_j a_{ij}x_j)$$

Exercise 4: What happens with Theorem 2 if the assumption on the uniqueness of the rest point is dropped?

9.3. The Lotka-Volterra equations for food chains

Let us investigate food chains with n members (chains with up to six members are found in nature). The first population is the prey for the second, which is the prey for the third etc. ... up to the n-th, which is at the top of the food pyramid. Taking competition within each species into account, and assuming constant interaction terms, we obtain

$$\begin{aligned} \dot{x}_1 &= x_1(r_1 - a_{11}x_1 - a_{12}x_2) \\ \dot{x}_j &= x_j(-r_j + a_{j,j-1}x_{j-1} - a_{jj}x_j - a_{j,j+1}x_{j+1}) \qquad j=2,\ldots,n-1 \\ \dot{x}_n &= x_n(-r_n + a_{n,n-1}x_{n-1} - a_{nn}x_n) \end{aligned} \tag{9.9}$$

with all r_j, $a_{ij}>0$. The case $n=2$ is just (7.15). We shall presently see that the general case leads to nothing new:

Theorem. *If* (9.9) *admits an interior equilibrium* $\mathbf{p}$, *then* $\mathbf{p}$ *is globally stable in the sense that all orbits in* int $\mathbf{R}^n_+$ *converge to* $\mathbf{p}$.

In order to prove this we write (9.9) as $\dot{x}_i = x_i w_i$ and try

$$V(x) = \sum c_i(x_i - p_i \log x_i) \tag{9.10}$$

for suitably chosen c_i, as a Ljapunov function in int $\mathbf{R}^n_+$. Clearly

$$\dot{V} = \sum c_i(\dot{x}_i - p_i\frac{\dot{x}_i}{x_i}) = \sum c_i(x_i w_i - p_i w_i) = \sum c_i(x_i - p_i)w_i\ . \tag{9.11}$$

Since $\mathbf{p}$ is an equilibrium, we have

$$r_j = a_{j,j-1}p_{j-1} - a_{jj}p_j - a_{j,j+1}p_{j+1}$$

for $j = 2,\ldots,n-1$, and similar equations for $j=1$ or n. This implies

$$w_j = a_{j,j-1}(x_{j-1} - p_{j-1}) - a_{jj}(x_j - p_j) - a_{j,j+1}(x_{j+1} - p_{j+1})$$

Writing $y_j = x_j - p_j$, we obtain from (9.11)

$$\dot{V} = -\sum_{j=1}^{n} c_j a_{jj} y_j^2 + \sum_{j=1}^{n-1} y_j y_{j+1}(-c_j a_{j,j+1} + c_{j+1}a_{j+1,j})\quad . \tag{9.12}$$

We are still free to choose the constants $c_j>0$. Let us do it in such a way that

$$\frac{c_{j+1}}{c_j} = \frac{a_{j,j+1}}{a_{j+1,j}} \tag{9.13}$$

holds for $j = 1,\ldots,n$. (9.12) then implies

$$\dot{V} = -\sum c_j a_{jj}(x_j - p_j)^2 \le 0\ . \tag{9.14}$$

By Ljapunov's theorem the ω-limit of every orbit in $\operatorname{int} \mathbf{R}^n_+$ consists of the equilibrium $\mathbf{p}$.

Exercise 1: Show that in the absence of competition within the predators ($a_{jj} = 0$ for $j = 2,\ldots,n$), $\mathbf{p}$ is still a global attractor in $\operatorname{int} \mathbf{R}^n_+$. (Hint: an ω-limit is an invariant set.)

Exercise 2: Discuss the different phase portraits for (9.9). Show that with increasing r_1 more and more predators can subsist. What are the *bifurcation values*?

Exercise 3: Study food chains with recycling

$$\dot{x}_1 = Q - a_{12}x_1x_2 + \sum_{k=2}^{n} b_k x_k$$

$$\dot{x}_j = x_j(-r_j + a_{j,j-1}x_{j-1} - a_{j,j+1}x_{j+1}) \quad j = 2,\ldots,n \ .$$

9.4. The exclusion principle

The *exclusion principle* states that if n populations depend linearly on m resources, with $m<n$, then at least one of the populations will vanish. In the long run, therefore, more populations than there are resources (or *ecological niches*, in another interpretation) cannot subsist. The assumption on the linear dependence of the resources is crucial. It means that the growth rate of the i-th population is of the form

$$\frac{\dot{x}_i}{x_i} = b_{i1}R_1 + \cdots + b_{im}R_m - \alpha_i \quad i = 1,\ldots,n \ . \tag{9.15}$$

The constant $\alpha_i>0$ indicates the rate of decline in the absence of any resource. R_k is the abundance of the k-th resource, and the coefficients b_{ik} describe the efficiency of the i-th species in making use of the k-th resource. The abundance of the resources depends of course on the population densities. If the dependence is linear, i.e. if

$$R_k = \bar{R}_k - \sum x_i a_{ki} \tag{9.16}$$

with positive constants $\bar{R}_k$ and a_{ki}, then (9.15) is a special case of the Lotka-Volterra equation (9.1). We don't need assumption (9.16), however. It is enough to postulate that the resources can be exhausted, i.e. that the densities x_i cannot grow to infinity.

Since $n > m$, the system of equations

$$\sum_{i=1}^{n} c_i b_{ij} = 0 \quad j = 1,\ldots,m$$

admits a nontrivial solution $(c_1,\ldots,c_n)$. Let

$$\alpha = \sum_{i=1}^{n} c_i \alpha_i \,.$$

We shall only consider the general case where $\alpha \neq 0$. By an appropriate choice of the c_i, we can assume $\alpha > 0$. (9.15) implies

$$\sum c_i (\log x_i)^{\cdot} = \sum c_i \frac{\dot{x}_i}{x_i} = -\alpha \,.$$

Integrating from 0 to T we obtain

$$\prod_{i=1}^{n} x_i(T)^{c_i} = Ce^{-\alpha T} \tag{9.17}$$

for some constant C. For $T \to +\infty$ the right hand side converges to 0. Since all $x_i(T)$ are bounded, there must be at least one index i such that $\liminf_{T\to\infty} x_i(T) = 0$, which spells extinction for the corresponding species.

9.5. A model for cyclic competition

Within the class of Lotka-Volterra equations, the existence of an interior rest point implies its stability for models of food chains. This is not so, however, for models of competitive interactions: as we have seen in **8.3**, competition between two species may lead to the extinction of one of them even if an interior rest point exists (in the case of bistability).

If three or more species compete, another and rather curious thing can occur. It may look for some time as if species 1 were bound to be the unique survivor; then, suddenly, its density drops, species 2 takes its place and seems to dominate the "ecosystem"; after some time, it in turn collapses, however, and leaves the field to species 3, which appears to be the ultimate winner; but then, species 1 suddenly rallies and outcompetes its rivals, and so another "round" starts. The species supersede each other in cyclic fashion: the time spans during which one species predominates grow larger and larger. An observer may get the impression that one species is better adapted and the other two doomed to extinction, until suddenly, without exterior cue, another revolution occurs.

We shall examine such a behaviour for the equation

$$\begin{aligned}\dot{x}_1 &= x_1(1-x_1-\alpha x_2-\beta x_3)\\ \dot{x}_2 &= x_2(1-\beta x_1-x_2-\alpha x_3)\\ \dot{x}_3 &= x_3(1-\alpha x_1-\beta x_2-x_3)\end{aligned} \tag{9.18}$$

with $0 < \beta < 1 < \alpha$ and $\alpha + \beta > 2$. Of course, the assumptions behind this equation are so artificial that we are never going to find them "in the field". But they do help with computations and display features which we should be prepared to meet in more general situations.

The special symmetry assumption behind the model is that of a cyclic interaction between the species: if we replace 1 by 2, 2 by 3 and 3 by 1, equation (9.18) will remain unchanged. This cyclic symmetry leads to a drastic simplification of some computations, which we shall use a few more times later on.

In most cases it is a difficult task to find the eigenvalues of a Jacobian. For cyclic symmetry, however, it is child's play. An $n \times n$-matrix is said to be *circulant* if it is of the form

$$\begin{bmatrix} c_0 & c_1 & c_2 & \cdots & c_{n-1} \\ c_{n-1} & c_0 & c_1 & \cdots & c_{n-2} \\ \cdot & \cdot & \cdot & & \cdot \\ \cdot & \cdot & \cdot & & \cdot \\ \cdot & \cdot & \cdot & & \cdot \\ c_1 & c_2 & c_3 & \cdots & c_0 \end{bmatrix} \tag{9.19}$$

where a cyclic permutation sends the elements of each row into those of the next one.

Exercise 1: Check that the eigenvalues of (9.19) are

$$\gamma_k = \sum_{j=0}^{n-1} c_j \lambda^{jk} \qquad k = 0,\ldots,n-1 \tag{9.20}$$

and the eigenvectors

$$\mathbf{y}_k = (1,\lambda^k,\lambda^{2k},\ldots,\lambda^{(n-1)k}) \tag{9.21}$$

where λ is the n-th root of unity

$$\lambda = \exp(2\pi i/n) \quad . \tag{9.22}$$

Now let us return to (9.18) and note that it admits a unique interior rest point **m** given by

$$m_1 = m_2 = m_3 = \frac{1}{1+\alpha+\beta} \,. \tag{9.23}$$

The Jacobian at the point **m** is

$$\frac{1}{1+\alpha+\beta} \begin{bmatrix} -1 & -\alpha & -\beta \\ -\beta & -1 & -\alpha \\ -\alpha & -\beta & -1 \end{bmatrix} \,. \tag{9.24}$$

This matrix is circulant. By (9.20) its eigenvalues are $\gamma_0 = -1$ (with eigenvector (1,1,1)) and

$$\gamma_1 = \bar{\gamma}_2 = \frac{1}{1+\alpha+\beta}(-1-\alpha e^{2\pi i/3}-\beta e^{4\pi i/3}) \quad .$$

The real part of γ_1 and γ_2 is thus

$$\frac{1}{1+\alpha+\beta}(-1+\frac{\alpha+\beta}{2}) \tag{9.25}$$

which is positive by assumption. Hence **m** is a saddle.

There are four further rest points, all on bd $\mathbf{R}^3_+$, namely **0** (which is a source) and the saddles $\mathbf{e}_1$,$\mathbf{e}_2$,$\mathbf{e}_3$ (the standard unit vectors).

The restriction of (9.18) to the face $x_3 = 0$ yields a competition equation for x_1 and x_2, which we have studied in **8.3**. In the absence of species 3, species 2 will outcompete species 1 (see Fig. 8.2). This implies that the stable manifold of $\mathbf{e}_2$ is the two-dimensional set $\{(x_1,x_2,x_3):x_1 \geq 0,x_2 > 0,x_3 = 0\}$, while the unstable manifold of $\mathbf{e}_1$ consists of a single orbit $\mathbf{o}_2$ converging to $\mathbf{e}_2$.

On the other boundary faces, the situation is similar: on the plane $x_1 = 0$, there is an orbit $\mathbf{o}_3$ with α-limit $\mathbf{e}_2$ and ω-limit $\mathbf{e}_3$, and on $x_2 = 0$ an orbit $\mathbf{o}_1$ from $\mathbf{e}_3$ to $\mathbf{e}_1$. We denote by F the set consisting of the three saddles $\mathbf{e}_1$,$\mathbf{e}_2$ and $\mathbf{e}_3$ and the three connecting orbits $\mathbf{o}_1$,$\mathbf{o}_2$, and $\mathbf{o}_3$ (see Fig. 9.2). We shall show that all orbits in int $\mathbf{R}^3_+$ (with the exception of those on the diagonal) have F as ω-limit. The state thus remains for a long time close to the rest point $\mathbf{e}_1$, then travels along $\mathbf{o}_2$ to the vicinity of the rest point $\mathbf{e}_2$, lingers there for a still longer time, then jumps over to the rest point $\mathbf{e}_3$ and so on, in cyclic fits and starts.

To verify this, we shall use the functions

$$S = x_1+x_2+x_3 \tag{9.26}$$

and

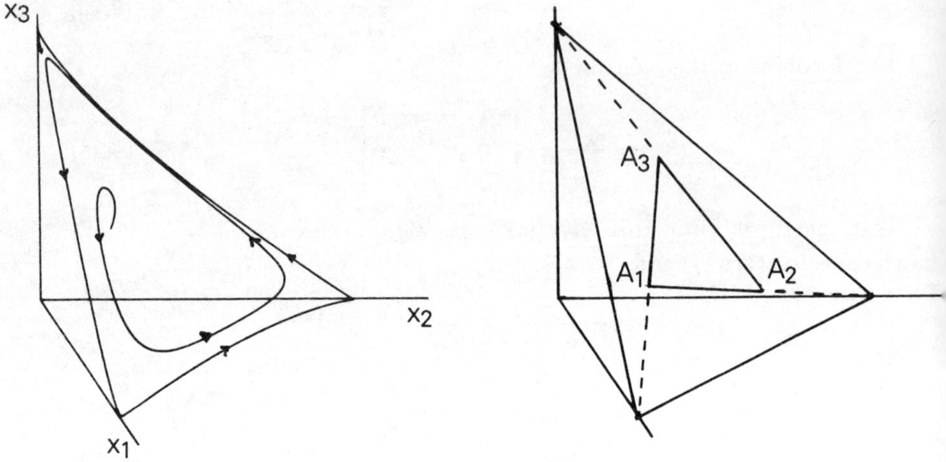

Figures 9.2 and 9.3

$$P = x_1 x_2 x_3 \ . \tag{9.27}$$

One has

$$\dot{S} = x_1+x_2+x_3-[x_1^2+x_2^2+x_3^2+(\alpha+\beta)(x_1x_2+x_2x_3+x_3x_1)] \tag{9.28}$$

and

$$\dot{P} = \dot{x}_1x_2x_3+x_1\dot{x}_2x_3+x_1x_2\dot{x}_3 = P(3-(1+\alpha+\beta)S) \ . \tag{9.29}$$

(9.28) implies $\dot{S} \leq S(1-S)$, so that no population can explode. A straightforward computation yields

$$\left(\frac{P}{S^3}\right)^{\cdot} = S^{-4}P\left(1-\frac{\alpha+\beta}{2}\right)[(x_1-x_2)^2+(x_2-x_3)^2+(x_3-x_1)^2] \leq 0 \ .$$

The theorem of Ljapunov then implies that any orbit which is not on the (invariant) diagonal $x_1 = x_2 = x_3$ converges to the boundary (the set where P vanishes). But we have already investigated the behaviour there: the only candidate for an ω-limit is the set F.

Of course such an ω-limit can never occur "in reality". Since $\liminf x_i(t) = 0$, one of the species will sooner or later vanish, and then one of the remaining species will outcompete the other one. Still, the model is of biological interest, as it suggests a surprising mechanism for sudden upheavals in ecological communities.

It is instructive to study the behaviour of the time averages for this system. We have seen in **9.2** that time averages converge to an interior rest point whenever the orbit stays away from the coordinate planes. In (9.18), however, orbits converge to bd $\mathbf{R}^3_+$: as we shall see in a moment, their time averages do not longer converge.

Since the orbits of (9.18) spend most of their time near the fixed points $\mathbf{e}_1$,$\mathbf{e}_2$ and $\mathbf{e}_3$, their time averages

$$\mathbf{z}(T) = \frac{1}{T}\int_0^T \mathbf{x}(t)\,dt$$

will converge to the plane spanned by these three points, which is given by

$$z_1+z_2+z_3 = 1\ . \tag{9.30}$$

Now consider again equation (9.7) for the time averages. Since $x_i(T)$ is bounded from above, every accumulation point of the left hand side is nonpositive. Thus every limit point $\mathbf{z}$ of the time averages $\mathbf{z}(T)$ satisfies

$$r_i+\sum_{j=1}^{n} a_{ij}z_j \leq 0\ . \tag{9.31}$$

To be more precise, consider a sequence $T_k \to +\infty$ such that $\mathbf{z}(T_k) \to \mathbf{z}$, and let $\bar{\mathbf{x}}$ be a limit point of $\mathbf{x}(T_k)$. There are two possibilities:
(a) $\bar{\mathbf{x}}$ lies on one of the three connecting orbits $\mathbf{o}_1$,$\mathbf{o}_2$ or $\mathbf{o}_3$. Then only one of the coordinates of $\bar{\mathbf{x}}$ is zero; say $\bar{x}_1 = 0, \bar{x}_2 > 0, \bar{x}_3 > 0$. (9.31) yields one inequality and two equations to be satisfied by $\mathbf{z}$:

$$\begin{aligned} 1-z_1-\alpha z_2-\beta z_3 &\leq 0 \\ 1-\beta z_1-z_2-\alpha z_3 &= 0 \\ 1-\alpha z_1-\beta z_2-z_3 &= 0\ . \end{aligned} \tag{9.32}$$

(9.32) determines a line segment between the interior fixed point $\mathbf{m}$ and the intersection of the x_2- and x_3-isoclines in the plane $x_1 = 0$. Together with (9.30) this determines the position of $\mathbf{z}$. Let us call this point A_1, and define A_2 and A_3 similarly. (See Fig. 9.3).

Exercise 2: Compute A_1,A_2 and A_3 explicitly.

(b) $\bar{\mathbf{x}}$ is one of the three boundary rest points, say $\mathbf{e}_1$. Then (9.31) yields one equality and two inequalities: this is just like (9.32) with equality and inequality signs interchanged. This shows that any such limit point $\mathbf{z}$ lies on the line segment between A_2 to A_3. Since the set of all limit points of $\mathbf{z}(T)$ is obviously connected, every point on this segment will be a limit point.

This shows that the set of all limit points of the time averages $\mathbf{z}(T)$ is just the boundary of the triangle $A_1A_2A_3$, which is given by the intersection of the plane (9.30) with the region (9.31).

Exercise 3: Show that the time intervals which a solution spends near the rest points $\mathbf{e}_1, \mathbf{e}_2, \mathbf{e}_3, \mathbf{e}_1, \ldots$ increase exponentially with exponent $\frac{\alpha-1}{1-\beta}>1$. Using this fact, find another (independent and even more general) proof of the above result.

Exercise 4: Analyse (9.18) for other values of α and β.

Exercise 5: Let $\hat{\mathbf{x}}$ be an interior fixed point of (9.1). Let $\mathbf{c}$ be a left eigenvector of the Jacobian at $\hat{\mathbf{x}}$, i.e.

$$\sum_{i=1}^{n} c_i \hat{x}_i a_{ij} = \lambda c_j \,, \quad \lambda \neq 0$$

and $\mathbf{y} = \mathbf{0}$ be the only solution of the system

$$\sum_{i=1}^{n} c_i y_i = 0 \quad \text{and} \quad \sum_{i,j} c_i a_{ij} y_i y_j = 0 \quad .$$

Show that under these assumptions, all ω-limits of solutions of (9.1) are either $\hat{\mathbf{x}}$ or contained in bd $\mathbf{R}^n_+$. (Hint:

$$V(x) = (\epsilon + \sum c_i(x_i - \hat{x}_i))^\epsilon \prod x_i^{-c_i\hat{x}_i} \tag{9.33}$$

is a Ljapunov function for small ϵ.) Apply this result to (9.18).

Exercise 6: Show that Volterra's Ljapunov function (9.10) arises as the limit case $\epsilon \to \infty$ of (9.33).

9.6. Notes

The model (9.18) for cyclic competition is due to May and Leonard (1975). It was discussed in subsequent papers by Chenciner (1977), which contains the result in exercise 9.5.5, Coste et al. (1979) and Schuster et al. (1979b). The interesting behaviour of the time averages was discovered by E.C. Zeeman (personal communication). The results on food chains are due to Harrison (1978), So (1979) and Gard and Hallam (1979). The "chaotic system" displayed in Fig. 9.1 was discovered by Arneodo et al. (1980). The exclusion principle is due to Volterra (1931). In Armstrong and McGehee (1976) one finds another proof, as well as an example showing that the principle need not be valid for nonlinear dependence on the resources. For Theorem 9.2.2 we refer to Coste et al.

(1978) and Schuster et al. (1980). The difference equation of exercise 9.2.3 was analysed in Hofbauer et al. (1979). The result on separating hyperplanes used in the proof of Theorem 9.2.1 can be found in Nikaido (1968). The ubiquity of Lotka-Volterra equations is stressed in Peschel and Mende (1984). For a discussion of stability properties, we refer to Chapter 21.

PART III: TEST TUBE EVOLUTION AND HYPERCYCLES: A PREBIOTIC PRIMER

10. Prebiotic Evolution

10.1. Polynucleotides

The spectacular progress of molecular biology has shed much light on the chemical activities within the cell, including metabolism, growth and division. The insights into the genetic machinery were of particular interest. In 1944, Avery discovered that the genetic information is stored in the form of *polynucleotides* as DNA or (for some viruses and bacteriophages) as RNA. These *nucleic acids* are long chainlike molecules consisting of a sugar-phosphate backbone along which the nucleotides are arranged. In DNA, there are four types of nucleotides, which differ in their bases, namely guanine (G), cytosine (C), adenine (A) and thymine (T). In RNA, thymine is replaced by uracil (U). The sequence of nucleotides specifies the information of the genome, just as the sequence of letters specifies the information of a book. The molecular messages are enormously long. The DNA molecules are the largest polymers in nature (3.10^9 nucleotides in human DNA, and we don't hold the record in this field).

The genetic information must get copied in order to get transmitted. The copying mechanism is based on the *complementarity* of the bases: because of their stereochemical configurations, A can form bonds with T (or U) and C with G. Along one strand of nucleotides - say AACCGATCG - a complementary strand, a kind of "negative" will be formed - TTGGCTAGC in this case. If the two strands separate, a copy of the original message will be built along the complementary strand. Molecules of DNA consist of two complementary strands twisting around each other in the form of a *double helix*. This structure was discovered by Watson and Crick in 1953. The essential step during cell division is the splitting of the two strands and the growth of a negative image along each one of them. This makes two look-alike DNA-molecules out of the original one (see Fig. 10.1). The precise chemical mechanism is exceedingly complicated, however, and not yet understood in every detail.

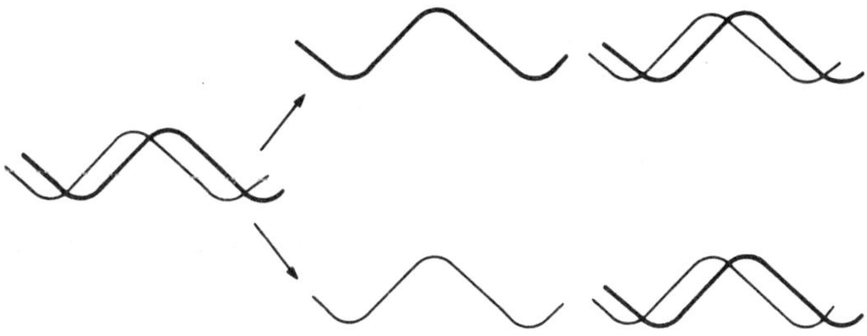

Figure 10.1

10.2. Polypeptides

Besides polynucleotides, there are lots of other molecules in a cell, of course. A particularly important class is that of *enzymes*, or biocatalysts - these are *proteins* which serve as machinery and in part also as building blocks of the cell. The enzymes are again polymers, namely chains of *amino acids*. Twenty types of amino acids occur. The length of enzymes, or polypeptides, is of the order of $10^2 - 10^4$. The primary structure of a polypeptide is defined by the sequence of its amino acids whilst the secondary structure, the so-called α-helix, stabilizes the molecule. The tertiary structure consists of one or several very complicated spatial arrangements; the foldings and twists are effected by hydrogen bonds and other couplings. The characteristic stereochemical shape of the enzyme is responsible for its very specific *catalytic action*. A quarternary structure, finally, may consist of several interlocking polypeptidic chains.

Each cell contains many thousands of enzymes.

10.3. The genetic code

How does the genome instruct the cell? There must be some way of translating polynucleotides into polypeptides, that is some code from an alphabet with four letters - the polynucleotides - into an alphabet of twenty letters - the amino acids. The *genetic code* was deciphered in 1968 by Kornberg and his coworkers. To each of the $4^3=64$ possible triplets of nucleotides it assigns one of the twenty amino acids: for instance, GGC leads to glycine, GCC to alanine etc... Amazingly enough, this code is *universal* (apart from a few marginal exceptions): it is the same one for all living beings on Earth.

The translation works by copying strings of the DNA-information contained in the nucleus into a *messenger-RNA* (m-RNA) which leaves the nucleus, becomes attached to a *ribosome* and gets linked via Watson-Crick bonding with *transfer-RNA* (t-RNA), thereby building up, step by step, the suitable protein (see Fig. 10.2).

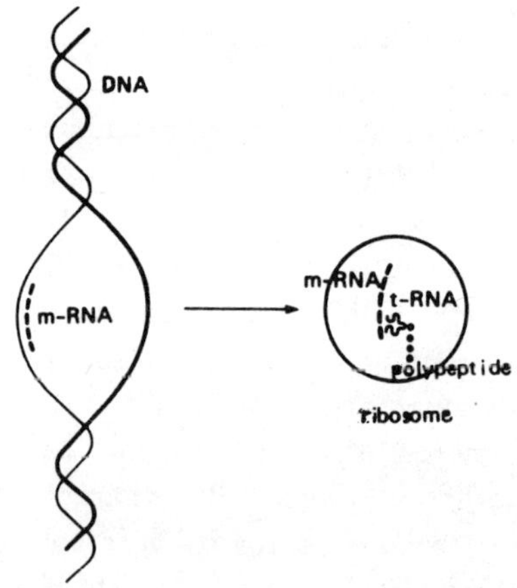

Figure 10.2

The instruction thus leads from polynucleotide to polypeptide. The so-called "dogma" of molecular genetics asserts that the converse direction is ruled out, i.e. that polypeptides are never translated into polynucleotides.

10.4. The problem of the origin of life

In the context of molecular biology, the problem of the origin of life can be split into two parts. First, how could something as incredibly complex as the molecular reproduction machinery ever come into being? And second, why did only one such mechanism evolve? Both the existence and the uniqueness of the genetic code are riddles. If one excludes sheer accident or supernatural creation, one is faced with the task of delineating the natural laws which governed the transformation from lifeless matter into living beings - i.e. into systems able to sustain and reproduce themselves, occasionally with inheritable variations. What one looks for are the principles of a prebiotic evolution admitting a selforganization of macromolecules up to the complexity of the cellular reproduction mechanism.

Let us stress here - in a crudely oversimplified way - a major difference between prebiotic evolution and the more familiar biological evolution in the sense of Darwin. In the latter case, the bifurcations of evolutionary lines led to the millionfold profusion of species populating today's Earth. Many evolutionary alternatives are able to survive side by side. Prebiotic evolution has been different. Since it led up to a single molecular translation machinery, its decisions must have been universal: there was no room for coexisting alternatives. The descendency tree did not branch out (cf. Fig. 10.3).

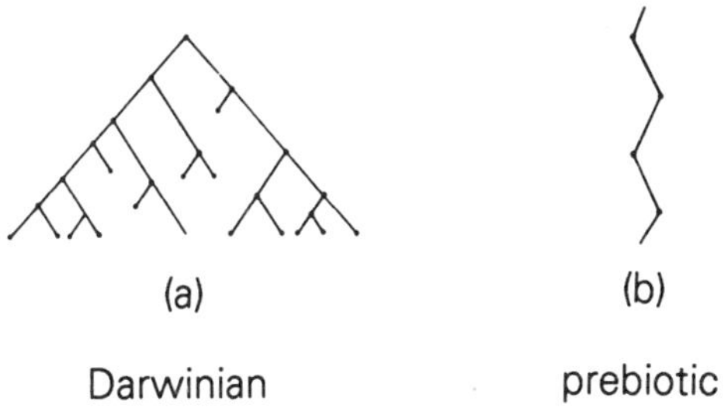

Figure 10.3

In principle, one can approach a theory of prebiotic evolution with two different standards of expectation. One can ask: how has it happened? This seems impossibly difficult to answer. There are no fossile traces dating back to the origin of life and the construction of descendency trees from actually existing macromolecules does not seem to lead all that far into the past either.

But one can also ask, in a more modest vein: in which way could it have happened? What theoretical possibilities are there to account for a spontaneous emergence of life? An answer to this question, although certainly still distant, does not seem altogether out of reach. The history of science knows of many breakthroughs achieved in fields which seemed hopelessly remote from direct observation.

10.5. The first steps

Life seems to have started some 4 billion years ago on our planet. On one hand, indeed, continents (and liquid water, for that matter) don't date much further back. On the other hand, fossile traces of microorganisms have been found in rocks aged 3.5 billion years.

The first steps in the evolution seem quite well established, and in some sense even subject to an experimental approach. The *primordial atmosphere* contained no free oxygen - which was produced only later by metabolism - but nitrogen, hydrogen and carbone, albeit in other compounds than in today's air. If these conditions are imitated in a laboratory, and some energy pumped into the mixture (electric discharges, for example, or ultraviolet radiation), then "organic" substances appear, like lipids, sugars, amino acids and even nucleotides. Those experiments, first undertaken by Miller and Urey in 1953, have been repeated since in many variations, and lead to the conclusion that the "building blocks" of life could evolve spontaneously under natural conditions. (Further proofs are the traces of amino acids found in meteorites.)

We can take it, therefore, that amino acids and nucleotides were present in the "primordial soup" and in a modest way able to polymerize. Especially under the catalytic influence of naturally occurring metallic ions, a huge variety of short chains could accumulate. But how could those which were "biologically correct" get singled out and developed further? We shall examine some aspects of a theory by Eigen and Schuster dealing with this problem and leading up to differential equations modelling the reaction kinetics of selfreplicating molecules.

10.6. Notes

A famous textbook on molecular genetics is due to Watson (1976). Concerning prebiotic evolution, we refer to Fox and Dose (1972), Miller and Orgel (1973), Schuster (1981), Eigen et al. (1981) and Maynard Smith (1986). A deep essay on the origin of life is Monod (1970). The hitherto accepted views on the composition of the primordial atmosphere have recently been challenged, see Thaxton et al. (1984).

11. Branching Processes and the Complexity Threshold

11.1. Branching processes and extinction probabilities

Let us consider, to begin with, a population with non-overlapping generations. We shall assume that all individuals reproduce independently of each other and according to the same probability law. This defines a so-called *branching process.*

We denote by $f_0, f_1, f_2, \ldots$ the probabilities that an individual has 0,1,2,... offspring in the next generation. Let Z_n denote the population size in the n-th generation and

$$P(Z_n = i \mid Z_m = j)$$

the conditional probability that there are i individuals in the n-th generation, provided that there were j individuals in the m-th generation. In particular,

$$P(Z_n = 0 \mid Z_0 = k)$$

is the probability that a population of size k vanishes within n generations. One has

$$P(Z_n = 0 \mid Z_0 = k) = [P(Z_n = 0 \mid Z_0 = 1)]^k \tag{11.1}$$

since the extinction of the population is the result of k independent events, viz. that the offspring of all k individuals originally present reaches extinction. Furthermore

$$P(Z_{n+1} = 0 \mid Z_1 = k) = P(Z_n = 0 \mid Z_0 = k) \ .$$

Let us assume now that $Z_0 = 1$ and define, for $s \in [0,1]$:

$$F(s) = \sum_{k=0}^{\infty} f_k s^k \quad \text{and} \quad F_n(s) = \sum_{k=0}^{\infty} P(Z_n = k) s^k \ . \tag{11.2}$$

Since $f_i = P(Z_1 = i)$ one has $F_1(s) = F(s)$. The probability of the extinction of the population within the first n generations is

$F_n(0) = P(Z_n = 0)$. The sequence $F_n(0)$ is monotonically increasing and converges to a limit q, which is the *probability of eventual extinction of the population.*

The event $Z_{n+1} = 0$ can be decomposed according to the number of offspring after one generation. This yields

$$P(Z_{n+1} = 0) = \sum_{k=0}^{\infty} P(Z_1 = k) P(Z_{n+1} = 0 \mid Z_1 = k)$$

$$= \sum_{k=0}^{\infty} f_k [P(Z_n = 0)]^k = \sum_{k=0}^{\infty} f_k [F_n(0)]^k$$

i.e.

$$F_{n+1}(0) = F(F_n(0)) \,. \tag{11.3}$$

From $F_n(0) \rightarrow q$ and (11.3) it follows that

$$q = F(q) \,. \tag{11.4}$$

Let $a > 0$ be a solution of (11.4). Since F is monotonic, one has $F(0) < F(a) = a$. Since $F_n(0) \leq a$ implies $F_{n+1}(0) = F(F_n(0)) \leq$ $\leq F(a) = a$, one has by induction $F_n(0) \leq a$ for all n and hence $q \leq a$. Thus q is the smallest positive root of $F(s) = s$.

We shall neglect the trivial case $f_0 + f_1 = 1$ (at most one offspring). The function F, thus, is strictly convex on $[0,1]$. It satisfies $F(0) = f_0 \geq 0$ and $F(1) = 1$. This implies that for $\frac{dF}{ds}(1) > 1$, the equation $F(s) = s$ has exactly one root in $(0,1)$, while for $\frac{dF}{ds}(1) \leq 1$, its smallest positive root is 1 (see Fig. 11.1). Now the *mean offspring* of one individual after one generation is given by

$$m = \sum_{k=0}^{\infty} k f_k = \frac{dF}{ds}(1) \,. \tag{11.5}$$

Hence we obtain:

(a) *if* $m \leq 1$, *the population reaches extinction with probability* 1;

(b) *if* $m > 1$, *the probability* q *of extinction is strictly smaller than* 1.

Branching processes were already studied in the last century in connection with the extinction of family names. It was only much later that the general relevance of such processes was recognized - for example in chain reactions. They also serve to describe populations of selfreproducing macromolecules.

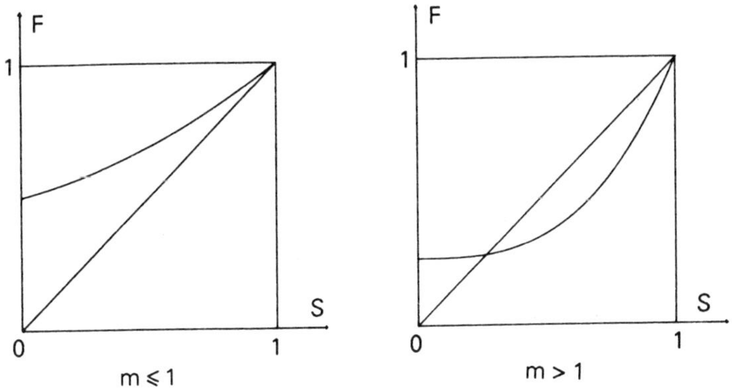

Figure 11.1

11.2. Error bound and complexity threshold

Let us consider a polymeric chain of ν nucleotides. If we assume that there is a fixed probability p of a single nucleotide being copied correctly, then the probability that a copy of the whole chain is exact is p^ν. Under appropriate boundary conditions, the replication time is fairly well defined: we shall view it as the length of a generation. If there is no "ageing", the survival probability of a molecule is a constant which we shall denote by w. During one generation step, then, the molecule either survives (with probability w) and produces a copy (which is accurate with probability p^ν) or it is hydrolysed (with probability $1-w$). A given molecule then yields 0,1 or 2 molecules of the same type after one unit of time: the probabilities are $1-w, w(1-p^\nu)$ and wp^ν, respectively, and the mean is $w(1+p^\nu)$.

The population of "correct" polynucleotides thus evolves according to a branching process. By the results of the previous section, extinction is certain if and only if

$$p^\nu \leq \frac{1-w}{w} \tag{11.6}$$

and hence there is a strictly positive probability of indefinite survival of the molecular population if and only if

$$\nu < \frac{\log\frac{1-w}{w}}{\log p} \sim \frac{1}{1-p}\log\frac{w}{1-w} \tag{11.7}$$

where the approximation holds when $1-p$ (the probability of a one-digit error) is small.

Exercise 1: The probability of extinction is $\min(1, \frac{1-w}{wp^\nu})$.

Exercise 2: As an alternative approach, let us consider the length of life of a polynucleotide as a generation span, and assume that it produces a mean number of σ (correct or incorrect) copies. Show that the survival of the molecular line is possible iff $\sigma p^\nu > 1$, and compare this with (11.7).

Relation (11.7) implies that the *error probability* $1-p$ defines a threshold for the length ν of the polynucleotide: if ν is larger than this threshold, then the molecular line has to reach extinction. Only for smaller ν is there a chance for long term survival. The maximal length of the molecule is (roughly) inversely proportional to the probability of a one digit error. This bound on the length of the message limits the amount of information: the error probability defines a *complexity threshold.*

There is no denying that this model is based on drastic simplifications. The probability p of copying a nucleotide correctly depends on the nucleotide and its neighbours; other errors than simple "misprints" occur; and back mutations from erroneous copies have not been taken into account. Still, we shall presently see that the main conclusion - the error bound relation (11.7) - seems to be verified in nature.

11.3. The information crisis

There are several ways in which polynucleotides get copied, with different error probabilities and consequently different complexity thresholds. It would seem as if the actual lengths, at least in their order of magnitude, do tally with these theoretical bounds.

The most primitive replication of RNA-molecules works without the help of enzymes. The error probability is some 5%, and the maximal length consists of 15 nucleotides or so. Such *oligonucleotides* occur in the "primordial soups" of lab experiments. The first genetic information, presumably a predecessor of present-day t-RNA, seems to have had a length of some 80 nucleotides, which more or less fits with this complexity threshold.

The next best copying mechanism uses one enzyme and is found in *phages* (which are, loosely speaking, viruses of bacteria). The error probability is down by a factor of 10^2 and the complexity threshold therefore 100 times larger. In fact, the RNA-genome of Q_β-phages consists of some 4500 nucleotides.

Still better copying can be found in *bacteria*: their DNA-molecules get replicated with the help of enzymes and a "proof-reading" mechanism: the one-digit accuracy increases by 10^2–10^3, and so does the complexity threshold. The DNA-molecule of E. Coli contains, accordingly, some 4.10^6 nucleotides.

The precision of the replication in the cells of higher organisms is again improved by several orders of magnitude. Estimates of the accuracy are yet lacking, but the length of the human genome is 3.10^9, and this is not the uppermost bound.

The complexity threshold relation seems to mean something, therefore. It can be viewed as an analogue, on the molecular level, of the allometric relations which set limits to the growth of organisms. For example, the height of trees is restricted by the strength of wood and the capacity for water transport of the trunk, the size of insects is restricted by the rate of oxygen transport through capillary diffusion, and the body weight of vertebrates is limited by the carrying capacity of the bones.

Let us turn back to the step from enzyme free replication to replication with enzymes. How could it possibly occur? This is quite a problem. In order to code an enzyme, a polynucleotide has to have some minimal length - and this length exceeds the bound given by error free replication! In other words: no better replication without a longer sequence, and no longer sequence without a better replication.

This modern version of the paradox of the hen and the egg suggests that there must have been, at an early stage, a "ganging up" of selfreplicating molecules. We shall, therefore, consider models for the coexistence of several types of polynucleotides.

11.4. Multitype branching

Let us consider a population consisting of n types of polynucleotides $I_1,\ldots,I_n$. Each molecule of type I_i can generate molecules of the same type or (by mutation) of different type. Replications will be homogeneous in time and mutually independent. For simplicity, we shall first assume discrete generations. In each generation, a polymer of type I_i produces r_1 polymers of type I_1,r_2 polymers of type I_2 ,..., and r_n polymers of type I_n with probability $p_i(r_1,\ldots,r_n)$. Let $Z_i(k)$ denote the total number of polymers of type I_i in generation k. The vector $\mathbf{Z}(k) = (Z_1(k),\ldots,Z_n(k))$ is a random variable.

Let m_{ij} be the mean number of polymers of type I_i derived from a polymer of type I_j within one generation. Since $m_{ij} \geq 0$, we shall be able to use the theory of *Perron-Frobenius*. *If M is a nonzero $n \times n$ - matrix with nonnegative elements, there exists a unique nonnegative eigenvector λ which is dominant in the sense that $|\mu| \leq \lambda$ for all other eigenvalues μ of M. There exist right and left eigenvectors $\mathbf{u} \geq \mathbf{0}$ (i.e. $u_i \geq 0$ for all i) and $\mathbf{v} \geq \mathbf{0}$ such that*

$$M\mathbf{u} = \lambda\mathbf{u} \qquad \mathbf{v}M = \lambda\mathbf{v} \,. \tag{11.8}$$

They can be chosen such that

$$\mathbf{u}\cdot\mathbf{v} = 1 \;\; and \;\; \mathbf{1}\cdot\mathbf{v} = \sum v_i = 1 \tag{11.9}$$

where

$$\mathbf{1} = (1,\ldots,1) \tag{11.10}$$

If M is irreducible (i.e. if for every pair of indices (i,j) there exists an integer $k = k(ij) > 0$ such that the (i,j)-th element of M^k is positive), then λ is simple and positive; $\mathbf{u}$ and $\mathbf{v}$ are uniquely defined. Furthermore $\mathbf{u}>\mathbf{0}$ (i.e. $u_i>0$ for all i) and $\mathbf{v}>\mathbf{0}$. No other eigenvalue of M is associated with an eigenvector whose components are all positive.

In this chapter we shall always assume M to be *primitive* (i.e. there exist a $k>0$ such that all elements of M^k are positive). In this case $|\mu|<\lambda$ for all other eigenvalues μ.

Exercise: Show that the matrix T with $t_{ij} = u_i v_j$ is idempotent ($T^2 = T$), that $TM = MT = \lambda T$ and (less easy!) that $\lambda^{-k}M^k \to T$ for $k \to \infty$.

Returning to the multitype branching process, let q_i denote the *probability of extinction* ($\mathbf{Z}(k) = \mathbf{0}$ for some k) given that the initial population consists of a single polymer of type I_i. (This is the condition $\mathbf{Z}(0) = \mathbf{e}_i$.) Then a result analogous to the single type branching case holds:

(a) *if $\lambda \leq 1$, extinction is certain: $q_i = 1$ for all i;*

(b) *if $\lambda > 1$, there is a positive probability of survival for infinite time, $q_i < 1$ for all i. If the population does not vanish, then it explodes almost surely (this means up to probability 0). The frequency of type I_i in generation k, given by*

$$X_i(k) = \frac{Z_i(k)}{Z_1(k)+\ldots+Z_n(k)}$$

satisfies almost surely

$$X_i(k) \to \frac{u_i}{u_1+\ldots+u_n} . \tag{11.11}$$

The vector $\mathbf{y}(k)$ of *expectation values* ($y_i(k) = EZ_i(k)$) satisfies the linear recurrence relation

$$\mathbf{y}(k+1) = M\mathbf{y}(k) \tag{11.12}$$

and hence is of the form $\mathbf{y}(k) = M^k\mathbf{y}(0)$. The vector $\mathbf{x}(k)$ of *normalized mean values*

$$x_i(k) = \frac{y_i(k)}{y_1(k)+\ldots+y_n(k)}$$

satisfies therefore

$$\mathbf{x}(k+1) = \frac{M\mathbf{x}(k)}{\mathbf{1}\cdot M\mathbf{x}(k)} . \tag{11.13}$$

Since

$$\lambda^{-k}\mathbf{y}(k) = \lambda^{-k}M^k\mathbf{y}(0) \to T\mathbf{y}(0) \tag{11.14}$$

one has

$$\lambda^{-k}y_i(k) \to \sum_j u_i v_j y_j(0) = u_i \mathbf{v}\cdot\mathbf{y}(0)$$

and hence

$$x_i(k) \to \frac{u_i}{u_1+\ldots+u_n} . \tag{11.15}$$

The recurrence relations (11.12) and (11.13) correspond to the discrete dynamical systems

$$\mathbf{y} \to M\mathbf{y} \tag{11.16}$$

on $\mathbf{R}^n_+$ and

$$\mathbf{x} \to \frac{M\mathbf{x}}{\mathbf{1}\cdot M\mathbf{x}} \tag{11.17}$$

on S_n, respectively. Both are related to the branching process, but the latter is more apposite to its description than the former. Indeed, it can be shown that the variance of the random vector $Z_i(k)$ increases at an exponential rate (if $\lambda>1$) so that the mean value $y_i(k)$ tells us rather little about $Z_i(k)$. On the other hand, if the population does not reach extinction, then the behaviour of the stochastic frequencies $X_i(k)$ is faithfully reflected by that of the mean frequencies $x_i(k)$, as shown by (11.11) and (11.15). The variance, so to speak, is all in the total population size, and not in the relative frequencies.

11.5. The continuous case

We have assumed discrete generations so far. It is obvious, however, that a continuous time version of multitype branching is a better description of polynucleotide replication. It is more complicated, but leads to essentially the same results as in the discrete case.

Let $\mathbf{Z}(t) = (Z_1(t),\ldots,Z_n(t))$ be the random vector of population numbers at time t and $\mathbf{y}(t) = (y_1(t),\ldots,y_n(t))$ the vector of expectation values. If $m_{ij}(t)$ denotes the mean number of polymers of type I_i produced by one polymer of type I_j in time t, and if $m_{ij}(t)>0$ for all i,j and some $t>0$ (for all $t>0$, therefore), then

$$\mathbf{y}(t+s) = M(t)\mathbf{y}(s) \tag{11.18}$$

so that $\mathbf{y}(t)$ is a solution of the linear differential equation

$$\dot{\mathbf{y}} = A\mathbf{y} \tag{11.19}$$

with

$$A = \lim_{\Delta t \to 0} \frac{M(\Delta t) - I}{\Delta t} . \tag{11.20}$$

It follows that

$$M(t) = e^{At} . \tag{11.21}$$

With $M = M(1)$ and $\lambda,\mathbf{u},\mathbf{v}$ as in the previous section, we obtain the same results:

(a) *extinction with probability one if* $\lambda \leq 1$

(b) *positive probability of indefinite survival if* $\lambda > 1$. *In this case, if the population does not vanish, then the frequency*

$$X_i(t) = \frac{Z_i(t)}{Z_1(t)+\ldots+Z_n(t)}$$

satisfies almost surely

$$X_i(t) \to \frac{u_i}{u_1+\ldots+u_n} . \tag{11.22}$$

The quotient rule and (11.19) imply that the normalized mean vector $\mathbf{x}(t)$ with components

$$x_i(t) = \frac{y_i(t)}{y_1(t)+...+y_n(t)}$$

satisfies

$$\dot{\mathbf{x}} = A\mathbf{x} - \mathbf{x}(\mathbf{1}\cdot A\mathbf{x}) \; , \tag{11.23}$$

that is

$$\dot{x}_i = \sum_j a_{ij}x_j - x_i(\sum_{k,j} a_{kj}x_j) \; .$$

Equation (11.23) is defined on $\mathbf{R}^n$, but since the x_i are frequencies, we are only interested in the restriction to the simplex S_n. This simplex is positively invariant: indeed,
(i) the sum $S = x_1+...+x_n$ satisfies

$$\dot{S} = (\mathbf{1}\cdot A\mathbf{x})(1-S) \tag{11.24}$$

which has $S(t) = 1$ as solution (thus if the solution of (11.23) starts on the plane $\sum x_i = 1$, it remains there); and
(ii) if $x_i = 0$ then $\dot{x}_i \geq 0$ so that the vector field on bdS_n points inward (no orbit starting in S_n can escape).
Exercise: (11.23) is obtained from the linear equation (11.19) by normalization. The converse passage is just as easy. Setting

$$\Phi(t) = \int_0^t \mathbf{1}\cdot A\mathbf{x}(u)du$$

one easily checks that if $\mathbf{x}(t)$ is a solution of (11.23), then $\mathbf{y}(t) = \mathbf{x}(t) \exp \Phi(t)$ is a solution of (11.19).

An explicit solution of (11.19), and hence also of (11.23), can be written down, but one does not need it to understand the qualitative behaviour of (11.23). Just as with the discrete case (11.14) one has $\lambda^{-t}\mathbf{y}(t) \to T\mathbf{y}(0)$ and so

$$\mathbf{x}(t) \to (\mathbf{1}\cdot\mathbf{u})^{-1}\mathbf{u} \; . \tag{11.25}$$

An equilibrium $\mathbf{x}$ of (11.23) must satisfy $A\mathbf{x} = \mathbf{x}(\mathbf{1}\cdot A\mathbf{x})$ and hence be a right eigenvector of A. There is only one such eigenvector in S_n, namely $(\mathbf{1}\cdot\mathbf{u})^{-1}\mathbf{u}$. This is a global attractor for all orbits in S_n.

In the limiting case of *replication without mutation*, $m_{ij}(t) = 0$ for $i \neq j$ and hence $M(t)$ and A are diagonal. (11.23) reduces to

$$\dot{x}_i = x_i(a_i - \bar{a}) \tag{11.26}$$

with

$$\bar{a} = \sum_{j=1}^{n} a_j x_j \,.$$

In general, one of the a_j(a_1, say) will be larger than the others. By the quotient rule

$$\left(\frac{x_1}{x_j}\right)^{\cdot} = \left(\frac{x_1}{x_j}\right)(a_1 - a_j) > 0$$

for $j \neq 1$, so that $\frac{x_1}{x_j} \rightarrow +\infty$ for $t \rightarrow +\infty$. Since the frequencies are bounded by 1, this implies $x_j(t) \rightarrow 0$ for $j \neq 1$ and hence $x_1(t) \rightarrow 1$. All orbits in S_n converge to $\mathbf{e}_1$. All but one of the types of macromolecules vanish.

If one takes mutations into account, the off-diagonal terms $m_{ij}(t)$ will be positive, but very small. The global attractor $(\mathbf{1} \cdot \mathbf{u})^{-1}\mathbf{u}$ will be a point in int S_n, but very close to $\mathbf{e}_1$. Again, one type of macromolecule will predominate: the others will subsist in tiny quantities as erroneous offshoots only.

As mentioned in **11.3** we are interested in the coexistence of several types of molecular information carriers. Independent replication, as we have just seen, is not the solution, since all but one of the molecular types will be present in minimal quantities only. What one needs is some coexistence of macromolecules. The simplest way to conceive this is through catalytic interaction: each type of polynucleotide should help with the reproduction of another.

In the sophisticated reproduction machinery of today's cells, the replication of polynucleotides is catalysed by specific polypeptides. Under more primitive conditions, however, some types of polynucleotides can do the job. This has been discovered in the last few years only. We shall be interested not in the actual chemical kinetics, but in the principle behind it, and investigate some simplified theoretical models. Probabilistic methods (like branching processes) seem to be unsuited for this kind of problems, where the assumption of independent replication fails. We shall try to use deterministic equations similar to (11.23).

11.6. Notes

The error threshold relation (11.7) was first derived by Eigen (1971) (see also Eigen and Schuster (1979), Schuster and Swetina (1982) and Küppers (1983)). The connection with branching processes was established in Schuster and Sigmund (1980a) and Demetrius, Schuster and Sigmund (1985). An excellent introduction to branching processes and the Perron-Frobenius theorem is given by Karlin (1966). Experimental results determining the accuracy of replication for Q_β- phages are due to Domingo et al. (1976). In Eigen et al. (1981) and Eigen and Schuster (1985), the "information crisis" is described as one of the main bottlenecks of prebiotic evolution. Equation (11.23) was derived by Eigen (1971) and discussed by Thompson and McBride (1974) and Jones et al. (1976). An account of the catalytic action of polynucleotides can be found in Cech (1986).

12. Catalytic Growth of Selfreproducing Molecules

12.1. The hypercycle

It is a tempting (but of course rather difficult) task to set up evolutionary experiments within a chemical reactor and watch selection operate on the level of selfreproducing macromolecules. Several types M_1 to M_n of polynucleotides would compete for energy rich molecular compounds (ATP, GTP etc...); the excess amount of polynucleotides, as well as the degradation products, would get eliminated by a steady flow (Fig. 12.1).

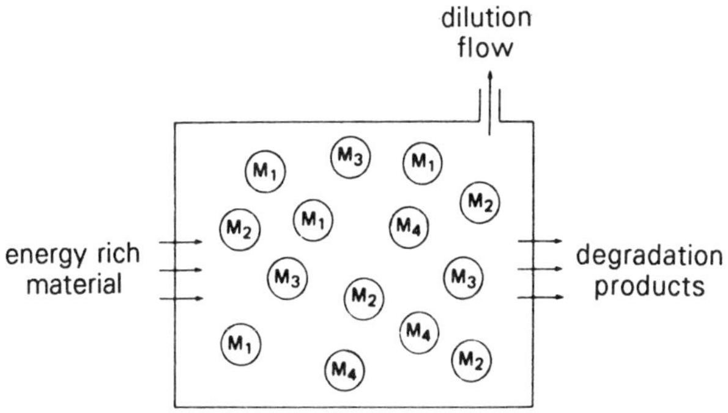

Figure 12.1

If y_i denotes the concentration of M_i, then $\frac{\dot{y}_i}{y_i}$ corresponds to its rate of growth. Depending on the chemical kinetics, many different expressions for this rate are conceivable. We only list a few of them: (i) independent, error-free replication would lead to a constant rate $a_i > 0$; (ii) an autocatalytic mechanism would correspond to a rate which increases with the concentration, for example an expression like $b_i y_i (b_i > 0)$; (iii) but we shall be mostly interested in a closed feedback loop where M_1 catalyses the replication of M_2, M_2 that of M_3 etc... until M_n, in its turn, catalyses the replication of M_1 (see Fig. 12.2). The rate of growth of M_i, then, is enhanced by the presence of its predecessor M_{i-1}: it could be, for example, an expression like $k_i y_{i-1}$, with $k_i > 0$ (we count indices cyclically modulo n, so that $y_0 = y_n$). Such a *hypercycle* has been suggested by Eigen as a solution to the "information crisis". It represents the simplest form of mutual help: every molecular type would benefit, directly or indirectly, from every other one.

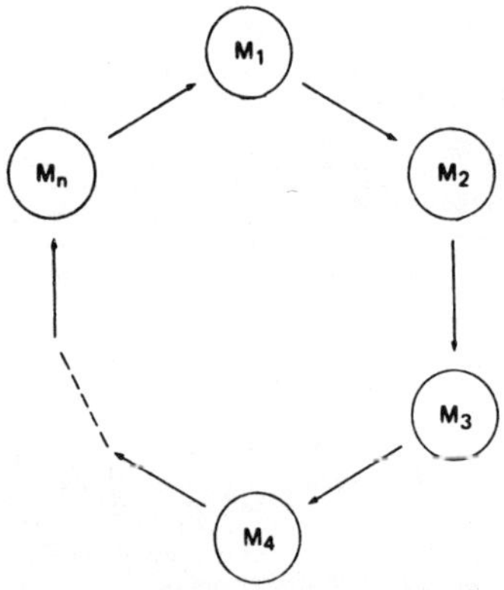

Figure 12.2

The three examples mentioned so far are:

(i) *independent replication*:

$$\dot{y}_i = a_i y_i \tag{12.1}$$

(ii) *autocatalytic replication*:

$$\dot{y}_i = b_i y_i^2 \tag{12.2}$$

(iii) *hypercyclic replication:*

$$\dot{y}_i = k_i y_i y_{i-1} \qquad (12.3)$$

12.2. Equations for the flow reactor

Obviously some supplementary mechanism must provide for the boundedness of the solutions. We shall assume that it is a *dilution flow* removing the excess concentration in the reactor so that the total concentration $\sum y_i$ remains equal to some constant c. This means that the growth rate $\frac{\dot{y}_i}{y_i}$, which is given by some function $f_i(\mathbf{y})$ in the undiluted system, is reduced by a term $\bar{f}$ which is unspecific (i.e. does not depend on i). Thus we obtain

$$\dot{y}_i = y_i(f_i(\mathbf{y}) - \bar{f}) \ . \qquad (12.4)$$

Clearly $\bar{f}$ has to be equal to the average growth rate

$$\bar{f} = \sum \frac{y_i}{c} f_i(\mathbf{y}) \ .$$

Exercise 1: Show that $S = \sum y_i$ satisfies $\dot{S} = (c - S)\bar{f}$ and hence that the simplex $\{\mathbf{x} \in \mathbf{R}^n_+ : S = c\}$ is invariant. Show that its boundary is also invariant (if $y_1(t)$ is zero for some time t, then it is zero for all times).

Exercise 2: If the f_i are homogeneous of degree s (i.e. if $f_i(\alpha y_1, \ldots, \alpha y_n) = \alpha^s f_i(y_1, \ldots, y_n)$ for $\alpha > 0$), then the relative concentrations $x_i = y_i / c$ satisfy

$$\dot{x}_i = x_i(f_i(\mathbf{x}) - \bar{f})c^s$$

with $\bar{f} = \sum x_i f_i(\mathbf{x})$. By the change in time scale $\tau = c^s t$, we get rid of the factor c^s and obtain (12.4) with $c = 1$.

If we apply the dilution flow to (12.1), (12.2) and (12.3) , we obtain in this way equations for the relative concentrations

$$x_i = \frac{y_i}{y_1 + \cdots + y_n} \ .$$

Independent replication (12.1) leads to

$$\dot{x}_i = x_i(a_i - \bar{f}) \qquad (12.5)$$

with

$$\bar{f} = \sum a_j x_j$$

an equation which we have already analysed in **11.5**. Autocatalytic replication (12.2) leads to

$$\dot{x}_i = x_i(b_i x_i - \bar{f}) \tag{12.6}$$

with with $\bar{f} = \sum b_j x_j^2$ and the hypercycle (12.3) to

$$\dot{x}_i = x_i(k_i x_{i-1} - \bar{f}) \tag{12.7}$$

with $\bar{f} = \sum k_j x_j x_{j-1}$. Equations (12.5) to (12.7) are just equations (12.1) to (12.3) modified by some term $-x_i\bar{f}$ which means that $\sum \dot{x}_i$, the sum of the right hand sides, is 0 on S_n. They leave the simplex S_n invariant, and we shall only consider this restriction to S_n. For a slightly different approach to these equations, we refer to exercise 12.4.3.

12.3. The competition of autocatalytic macromolecules

Equation (12.5) was analyzed in Section **11.5**. Equation (12.6) is just as easy to deal with. By the quotient rule

$$\left(\frac{x_i}{x_j}\right)^{\cdot} = \left(\frac{x_i}{x_j}\right)(b_i x_i - b_j x_j) = \left(\frac{x_i}{x_j}\right)\left(\frac{x_i}{x_j} - \frac{b_j}{b_i}\right) b_i x_j \, .$$

Clearly if the ratio $\frac{x_i}{x_j}$ is equal to $\frac{b_j}{b_i}$, then it will remain constant. If it is larger, it will increase, and if it is smaller, it will decrease further. This implies that the final outcome depends on the initial values $b_1 x_1(0), \ldots, b_n x_n(0)$. In general one of these values will be larger than the others and the corresponding molecular species will outcompete the others. If several of the initial values happen to be equal, then the ratio of the corresponding frequencies remains constant (see Fig. 12.3).

Just as for constant rates of growth (12.5), one molecular species will win the competition at the expense of the others. However, the selection depends on the initial conditions: in a competition of autocatalytic species, molecular types which are present in small quantities only are doomed to extinction.

Exercise 1: Find a Ljapunov function for (12.6).

Exercise 2: Compute the eigenvalues for the rest point in int S_n.

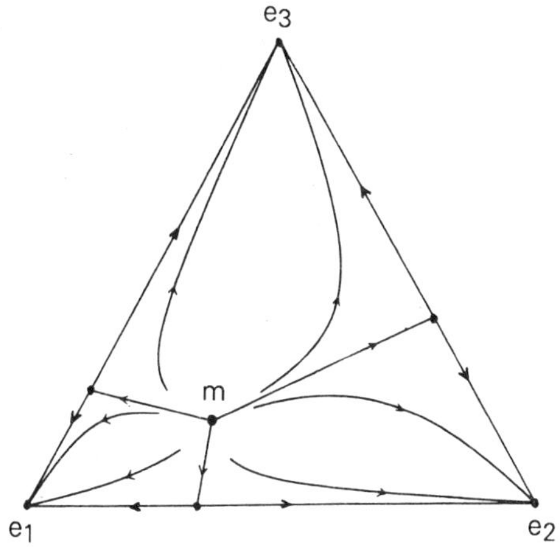

Figure 12.3

12.4. A barycentric transformation for the hypercycle

Let us now investigate the hypercycle equation (12.7). It admits a unique equilibrium $\mathbf{p}$ in int S_n, given by the equations

$$k_1 x_n = k_2 x_1 = \cdots = k_n x_{n-1} \quad (= \bar{f})$$

and

$$x_1 + x_2 + \cdots + x_n = 1 .$$

This yields

$$p_i = \frac{k_{i+1}^{-1}}{\sum_j k_{j+1}^{-1}} \qquad i = 1,\ldots,n . \tag{12.8}$$

The *stability analysis* of $\mathbf{p}$ is considerably simplified by a change in coordinates which transforms $\mathbf{p}$ into the central point $\mathbf{m} = \frac{1}{n}\mathbf{1}$ of S_n. Thus we set

$$y_i = \frac{k_{i+1} x_i}{\sum_{j=1}^{n} k_{j+1} x_j} . \tag{12.9}$$

The map $\mathbf{x} \rightarrow \mathbf{y}$ (which sends $\mathbf{p}$ into $\mathbf{m}$) is a differentiable map from S_n onto itself. Its inverse is given by

$$x_i = \frac{k_{i+1}^{-1} y_i}{\sum_s k_{s+1}^{-1} y_s} . \tag{12.10}$$

The hypercycle equation (12.7) is transformed into

$$\dot{y}_i = \frac{1}{[\sum k_{j+1} x_j]^2} [(\sum k_{j+1} x_j) k_{i+1} \dot{x}_i - k_{i+1} x_i (\sum k_{j+1} \dot{x}_j)]$$

which, as the terms with $\bar{f}$ cancel, reduces to

$$\begin{aligned} \dot{y}_i &= (\sum k_{j+1} x_j) y_i y_{i-1} - k_{i+1} x_i \sum y_j y_{j-1} \\ &= \frac{(\sum y_j) y_i y_{i-1}}{\sum_s k_{s+1}^{-1} y_s} - \frac{y_i}{\sum_s k_{s+1}^{-1} y_s} \sum_j y_j y_{j-1} , \end{aligned}$$

that is,

$$\dot{y}_i = y_i (y_{i-1} - \sum y_j y_{j-1}) M(y_1, \ldots, y_n) \tag{12.11}$$

with

$$M(y_1, \ldots, y_n) = (\sum_s k_{s+1}^{-1} y_s)^{-1} > 0 \tag{12.12}$$

on S_n. Now we shall use the following:

Theorem. *If two differential equations of the form*

$$\dot{x}_i = f_i(x_1, \ldots, x_n) \tag{12.13}$$

and

$$\dot{x}_i = f_i(x_1, \ldots, x_n) W(x_1, \ldots, x_n) \tag{12.14}$$

differ only by a positive factor W which does not depend on i, then these equations admit the same orbits.

Indeed, if $\mathbf{x}$ is a rest point of (12.13), it is also a rest point of (12.14), and vice versa. If $\mathbf{x}$ is not a rest point, on the other hand, then some component of the vector field - say $\dot{x}_n$, for example - is different from 0. In a sufficiently small neighbourhood of $\mathbf{x}$, the function $x_n(t)$ is invertible, i.e. t can be expressed as a function of x_n and so can $x_i = x_i(x_n)$. The relations

$$\frac{dx_i}{dx_n} = \frac{\frac{dx_i}{dt}}{\frac{dx_n}{dt}} = \frac{f_i}{f_n} = \frac{f_i W}{f_n W} \quad (1 \le i \le n-1) \tag{12.15}$$

imply that (12.13) and (12.14) lead both to the same functions x_i (of x_n). In other words, the vector fields (12.13) and (12.14) always point into the same direction and differ only by the ratio $W(x_1,\dots,x_n)$ of their magnitudes. The orbits in the phase-space, therefore, are the same: the factor W corresponds to a *change in velocity*.

Exercise 1: Give a different proof of this: Assume that the solutions $\mathbf{x}(t)$ of (12.14) are given, and show that for a suitable (strictly monotonic) change in time-scale $\tau = \varphi(t)$, $\mathbf{x}(\tau)$ is a solution of (12.13).

Returning to the hypercycle equation (12.11), we can drop the factor $M(y_1,\dots,y_n)$ without changing the orbits. Writing x_i for y_i again, we obtain

$$\dot{x}_i = x_i(x_{i-1} - \bar{f}) \tag{12.16}$$

with

$$\bar{f} = \sum_j x_j x_{j-1} \tag{12.17}$$

which is just (12.7) with $k_1 = k_2 = \cdots = k_n = 1$. Thus a change in coordinates followed by a change in velocity allows one to get rid of the coefficients k_i. As we shall presently see, this simplifies the computations a lot.

Exercise 2: Apply this barycentric transformation also to (12.6).

Exercise 3: Assume that the right hand side of (12.2) or (12.3) is modified by an unspecific dilution term $-y_i\Phi(\mathbf{y})$ which keeps $\sum y_i$ within some bounds (but not necessarily constant). Show that the relative concentrations obey (12.6) resp. (12.7), up to a change in velocity.

12.5. The computation of the eigenvalues

The Jacobian of (12.16) at the point $\mathbf{m}$, which is its unique equilibrium in int S_n, is given by

$$\begin{aligned}\frac{\partial \dot{x}_i}{\partial x_j} &= \frac{\partial}{\partial x_j}[x_i(x_{i-1} - \sum_{s=1}^{n} x_s x_{s-1})] \\ &= \frac{\partial x_i}{\partial x_j}(x_{i-1} - \sum_{s=1}^{n} x_s x_{s-1}) + x_i(\frac{\partial x_{i-1}}{\partial x_j} - (x_{j-1} + x_{j+1}))\end{aligned}$$

The first term vanishes at $\mathbf{m}$, and we are left with

$$\frac{\partial \dot{x}_i}{\partial x_j} = \frac{1}{n}(1-\frac{2}{n}) \text{ for } j=i-1 \tag{12.18}$$

$$= -\frac{2}{n^2} \text{ otherwise .} \tag{12.19}$$

The Jacobian is *circulant* (see **9.5**). The elements in the first row are

$$-\frac{2}{n^2}, -\frac{2}{n^2}, \ldots, -\frac{2}{n^2}, \frac{1}{n}-\frac{2}{n^2}$$

and the other rows are obtained by cyclic permutation. The eigenvalues are given by (9.20). This yields

$$\gamma_0 = \frac{1}{n} - \frac{2n}{n^2} = -\frac{1}{n} \tag{12.20}$$

and, for $j=1,..,n-1$

$$\gamma_j = \sum_{k=0}^{n-1}(-\frac{2}{n^2})\lambda^{kj} + \frac{1}{n}\lambda^{(n-1)j} = \frac{\lambda^{-j}}{n} \tag{12.21}$$

with $\lambda = e^{2\pi i/n}$. The eigenvalue γ_0 corresponds to the eigenvector $\mathbf{1}$ which is orthogonal to the simplex S_n. Since we are only interested in the restriction of (12.16) to S_n, we need not consider it any further. For the $(n-1)$-dimensional dynamical system on S_n, we get:

Theorem. *The eigenvalues of the Jacobian of the hypercycle equation* (12.16) *at the point* $\mathbf{m}$ *are given by*

$$\gamma_j = \frac{1}{n}e^{2\pi ij/n} \quad j = 1,\ldots,n-1 \,. \tag{12.22}$$

For $n=2$ and $n=3$, these eigenvalues have negative real parts. By the theorem of Hartman and Grobman, $\mathbf{m}$ is asymptotically stable, therefore. For $n=4$, two complex conjugate eigenvalues lie on the imaginary axis, so that we cannot apply the theorem of Hartman and Grobman. We shall see in the next section, however, that $\mathbf{m}$ is still asymptotically stable. For $n \geq 5$, finally, $\mathbf{m}$ is always unstable, since some of the eigenvalues have positive real parts.

Exercise: Compute the eigenvalues of (12.7) at $\mathbf{p}$.

12.6. A Ljapunov function for short hypercycles

The function

$$P(\mathbf{x}) = x_1 x_2 \cdots x_n \tag{12.23}$$

vanishes on bd S_n (where $x_i = 0$ for at least one i) and is strictly positive in int S_n. It attains its maximal value (in S_n) at the point $\mathbf{m}$. The time derivative of the function $t \to \log P(\mathbf{x}(t))$ (where $\mathbf{x}(t)$ is a solution of (12.16)) is given by

$$(\log P)^{\cdot} = \sum_{i=1}^{n} \frac{\dot{x}_i}{x_i} = \sum_{i=1}^{n} (x_{i-1} - \bar{f}) = \Psi(\mathbf{x}) \tag{12.24}$$

where

$$\Psi(\mathbf{x}) = (x_1 + \ldots + x_n)^2 - n \sum_{j=1}^{n} x_j x_{j-1} \tag{12.25}$$

(here we have used the fact that $x_1 + \ldots + x_n = 1$ on S_n).

For $n=2$

$$\Psi(\mathbf{x}) = (x_1 + x_2)^2 - 2(x_1 x_2 + x_2 x_1) = (x_1 - x_2)^2 \, .$$

Hence $\Psi(\mathbf{x}) \geq 0$, with $\Psi(\mathbf{x}) = 0$ iff $x_1 = x_2$, i.e. iff $\mathbf{x} = \mathbf{m}$. By Ljapunov's theorem every orbit in int S_2 converges to $\mathbf{m}$. A similar result holds for $n=3$, where

$$\begin{aligned}\Psi(\mathbf{x}) &= (x_1 + x_2 + x_3)^2 - 3(x_1 x_2 + x_2 x_3 + x_3 x_1) \\ &= x_1^2 + x_2^2 + x_3^2 - (x_1 x_2 + x_2 x_3 + x_3 x_1) \\ &= \frac{1}{2}[(x_1 - x_2)^2 + (x_2 - x_3)^2 + (x_3 - x_1)^2] \quad .\end{aligned}$$

Again, $\Psi(\mathbf{x}) \geq 0$ with equality iff $x_1 = x_2 = x_3$, i.e. iff $\mathbf{x} = \mathbf{m}$. Thus every orbit in int S_3 converges to $\mathbf{m}$.

For $n=4$ the situation is slightly more complicated. One has

$$\begin{aligned}\Psi(\mathbf{x}) &= (x_1 + \ldots + x_4)^2 - 4(x_1 x_2 + x_2 x_3 + x_3 x_4 + x_4 x_1) \\ &= [(x_1 + x_3) - (x_2 + x_4)]^2 \quad .\end{aligned}$$

Again $\Psi(\mathbf{x}) \geq 0$, but this time the set where $\Psi(\mathbf{x}) = 0$, i.e. the set

$$\{\mathbf{x} \in S_4 : x_1 + x_3 = x_2 + x_4\}$$

does not reduce to the point $\mathbf{m}$. The theorem of Ljapunov states that every orbit in int S_4 converges to the maximal invariant subset M of this set. What does M look like? Every point in M must additionally satisfy

$$(x_1+x_3)^{\cdot} = (x_2+x_4)^{\cdot}$$

that is,

$$x_1x_4+x_3x_2-(x_1+x_3)\bar{f} = x_2x_1+x_4x_3-(x_2+x_4)\bar{f}$$

or

$$(x_1-x_3)(x_4-x_2) = 0 \quad .$$

Thus M is contained in the set where $x_1 = x_3$ or $x_2 = x_4$. The invariance of M implies, however, that whenever $x_1 = x_3$, then $\dot{x}_1 = \dot{x}_3$, that is $x_4-\bar{f} = x_2-\bar{f}$ or $x_4 = x_2$ and similarly, whenever $x_2 = x_4$, then $x_1 = x_3$. Hence M consists of the rest point $\mathbf{m}$ only. Again every orbit in int S_4 converges to $\mathbf{m}$.

Theorem. *For "short hypercycles"* ($n = 2,3,4$) *the inner equilibrium is globally stable.*

For $n \geq 5$, *however, the inner equilibrium is unstable*, and one is left with the question whether a permanent coexistence of the different types of polynucleotides is possible.

Exercise 1: An alternative way of proving the global stability of $\mathbf{m}$ for $n = 3$ would be to use other Ljapunov functions than (12.23). Try e.g. $x_1^{-p} + x_2^{-p} + x_3^{-p}$.

Exercise 2: Write down a global Ljapunov function for (12.7) if $n \leq 4$. Does $P = \prod x_i^{p_i}$, with $\mathbf{p}$ the interior fixed point (12.8), always work?

12.7. Notes

The concept of an "evolution reactor" is due to Eigen (1971). A discrete version of it was used in the "serial transfer" experiments of Spiegelman (1971). The "hypercycle equation" (12.7) was first discussed in Eigen (1971). The local stability analysis for (12.16) was carried out in Eigen and Schuster (1979). The Ljapunov function (12.23) was used in Schuster et al. (1978) to discuss short hypercycles. Schuster et al. (1980) use the barycentric change in coordinates (12.9). In Eigen et al. (1980) it is shown that (12.7) can indeed be used as approximation for the kinetics of a realistic hypercycle. Higher order hypercycles have been studied by Lyubich (1986). Catalytic RNA is discussed in Cech (1986).

13. Permanence and the Evolution of Hypercycles

13.1. Permanence

As we have seen in the last chapter, hypercycles do not reach an equilibrium for $n \geq 5$. Numerical simulation shows that there is a limit cycle towards which almost all solutions in int S_n converge. This has not been rigorously proved, however. Our knowledge of the asymptotic regimes of hypercycles is still deficient. In spite of this, the situation is not all that bad. The "purpose" of the hypercycle is to allow for a coexistence of several types of selfreproducing macromolecules: whether the concentrations converge or oscillate in a regular or irregular way is of secondary importance. The main thing is that they don't vanish: more precisely, that there exists a *threshold value* $\delta > 0$ such that every solution in int S_n satisfies $x_i(t) > \delta$ for $i=1,..,n$ whenever t is large enough. This implies that if initially all species are present, even if only in tiny quantities, then after some time some sizeable amount of each will be present. No perturbation which is smaller than δ could wipe out a molecular species (and its information).

This property, which is important in many other contexts, deserves a name. A dynamical system defined on S_n will be called *permanent* if there exists a $\delta > 0$ such that $x_i(0) > 0$ for $i=1,\ldots,n$ implies

$$\liminf_{t \to +\infty} x_i(t) > \delta \tag{13.1}$$

for $i=1,\ldots,n$.

Let us stress that δ does *not* depend on the initial values $x_i(0)$. Permanence means more than just that no component will vanish. If every state is a rest point, for example, the system is not permanent. Even if initially the concentrations were abundant, a sequence of tiny perturbations could lead to the extinction of a species. For a permanent system, on the other hand, perturbations which are sufficiently small and rare cannot lead to extinction. The boundary of the state space S_n acts as a *repellor*.

The proof that the hypercycle is permanent is not obvious. We shall start with a more general theorem giving conditions for permanence which will be useful in many other situations too.

13.2. Average Ljapunov functions

Theorem 1. *Let us consider a dynamical system on S_n leaving the boundary invariant. Let $P:S_n \to \mathbb{R}$ be a differentiable function vanishing on* bdS_n *and strictly positive in* int S_n. *If there exists a continuous function Ψ on S_n such that the following two conditions hold:*

$$\text{for } \mathbf{x} \in \text{int } S_n, \quad \frac{\dot{P}(\mathbf{x})}{P(\mathbf{x})} = \Psi(\mathbf{x}) \tag{13.2}$$

$$\text{for } \mathbf{x} \in \text{bd}S_n, \quad \int_0^T \Psi(\mathbf{x}(t))\,dt > 0 \text{ for some } T > 0 , \tag{13.3}$$

then the dynamical system is permanent.

The value $P(\mathbf{x})$ measures the distance from $\mathbf{x}$ to the boundary. If one had $\Psi>0$ on bd S_n - a condition implying (13.3) - then $\dot{P}(\mathbf{x})>0$ for any $\mathbf{x} \in$ int S_n near the boundary, and so P would increase, i.e. the orbit would be repelled from bd S_n. In such a case, P would act like a Ljapunov function. Quite often, however, one cannot find a function P of this type. The weaker version defined above is said to be an *average Ljapunov function*: its time average acts like a Ljapunov function.

Now to the proof. The $T>0$ in (13.3) can obviously be chosen locally as a continuous function $T(\mathbf{x})$. Its infimum τ is positive, since bd S_n is compact. For $h>0$, we define

$$U_h = \{\mathbf{x} \in S_n: \text{there is a } T>\tau \text{ such that } \frac{1}{T}\int_0^T \Psi(\mathbf{x}(t))\,dt > h\} .$$

For $\mathbf{x} \in U_h$ we set

$$T_h(\mathbf{x}) = \inf\{T>\tau: \frac{1}{T}\int_0^T \Psi(\mathbf{x}(t))\,dt > h\} .$$

We show first that U_h is open and T_h upper semicontinuous: in other words, if $\mathbf{x} \in U_h$ and $\alpha>0$ are given, then for $\mathbf{y} \in S_n$ sufficiently close to $\mathbf{x}$, one has

$$\mathbf{y} \in U_h \quad \text{and} \quad T_h(\mathbf{y}) < T_h(\mathbf{x}) + \alpha . \tag{13.4}$$

Indeed, for α and $\mathbf{x}$ there is a $T \in [\tau, T_h(x)+\alpha)$ such that

$$\epsilon = \frac{1}{T}\int_0^T \Psi(\mathbf{x}(t))\,dt - h > 0 .$$

Since the solutions of ODE's depend continuously on the initial values,

$\mathbf{x}(t)$ and $\mathbf{y}(t)$ are near each other, for all $t \in [0,T]$, if $\mathbf{x}$ and $\mathbf{y}$ are sufficiently close. The uniform continuity of Ψ implies

$$|\Psi(\mathbf{x}(t)) - \Psi(\mathbf{y}(t))| < \epsilon \quad \text{for all } t \in [0,T]$$

and hence

$$\frac{1}{T}\int_0^T \Psi(\mathbf{y}(t))\,dt > \frac{1}{T}\int_0^T \Psi(\mathbf{x}(t))\,dt - \epsilon = h$$

from which (13.4) follows.

By (13.3) the family of nested sets U_h $(h>0)$ is an open covering of the compact set $\operatorname{bd} S_n$. There exists *one* $h>0$, then, such that U_h is an open neighbourhood of $\operatorname{bd} S_n$ (in S_n). Since $S_n \setminus U_h$ is also compact, P attains its minimum on this set. If we choose $p>0$ smaller than this minimum, then the set

$$I(p) = \{\mathbf{x} \in S_n : 0 < P(\mathbf{x}) \leq p\}$$

is contained in U_h. $I(p)$ is a "boundary layer" which is very thin if p is small.

We shall show that if $\mathbf{x} \in I(p)$, then there exists a $t>0$ such that $\mathbf{x}(t) \notin I(p)$. Indeed, otherwise $\mathbf{x}(t)$ would have to be in U_h for all $t>0$. For every such $\mathbf{x}(t)$, there exists a $T \geq \tau$ such that

$$\frac{1}{T}\int_t^{T+t} \Psi(\mathbf{x}(s))\,ds > h\ .$$

But since $\Psi = (\log P)^{\cdot}$ holds in $\operatorname{int} S_n$, this implies

$$h < \frac{1}{T}\int_t^{T+t} (\log P)^{\cdot}(\mathbf{x}(s))\,ds = \frac{1}{T}[\log P(\mathbf{x}(T+t)) - \log P(\mathbf{x}(t))]\ ,$$

that is,

$$P(x(T+t)) > P(\mathbf{x}(t))e^{hT} \geq P(\mathbf{x}(t))e^{h\tau}\ .$$

Hence there would exist a sequence t_n for which $P(\mathbf{x}(t_n))$ tends to $+\infty$, in contradiction to the boundedness of P.

Let us denote by $\bar{I}(p)$ the union of $I(p)$ with $\operatorname{bd} S_n$. All that needs to be shown is that there exists a $q \in (0,p)$ such that $\mathbf{x}(0) \notin \bar{I}(p)$ implies $\mathbf{x}(t) \notin I(q)$ for all $t \geq 0$.

The upper semicontinuous function T_h admits an upper bound $\bar{T}$ on the compact set $\bar{I}(p)$. Let t_0 be the time when $\mathbf{x}(t)$ reaches $\bar{I}(p)$:

$$t_0 = \min\{t>0 : \mathbf{x}(t) \in \bar{I}(p)\}$$

and let $\mathbf{x}(t_0)=\mathbf{y}$. Obviously $P(\mathbf{y})=p$. Let m be the minimum of Ψ on S_n. For $m \geq 0$, all is clear, since P never decreases. In the case $m<0$ we set $q=pe^{m\bar{T}}$. For $t \in (0,\bar{T})$

$$\frac{1}{t}\int_0^t \Psi(\mathbf{y}(s))\,ds \geq m$$

and hence, just as above

$$P(\mathbf{y}(t)) \geq P(\mathbf{y})e^{mt} > pe^{m\bar{T}} = q\ .$$

Hence the solution does not reach $I(q)$ for $t \in (0,\bar{T})$. Furthermore, since $\mathbf{y} \in I(p)$, there is a time $T \in [\tau,\bar{T})$ such that

$$P(\mathbf{y}(T)) \geq pe^{h\tau} > p\ .$$

At time $t+T$, thus, the orbit of $\mathbf{x}$ has left $I(p)$ without having reached $I(q)$. Repeating this argument, one sees that the orbit can never reach $I(q)$.

Theorem 2. *It is sufficient to verify* (13.3) *for all* $\mathbf{x}$ *in the* ω*-limits of boundary orbits.*

Proof: There exists, as before, an $h>0$ such that U_h is an open neighbourhood of $\omega(\mathbf{x})$ (in S_n) for all $\mathbf{x} \in \mathrm{bd}\, S_n$. Since $\mathbf{x}(t)$ converges to $\omega(\mathbf{x})$, there is a t_1 such that $\mathbf{x}(t) \in U_h$ for all $t \geq t_1$. There is a $t_2 \geq t_1+\tau$ such that

$$\frac{1}{t_2-t_1}\int_{t_1}^{t_2} \Psi(\mathbf{x}(t))\,dt > h\ ,$$

similarly a $t_3 \geq t_2+\tau$ such that

$$\frac{1}{t_3-t_2}\int_{t_2}^{t_3} \Psi(\mathbf{x}(t))\,dt > h$$

etc. We obtain a sequence $t_1, t_2, t_3, \ldots$ satisfying

$$\frac{1}{t_k-t_1}\int_{t_1}^{t_k} \Psi(\mathbf{x}(t))\,dt > h\ .$$

If k is sufficiently large, the time average

$$\frac{1}{t_k}\int_0^{t_k} \Psi(\mathbf{x}(t))\,dt$$

is close to the previous expression and hence positive.

Exercise: Show that the theorem is valid if Ψ is only assumed to be lower semicontinuous. This can always be achieved by defining, for $\mathbf{x} \in \mathrm{bd} S_n$

$$\Psi(\mathbf{x}) = \liminf_{\substack{\mathbf{y}\to\mathbf{x} \\ \mathbf{y}\in \mathrm{int}\, S_n}} \frac{\dot{P}}{P}(\mathbf{y}) \quad .$$

13.3. The permanence of the hypercycle

We shall use Theorem 13.2.1 to prove that *the hypercycle*

$$\dot{x}_i = x_i(k_i x_{i-1} - \bar{f}) \tag{13.5}$$

is permanent.

For $P(\mathbf{x})$ we choose the product $x_1 x_2 \cdots x_n$ which stood us already in good stead in **12.6**. We have $\dot{P} = P\Psi$ with $\Psi = \sum k_i x_{i-1} - n\bar{f}$. In order to show that P is an average Ljapunov function, it remains to verify condition (13.3). Thus we have to show that for every $\mathbf{x} \in \mathrm{bd}\ S_n$, there exists a $T{>}0$ such that

$$\frac{1}{T}\int_0^T \sum_{i=1}^n (k_i x_{i-1} - n\bar{f})\,dt > 0 \tag{13.6}$$

i.e. such that

$$\frac{1}{T}\int_0^T \bar{f}(\mathbf{x})\,dt < \frac{1}{Tn}\int_0^T \sum k_i x_{i-1} dt \ . \tag{13.7}$$

Since

$$k := \min k_i > 0 \tag{13.8}$$

and $\sum k_i x_{i-1} \geq k$ for all $\mathbf{x}{\in}S_n$, the right hand side of (13.7) is larger than $\frac{k}{n}$. It is enough, therefore, to show that there is no $\mathbf{x} \in \mathrm{bd}\ S_n$ such that for all $T{>}0$

$$\frac{1}{T}\int_0^T \bar{f}(\mathbf{x}(t))\,dt > \frac{k}{n} \ . \tag{13.9}$$

Let us proceed indirectly and assume that there is such an $\mathbf{x}{\in}S_n$. We shall show by induction that

$$\lim_{t\to+\infty} x_i(t) = 0 \tag{13.10}$$

for $i=1,\ldots,n$. Since $\mathbf{x} \in \text{bd } S_n$, there exists an index i_0 such that $x_{i_0}(t) \equiv 0$. Now if $x_i(t)$ converges to 0, then so does x_{i+1}; indeed, if $x_{i+1}(t)>0$, one obtains from

$$(\log x_{i+1})^{\cdot} = \frac{\dot{x}_{i+1}}{x_{i+1}} = k_{i+1}x_i - \bar{f}$$

by integrating from 0 to T and dividing by T

$$\frac{\log x_{i+1}(T) - \log x_{i+1}(0)}{T} = \frac{1}{T}\int_0^T (\log x_{i+1})^{\cdot} dt$$

$$= \frac{1}{T}\int_0^T k_{i+1}x_i(t)\,dt - \frac{1}{T}\int_0^T \bar{f}(\mathbf{x}(t))\,dt .$$

From $x_i(t) \to 0$ it follows that

$$\frac{1}{T}\int_0^T k_{i+1}x_i(t)\,dt < \frac{k}{2n}$$

for all sufficiently large T. Together with (13.9) this implies

$$\log x_{i+1}(T) - \log x_{i+1}(0) < -\frac{kT}{2n}$$

or

$$x_{i+1}(T) < x_{i+1}(0)\exp(-\frac{kT}{2n}) . \tag{13.11}$$

Hence $x_{i+1}(t) \to 0$ and (13.10) must hold. However, this contradicts the relation $\sum x_i = 1$. Hence P is an average Ljapunov function and (13.5) is permanent.

Let us mention that the same proof shows that the considerably more general "hypercycle equation"

$$\dot{x}_i = x_i(F_i(\mathbf{x})x_{i-1} - \bar{f})$$

is permanent whenever the functions $F_i(\mathbf{x})$ are all positive on S_n. Such equations describe the reaction kinetics for more realistic hypercycle models.

13.4. The competition of disjoint hypercycles

Let us assume now that the evolution reactor contains n types $M_1,\ldots,M_n$ of selfreplicating molecules organized into *several* disjoint hypercycles. This can be described by a permutation π of the et $\{1,\ldots,n\}$. Every permutation can be decomposed into *elementary cycles* $\Gamma_1,\ldots,\Gamma_s$: these correspond to hypercycles (see Fig. 13.1). The dynamics are given by

$$\dot{x}_i = x_i(k_i x_{\pi(i)} - \bar{f}) \tag{13.12}$$

with $k_i>0$ for $i=1,\ldots,n$.

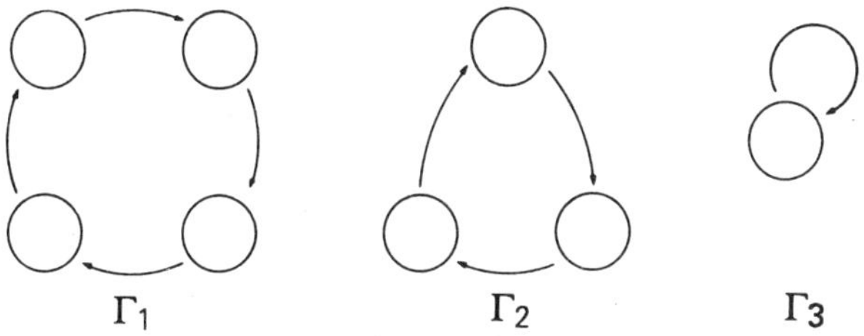

Figure 13.1

If π consists of a unique cycle, we obtain - up to a reordering of the indices - the familiar hypercycle equation (13.5). If the elementary cycle Γ_j consists of a unique element i (a rest point of the permutation π) then M_i is an autocatalytic molecular type.

As in **12.4** we may perform a transformation

$$y_i = \frac{k_{\tau(i)} x_i}{\sum_j k_{\tau(j)} x_j} \tag{13.13}$$

(with $\tau = \pi^{-1}$) and obtain from (13.12), up to a change in velocity,

$$\dot{x}_i = x_i(x_{\pi(i)} - \bar{f}) \ . \tag{13.14}$$

Again there is a unique fixed point in int S_n, namely the center $\mathbf{m} = \frac{1}{n}\mathbf{1}$.

We shall show first that *short hypercycles go to equilibrium: if the length* $|\Gamma_k|$ *of* Γ_k *is smaller than* 5 *and if* i *and* j *belong to* Γ_k*, then*

$$\frac{x_i}{x_j} \rightarrow 1 \tag{13.15}$$

for $t \rightarrow +\infty$. The proof is analogous to that in **12.6** for single hypercycles. We may assume that Γ_k consists of the indices $1,...,m$ (with $m \leq 4$). The set

$$Z = \{\mathbf{x} \in S_n : x_1 = \cdots = x_m\} \tag{13.16}$$

is invariant. Let us define

$$V = PS^{-m} \tag{13.17}$$

with $P = x_1 ... x_m$ and $S = x_1 + ... + x_m$. Then the function $t \rightarrow V(\mathbf{x}(t))$ satisfies

$$\dot{V} = P\Psi S^{-(m+1)} \tag{13.18}$$

with Ψ given as in (12.25), but with m instead of n. As in **12.6** this implies that every orbit in int S_n converges to Z, and hence that (13.15) holds.

If the evolution reactor contains only short hypercycles, i.e. if $|\Gamma_j| \leq 4$ for $j=1,\ldots,s$, then all orbits in int S_n converge to the set $W = \{\mathbf{x} \in S_n : x_i = x_j$ whenever i and j belong to the same cycle$\}$.

W is a simplex with s edges which is invariant under (13.14). If we introduce the coordinates

$$z_j = \sum_{i \in \Gamma_j} x_i \quad (j = 1,\ldots,s)$$

then we obtain as the restriction of (13.14) to W the equation

$$\dot{z}_j = z_j\left(\frac{1}{|\Gamma_j|} z_j - \bar{f}\right) . \tag{13.19}$$

This is an equation of the form (12.6) which describes the competition of autocatalytic molecular types. We know that in this case all but one of the types will vanish (except for some initial conditions of probability zero). The same is valid, then, for the competition of short hypercycles: all but one will vanish. Which one outcompetes the others depends on the initial conditions.

The same result is probably true for the competition of longer hypercycles, although this has not yet been proved. But is is easy to show that *as soon as there are more than one hypercycle in the reactor, the interior equilibrium* $\mathbf{m}$ *is unstable and the system not permanent.*

Indeed, let us consider two elementary cycles (Γ_1 and Γ_2, say) and the set

$$A = \{\mathbf{x} \in S_n : x_i < x_j \text{ for all } i \in \Gamma_2, j \in \Gamma_1\} . \tag{13.20}$$

The quotient rule implies that in A, one has

$$\left(\frac{x_i}{x_j}\right)^{\cdot} = \left(\frac{x_i}{x_j}\right) (x_{\pi(i)} - x_{\pi(j)}) < 0 \tag{13.21}$$

because $i \in \Gamma_2$ implies $\pi(i) \in \Gamma_2$, and $j \in \Gamma_1$ implies $\pi(j) \in \Gamma_1$. A is thus forward invariant, and every orbit in A converges to bd S_n. This shows that the dynamical system is not permanent, and that the equilibrium $\mathbf{m}$ (each of whose neighbourhoods intersects A) is unstable.

13.5. The evolution of hypercycles

We are now in a position to describe the evolution of hypercycles in a chemical reactor - ideally, in the "primordial soup". The different types of selfreplicating macromolecules are introduced step by step through random mutations. Initially, none of the possible hypercycles will be completed. The system, at this stage, consists of one or more open catalytic chains and is certainly not permanent, since molecular types without predecessors will tend to vanish.

Exercise: For the catalytic chain described by (13.5) with $k_1 = 0$, $k_2, \ldots, k_n > 0$, all but the "end species" M_n will tend to extinction (Hint: show by induction, for $1 \leq i \leq n-1$, that $x_i \to 0$ and x_i / x_{i+1} converges in an ultimately monotonic way to some limit. Alternatively, use Theorem 16.7.1).

Now if a mutation introduces the last "missing" type of a hypercycle, then the feedback loop starts operating. Let us assume that the first hypercycle thus completed is Γ_1. If Γ_1 is short, it goes to equilibrium; if it is larger, it pulsates; in any case, the concentration of each molecular type belonging to Γ_1 will be cushioned away from 0, and the integrated molecular message will be safely protected against small perturbations.

This implies that the average rate of growth $\bar{f}$ will be larger than some positive constant μ. From this follows, in turn, that all the concentrations of molecular types not belonging to Γ_1 will converge to 0. For a type M_p without predecessor, this is obvious: $\dot{x}_p = x_p(-\bar{f})$, so the rate of growth of x_p is smaller than $-\mu$. Once x_p is sufficiently small, the rate of growth of the type catalysed by M_p, say M_{p+1}, which is given by

$k_{p+1}x_p - \bar{f}$, will be smaller than $-\frac{\mu}{2}$ and so x_{p+1} will converge to 0 in its turn, etc. ... Even if a sequence of mutations should complete another, disjoint hypercycle - say Γ_2 - this newcomer is doomed to extinction. Indeed, all the concentrations of Γ_2 will initially be smaller than all the concentrations of Γ_1, so that the state is in the region A given by (13.20). In this case, $x_i \to 0$ for all $i \in \Gamma_2$.

At first glance this may look as if there were no evolution of hypercycles at all: the first one which is completed prevails. However, this is not quite the case. We have assumed so far that the hypercycles Γ_1 and Γ_2 are disjoint. If they have some species M in common, the outcome will be different. Let M_1 be the successor of M in Γ_1, and M_2 in Γ_2 (see Fig. 13.2). If x denotes the concentration of M, then

$$\dot{x}_1 = x_1(k_1 x - \bar{f})$$

$$\dot{x}_2 = x_2(k_2 x - \bar{f}) \tag{13.22}$$

and thus by the quotient rule

$$\left(\frac{x_1}{x_2}\right)^{\cdot} = \left(\frac{x_1}{x_2}\right)(k_1 - k_2)x \ . \tag{13.23}$$

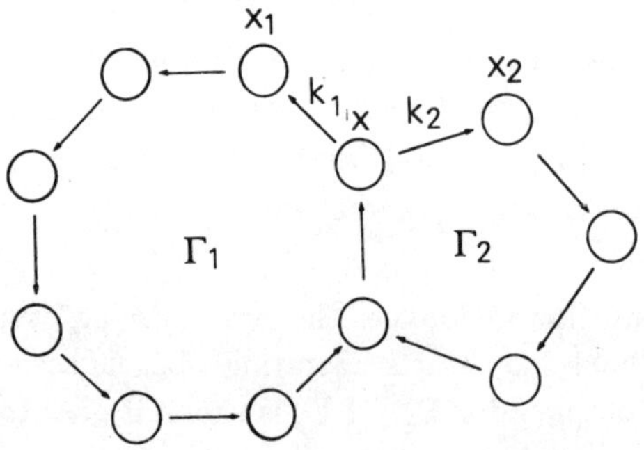

Figure 13.2

Let us forget about the highly improbable case that $k_1 = k_2$. Then either M_1 or M_2 will prevail at the cost of the other; the outcome depends on the rate constants k_1 and k_2. In particular, if $k_2 > k_1$, the new hypercycle Γ_2 will outcompete Γ_1.

This model for the evolution of hypercycles is obviously rather simpleminded. However, it leads to a plausible scenario: two hypercycles can never coexist, but one hypercycle can succeed another: in this case it will inherit parts of its predecessor. This "evolution" corresponds to a descendency tree which does not branch out, and agrees with the speculation on prebiotic evolution which was sketched in **10.4**.

13.6. Notes

Permanence was introduced in Schuster et al. (1979a). Permanence of a general "hypercycle equation" was shown in Hofbauer et al. (1981). Theorem 13.2.1 on average Ljapunov functions and its application to hypercycles is from Hofbauer (1981b), while Theorem 13.3.2 is due to Hutson (1984), who set it up in a considerably more general situation. In Chapter 19, permanence will be revisited. The competition and evolution of hypercycles is discussed in Schuster et al. (1979a). A stochastic approach to the evolution of hypercycles is described in Ebeling and Feistel (1982). On the role of hypercycles in prebiotic selforganization and on the necessity for a spatial division (compartmentation), we refer to Eigen and Schuster (1982). Recently, is has been shown that (13.5) admits a periodic orbit for $n \geq 5$; see Mallet-Paret and Smith (1988) and Hofbauer and Smith (in preparation).

PART IV: STRATEGIES AND STABILITY: AN OPENING IN GAME DYNAMICS

14. Some Aspects of Sociobiology

14.1. Sociobiology and genetics

Sociobiology is defined as the systematic investigation of the biological basis of social behaviour in humans and other animals. It is still a controversial science, especially in its applications to men, which are bound to be highly speculative since experiments on a large scale are impossible. (One of the few attempts seems to have been undertaken by an Egyptian ruler who let a group of newborn children be raised by a team of deaf and dumb slaves in the vain hope of finding the original language of mankind). Here we shall deal exclusively with animal behaviour. This leaves plenty of material for debate.

Behavioural traits in animals may be *acquired* or *innate.* A lion, for example, requires three years of coaching by his pride in order to master the hunting technique necessary for its survival. A swallow, on the other hand, does not need to be shown how to fly. Even if its wings were immobilized from its birth onwards, it will swoop through the air right after its release with perfect confidence and skill. A striking example of an innate behavioural program is displayed by the cuckoo, who never meets its parents.

The distinction between acquired and innate traits is often quite tricky. In one way or another, however, every type of behaviour rests on some genetic predispositions. In a trivial sense, this just means that it presupposes the right set of nerves and muscles. In another sense, one can argue that if some trait has evolved through natural selection (which often is hard to dispute), then there must have been a genetic variation for such a trait.

There is more direct evidence for *genes influencing behaviour.* One classical example is furnished by honeybees. Some bees are *hygienic*, meaning that they remove diseased larvae from the hive. The assumption that uncapping the cells is caused by a recessive allele u and removing the larvae by a recessive allele r (on another locus) is in agreement with the following experiment. Crossing a hygienic strain (uu/rr) with a non-hygienic strain (UU/RR) leads to a non-hygienic daughter generation (Uu/Rr). Backcrossing this with the pure hygienic strain produces four types of offspring: one quarter is hygienic (uu/rr) and another one non-hygienic (Uu/Rr). A further quarter uncaps the cells but does not remove the larvae (uu/Rr), and the last one does not uncap, but removes the larvae if the experimenter does the uncapping (Uu/rr).

Of course the complex pathways from the genetic program to the behavioural phenotype (via enzymes, hormones and neuromuscular developments) are unknown; but this holds just as well for more conventional traits like the form and the colour of the seeds of Mendelian peas. It must also be stressed that no complex behaviour is likely to be reducible to the action of a single gene, but is rather the outcome of a complex interaction of genetic and environmental factors. A "gene" for this or that trait means a genetic difference inducing a difference in behaviour (not necessarily in a deterministic way, but by affecting the probability of its occurence).

All behavioural traits which will be discussed in the following will be assumed to have evolved by natural selection, and hence to rest on a genetic basis.

14.2. Altruistic behaviour: gene selection versus group selection

Altruism is a particular challenge to the theory of natural selection. An animal behaves in an *altruistic* way if it promotes the welfare of another at the expense of its own. This seems difficult to reconcile with the notion of survival value, but it occurs nevertheless, and very frequently indeed. The fearless defence of nestlings by a bird mother, the self-immolation of soldier ants, the alarm call which warns the flock but attracts the attention of the predator upon the caller, the restraint of a wolf sparing its defeated rival are but a few examples of such acts.

A group may benefit from altruistic traits among its members, but this is not, in general, sufficient as an explanation. It reduces the probability of extinction of the group, but selection on group level is, as a rule, slow and much less effective than on the level of the individual. *Group selection* arguments, while not always unworkable as has been claimed, have been discredited in many cases. If a mutant gene (or rather gene complex) promoting altruistic behaviour reduces the fitness of its carrier, it will tend to be eliminated in spite of the boost it provides for the group. A basic task for sociobiology consists in finding explanations in terms of *gene selection* and individual survival values.

These explanations can be classified, broadly, into four categories. One of them is *manipulation.* The foster parents of a cuckoo are "cheated" by the intruder. Similarly, the queens of some species of ants invade the nests of other species, kill the host queen and use the host workers to bring up their own young. In some situations, the victims can be expected to develop countermeasures (but not in the last example. Why?); this may lead to "arms races" as for other host-parasite interactions. Manipulations can also be seen within one species. A young bird calling for more than its share of food can be said to manipulate its parents. An intriguing viewpoint is that *all* signals are manipulations.

An altruistic act caused by manipulation has a survival value, but not for the performer of the act. This is in contrast to the other three classes of sociobiological explanations for altruism, which we denote for short as the genetic, the strategic and the economic one:

(1) The genetic explanation is based on the notion of *kin selection.* A gene complex programming altruistic acts which benefit relatives may spread because it occurs, with a certain probability, within relatives whose reproductive success is increased. This will be discussed in the following two sections.

(2) The strategic explanation relies on the notion of *frequency-dependent fitness.* The fitness of an individual may depend on what the others are doing. An act which in certain situations would be called altruistic need not, in other situations, decrease the reproductive success of the individual performing it. Game theoretical considerations show that conflicting interests can lead to the evolution of stable and apparently cooperative traits of behaviour. This will be discussed in more detail in the next chapters.

(3) The economic explanation, finally, relies on the notion of *reciprocal altruism.* If the altruistic act benefits the recipient more than it costs the actor, and if it is likely to be returned at some later occasion, then such behaviour may become established, especially so if there is individual recognition between members of the population preventing "cheaters" from exploiting the others. As an example, we mention unrelated young male baboons teaming up: while one of them mounts a female, the other one fights off its consort. The roles are reversed on a later occasion.

14.3. Kin selection

A gene (or gene complex) can increase its probability of survival if it programs the mother to sacrifice herself, if necessary, in order to save her children. Indeed, the children will possess, with probability 1/2, copies of this gene. For the gene, an individual is a vehicle, a "survival machine" to guarantee the continuation of the germ line.

Thus every behaviour which favours relatives can be viewed as a kind of *selfishness*, namely selfishness of the corresponding gene which occurs – with more or less probability, depending on the relatedness – in the other individual as well. Thus the gene can be of benefit to its copies. The closer the relatedness of the recipient of an altruistic act, and the higher its potential fecundity, the more such "selflessness" makes biological sense. This leads to kin selection, i.e. the selection of genes causing individuals to assist close relatives: it works because the kin share those genes with high probability.

That the welfare of the group counts little, compared to the welfare of descendants, is illustrated by the sociobiology of lions. It often occurs that young males, when taking over a (usually unrelated) pride, kill the cubs right away. For the benefit of the species this is devastating, of course, but the fitness of the new males is increased thereby. Indeed, the time needed by females to come into reproductive condition again is shortened by a year or more if they lose their cub; and since the male is likely to last only two or three years as head of the pride, he must hurry to produce his own offspring.

All the more astonishing is the behaviour of social insects where *castes* of workers are completely sterile and devote their adult life to rearing the young of others. Such *eusociality* seems to have developed eleven times in Hymenoptera (ants, bees and wasps) and once in Isoptera (termites). Now Hymenoptera have strange *haplodiploid* genetics, and kin selection offers an explanation why this renders sterile castes more likely.

We shall need the notion of *degree of relatedness* $G(U,V)$ of the individual U to the individual V. This is defined, roughly, as the probability that a randomly chosen gene of U does also occur in V. (Of course, many genes are fixed in the population. We shall only consider "rare" genes and their likelihood to spread. We shall also neglect the possibility of inbreeding.)

In diploid species, the degree of relatedness of a child to its father (or mother) is 1/2, to a grandparent 1/4, to a brother (or sister) 1/2, to an uncle (or aunt) 1/4. One always has $G(U,V)=G(V,U)$.

In haplodiploid species, the case is different. Males develop from unfertilized eggs and are haploid, females from fertilized ones and are diploid (see Fig. 14.1). This makes for strange family ties. The degree of relatedness from father to son is 0, just like that from son to father. The degree of relatedness from mother to son is 1/2, that from son to mother is 1 (thus $G(U,V)$ is not always equal to $G(V,U)$!). The degree of relatedness from mother to daughter is 1/2, just like that from daughter to mother; from brother to sister or from brother to brother, it is 1/2; but the degree of relatedness from sister to brother is 1/4, and that from sister to sister 3/4.

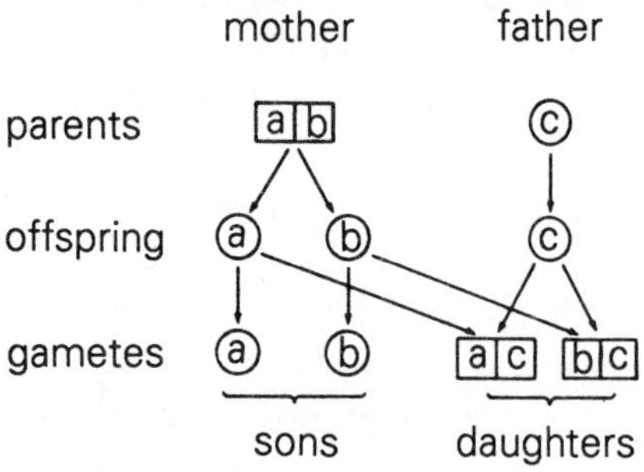

Figure 14.1

A haplodiploid female rearing a reproductive sister rather than producing a daughter is much better off than her diploid counterpart. There seems to be a conflict of interest, between workers and queen, concerning the sex ratio of the tribe. Queens are related equally to both male and female offspring and should, therefore, invest equally in the two sexes. Workers are three times less closely related to brothers than to sisters. For them, a 1 : 3 ratio would be favourable. They feed the larvae and have, therefore, a good opportunity for manipulation. In some cases, it looks indeed as if they have succeeded in twisting the sex ratio to their advantage.

14.4. Notes

Excellent introductions to sociobiology are by Dawkins (1976), Krebs and Davies (1984) and Trivers (1985). (The "bible", of course, is Wilson (1975).) The notion of kin selection is due to Hamilton (1964) and Maynard Smith (1964); its application to social insects is developed in Hamilton (1964) and Trivers and Hare (1976): we refer also to Dawkins (1982) for a critical review. For discussions of group selection, we refer to Williams (1966) and Eshel (1972).

15. Evolutionarily Stable Strategies

15.1. Conventional fighting

A long-standing theme of ethology is the prevalence of *conventional fights*. Conflicts among animals (especially within heavily armed species) are often settled by displays rather than all out fighting. Birds fluff out their feathers, frogs croak, beetles and bisons engage in pushing matches. A whole gamut of threatening signals and harmless assessments of strength serves to settle contests for food, territory and mates, so that escalated fights leading to injury or death are relatively rare. Stags, for example, engage in roaring matches, starting slowly at first and increasing the rate. If the intruder does not quit at that stage, this is followed by a parallel walk. This in turn may end with a retreat, or with a direct contest in strength: the stags clash head on, as if on a cue, interlock antlers and push against each other. If one stag turns earlier, and faces the flank of his opponent with the lethal points of his antlers, he halts his attack and resumes the parallel walk. Only a relatively small number of contests lead to serious injuries.

Such conventional fights have been compared with boxing tournaments, and the restrained nature of animal agression has often been stressed. Wolves, for example, refrain from dealing the killing bite if their opponent offers his throat in a gesture of surrender. This is obviously all to the good of the species, but needs an explanation from an evolutionary point of view. A stag ruthlessly killing his rivals should inherit their harems and increase his offspring. Its sons should inherit his fighting behaviour, and multiply faster than stags sticking to conventional displays. Why, then, do escalated contests remain rare?

15.2. A thought experiment: "hawks" and "doves"

The biologist Maynard Smith used game theory to explain the high frequency of conventional contests. It takes the form of a thought experiment: suppose there are only two possible phenotypes: one escalates the conflict until injury or the flight of the opponent settles the issue; the other one sticks to displays and retreats if the opponent escalates. These two phenotypes are usually described as "hawks" and "doves", although this is somewhat misleading. The conflicts, after all, are supposed to take place within one species and not between two; furthermore, real doves do escalate - we shall return to this in a moment.

The contests may take place over a morsel of food, the boundary line between territories or a potential mate. The prize corresponds to a gain in fitness G, while an injury reduces fitness by C.

If two "doves" meet, they posture, glare at each other, swell up, change colour or what not: but eventually, one of them retreats. The winner obtains G, the loser gets nothing, so that the average increase in fitness, for a "dove" meeting another "dove", is $\frac{G}{2}$. A "dove" meeting a "hawk" flees and its fitness remains unchanged, while that of the "hawk" increases by G. If a "hawk" meets a "hawk", finally, they escalate until one of the two gets knocked out. The fitness of the winner is increased by G, that of the loser reduced by C, so that the average increase in fitness is $\frac{G-C}{2}$, which is negative, if the cost of the injury exceeds the prize of the fight (as we shall always assume). One encapsulates this in a *payoff matrix*:

	hawk	dove
hawk	$\frac{G-C}{2}$	G
dove	0	$\frac{G}{2}$

(15.1)

In a population consisting mostly of "doves", "hawks" will spread, for they are likely to meet only "doves" and get a gain G out of each contest, while a "dove" will only get $\frac{G}{2}$. But in a population of mostly "hawks", it is the "dove" which is better off: it avoids every fight and keeps its fitness unchanged, while the "hawks" are at each others throat with an average loss in fitness of $\frac{G-C}{2}$. No phenotype is "better" than the other one: their fitness depends on the composition of the population. If the frequency of "hawks" is p, and that of doves correspondingly $1-p$, then the average increase in fitness will be

$$p\frac{G-C}{2} + (1-p)G$$

for the "hawks" and

$$(1-p)\frac{G}{2}$$

for the "doves". Equality holds iff $p = \frac{G}{C}$. If the frequency of "hawks" is less than $\frac{G}{C}$, they fare better than the "doves" and should spread, therefore; if their frequency is larger, they do less well and should diminish. Evolution should lead to a stable equilibrium with $\frac{G}{C}$ as the frequency of the "hawks".

In particular, if the cost of injury C is very large, the "hawk" frequency will be small. This results nearly in the "best for the species" (no hawks at all), but it is not the outcome of group selection: it follows from the frequency dependence of individual fitnesses.

It is well supported by observations that the "gloved fist" type of contest prevails especially among the heavily armed species. Apparently harmless animals have not developped traits for avoiding escalation. Real doves, for example, which under natural conditions cannot inflict serious injuries, will fight each other slowly to death when confined in a cage.

Exercise 1: If the display implies a decrease of fitness E (through loss of time and energy), compute the payoff matrix, the stable equilibrium and the frequency of "hawks" which maximizes the group benefit.

Exercise 2: Extend the previous analysis by allowing two other phenotypes: (a) "retaliators" which stick to displays unless the opponent starts escalating, in which case they escalate too; (b) "bullies" which fake an escalation, but run off if their opponent escalates too.

15.3. Game theory and biological evolution

The basic idea behind the "hawk-dove" model was to describe evolutionary stability in terms of game theory. This theory had been applied, previously, to all kinds of conflict in human societies, but not to animal societies. With the benefit of hindsight, it seems astonishing that biological "games" were neglected for so long in favour of political, military or economic ones. There are at least two reasons which make human conflicts harder to analyse than their "natural" counterparts in biology.

(1) The *payoff* in human systems is often doubtful, since it is hard to evaluate money, social prestige, health and other factors on a single utility scale. In the evolutionarily weighted contests in biology, the only relevant quantity is Darwinian fitness, i.e. the number of offspring or - in subtler situations - the number of copies of genes transmitted to the next generations. Although this notion presents considerable difficulties for theoretical analysis and empirical measurement, it provides, in principle at least, a scalar estimate for reproductive success.

(2) The *rationality axiom* frequently sets empirical and theoretical game theorists at odds. The investigation of animal behaviour is unhindered by such an assumption. Besides, biological conflicts tend to be fairly straightforward, compared with all but the simplest human conflicts of any real interest. The possibilities for conspiracy, fraud, lunacy, etc.... are vastly reduced.

The shift in the application of game theory from human to animal conflicts entails a shift in the meaning of "payoff" and also in the meaning of "strategy". Strategy, in the context of chess, or war, suggests a nicely calculated sequence of moves. This aspect of plotting and scheming was replaced by the notion of an innate trait in fighting behaviour, but it was soon realized that the game theoretical viewpoint was useful not only in the analysis of aggressive encounters, but in the modelling of *any* clash of interest: in the parent-offspring conflict, for example, concerning the duration of the weaning period, or in the male - female conflict about the respective share in parental investment. A strategy, thus, became a genetically programmed way of behaviour in pairwise contests. But such contests form only a small fraction of "subgames" in the struggle for life. Differences in resource allocation, size and sex ratio of the litter, dispersal rate etc... affect the fitness of an individual in quite another way. In such cases, the success is determined, not through playing against a particular opponent, or a series of opponents, but through *playing the field.*

Thus any behavioural phenotype, in fact any phenotype at all, may be viewed as a strategy, as long as we wish to analyse it by game theoretical means or, to put it less subjectively, as long as its fitness is frequency dependent (which means that it depends on the distribution of phenotypes in the population).

Selection is often viewed as an *optimization* process. In elementary situations this may indeed be the case, but in more complex situations, there is no guarantee for a steady improvement. Recombination, for example, may cause fitness to decrease (see **4.5**). The same holds for other genetic constraints, which will be discussed in chapters 28 and 29. Apart from this, there may also be strategic obstacles to an optimization of the *good of the species* (e.g. the mean fitness of a population), as the "hawk-dove" game shows.

Even in human games allowing forethought and deliberation, the self-interest of two players may lead to an outcome which is disastrous for both. This is best shown by the well known *prisoners' dilemma*. Two prisoners kept in separate cells are asked to confess a joint crime. If both confess, both will remain in jail for seven years. If only one confesses, he will instantly be freed (as witness of the prosecution), while the other gets sentenced to ten years. If none of them confesses, both will be detained for one year, pending investigation. This latter outcome is obviously the best, but will not be obtained. Each prisoner, indeed, should confess: it is the better option, no matter what the other one is going to decide. As a result, both will have to spend seven years in jail. Similar results hold for games where an individual is "playing the field". The best known example, in this respect, is the "tragedy of the commons", where the common meadow is ruined by over-exploitation, each villager trying to increase his gain.

Game theory can be used when classical optimization arguments fail.

15.4. Evolutionarily stable phenotypes

If adaptation leads to an equilibrium at all, it will be characterized by stability rather than optimality. An ***evolutionarily stable phenotype***, in Maynard Smith's definition, is a phenotype with the property that if all members of the population share it, no mutant phenotype could invade the population under the influence of natural selection.

In order to formalize this, let us denote by $W(I,Q)$ the fitness of an individual I-phenotype in a population whose composition is Q, and by $pI+(1-p)J$ the mixed population where p is the frequency of I-phenotypes and $1-p$ that of J-phenotypes. A population of I-phenotypes will be ***evolutionarily stable*** if, whenever a small amount of deviant J-phenotypes is introduced, the old phenotype I fares better than the newcomers J. This means that for all $J \neq I$,

$$W(J,\epsilon J+(1-\epsilon)I) < W(I,\epsilon J+(1-\epsilon)I) \tag{15.2}$$

for all $\epsilon>0$ which are sufficiently small.

We may obviously assume that $W(I,Q)$ is continuous in the second variable: a small change in the composition of the population will have only a slight effect on the fitness of I. By letting $\epsilon\to 0$, we obtain from (15.2) that

$$W(J,I) \leq W(I,I) \tag{15.3}$$

for all J, which means that no phenotype fares better against a population of I-phenotypes, than the I-phenotype itself. (The converse is not true in general: (15.3) does not imply (15.2).) In the "hawk-dove" game, none of the two phenotypes is evolutionarily stable: each can be invaded by the other one.

As a further example, let us consider the "sex ratio game": here, different phenotypes correspond to different sex ratios (i.e. frequencies of males among offspring). Already Darwin was puzzled by the prevalence of the sex ratio 1/2 in animal populations. It is not, as we may first think, an immediate result of the sex determination through X- and Y-chromosomes. In fact, at conception the ratio may be quite different, and subsequently change to yield a value near 1/2 at birth. What is the evolutionary reason of this? After all, as animal breeders know, a female biased sex ratio leads to a higher overall growth rate. Why are there so many males around?

We shall presently see that while the number of children is not affected by the sex ratio, yet the number of grandchildren is. Roughly speaking, if there were more females, then males would have good prospects. Since the same holds vice versa, this should lead to a sex ratio of 1/2.

To check this, let us denote by p the sex ratio of a given individual, and by m the average sex ratio in the population. Let N_1 be the population number in the daughter generation F_1 (of which mN_1 will be male and $(1-m)N_1$ female) and N_2 the number in the following generation F_2. Each member of F_2 has one mother and one father: the probability that a given male in the F_1 generation is its father is $\frac{1}{mN_1}$, and the expected number of children produced by a male in the F_1 generation is therefore $\frac{N_2}{mN_1}$ (assuming random mating). Similarly, a female in the F_1 generation contributes an average of $\frac{N_2}{(1-m)N_1}$ children. Since a p-

phenotype produces male and female children in the ratio p to $1-p$, its expected number of grandchildren will be proportional to

$$p\frac{N_2}{mN_1}+(1-p)\frac{N_2}{(1-m)N_1}$$

i.e., its fitness is proportional to

$$w(p,m) = \frac{p}{m} + \frac{1-p}{1-m}. \tag{15.4}$$

(We may clearly exclude the cases $m=0$ and $m=1$ which lead to immediate extinction.) For given $m\in(0,1)$ the function $p\to w(p,m)$ is affine linear, increasing for $m<1/2$, decreasing for $m>1/2$ and constant for $m=1/2$.

Let us now consider a phenotype with sex ratio q, and ask whether it is evolutionarily stable in the sense that no other phenotype with sex ratio p can invade. If such a deviant phenotype is introduced in a small proportion ϵ, the average sex ratio of the population is $r=\epsilon p+(1-\epsilon)q$. The q-phenotype fares better than the p-phenotype if and only if

$$w(p,r) < w(q,r)\ . \tag{15.5}$$

This is obviously the case for every p when $q=1/2$ (it is enough to note that p and r are either both smaller, or both larger than $1/2$). For $q<1/2$, a sex ratio $p>q$ will do better, however, and consequently spread; similarly, any $q>1/2$ can be invaded by a smaller p. Thus $1/2$ is the unique evolutionarily stable sex ratio in the sense of (15.2).

15.5. Evolutionarily stable strategies

For any given phenotype, we can conceive one which could supplant it - by adding properties like breathing fire, flying at supersonic speed etc. One of the first prerequisites for a game theoretical modelling is therefore the careful delimitation of those variations which are likely to occur in reality. This yields some well defined alternatives which we call "pure strategies". This could be the "hawk" and the "dove" strategy, or - in the context of the sex ratio game - "produce only sons" and "produce only daughters". In general, we shall assume that there are n such pure strategies R_1 to R_n. It will be useful to take into account "mixed strategies" too: these will consist in playing the pure strategies R_1 to R_n with some pre-assigned probabilities p_1 to p_n. (In the "hawk-dove" game, this would mean escalating with a certain probability; in the "sex-ratio" game, mixing male and female offspring with some given ratio). Since $p_i \geq 0$ and $\sum p_i = 1$, a strategy corresponds to a point $\mathbf{p} \in S_n$. The corners $\mathbf{e}_i$ are the pure strategies R_i.

Does one find in nature phenotypes realizing mixed strategies? For games describing pairwise contests they seem to be rather rare. On the other hand, they seem to prevail when "playing the field". The sex ratio is but one example.

The payoff F_i for strategy R_i depends on the state of the population: escalating a conflict in a population of doves is more rewarding and less risky than in a population consisting of hawks. The F_i will thus be assumed to be functions of the frequencies of the strategies R_j in the population ($j=1,\dots,n$). The strategy mix of the population is given by a point $\mathbf{m}$ which also belongs to S_n. Since a $\mathbf{p}$-strategist plays R_i with probability p_i, its payoff will be given by

$$\sum p_i F_i(\mathbf{m}) = \mathbf{p} \cdot \mathbf{F}(\mathbf{m}) \,. \tag{15.6}$$

For expository reasons, let us assume that $F_i(\mathbf{m})$ is linear in $\mathbf{m}$, i.e. of the form

$$F_i(\mathbf{m}) = (A\mathbf{m})_i = \sum_{j=1}^{n} a_{ij} m_j \tag{15.7}$$

where the $n \times n$ matrix $A = (a_{ij})$ is the so-called payoff matrix. This linearity assumption is valid, for example, if the game consists of one or repeated encounters with a randomly chosen opponent: the probability of meeting an R_j strategist is m_j and the payoff (to the R_i-strategist) is then a_{ij}. In the "hawk-dove" game this assumption is satisfied and A is given by (15.1). In the sex-ratio game, however, it is not valid, as indicated by (15.4).

Under the linearity assumption, the payoff for a $\mathbf{p}$-strategist in a population whose strategy mixture is $\mathbf{m}$ is therefore given by

$$\mathbf{p} \cdot A\mathbf{m} = \sum_{i,j} p_i a_{ij} m_j \,. \tag{15.8}$$

Let us now recall the definition of evolutionary stability given in (15.2). The two phenotypes I and J correspond to two strategies $\mathbf{p}$ and $\mathbf{q}$. The population $\epsilon J + (1-\epsilon) I$ has the strategy mix $\epsilon \mathbf{q} + (1-\epsilon)\mathbf{p}$. Thus the strategy $\mathbf{p} \in S_n$ will be said to be *evolutionarily stable* if for all $\mathbf{q} \in S_n$ with $\mathbf{q} \neq \mathbf{p}$, the inequality

$$\mathbf{q} \cdot A(\epsilon \mathbf{q} + (1-\epsilon)\mathbf{p}) < \mathbf{p} \cdot A(\epsilon \mathbf{q} + (1-\epsilon)\mathbf{p}) \tag{15.9}$$

holds for all $\epsilon > 0$ which are sufficiently small, i.e. smaller than some appropriate $\bar{\epsilon}(\mathbf{q})$.

Equation (15.9) may be written as

$$(1-\epsilon)(\mathbf{p}\cdot A\mathbf{p}-\mathbf{q}\cdot A\mathbf{p})+\epsilon(\mathbf{p}\cdot A\mathbf{q}-\mathbf{q}\cdot A\mathbf{q}) > 0 \,. \tag{15.10}$$

This implies that $\mathbf{p}$ is an evolutionarily stable strategy (or ESS) if and only if the following two conditions are satisfied:

(a) *equilibrium condition*

$$\mathbf{q}\cdot A\mathbf{p} \le \mathbf{p}\cdot A\mathbf{p} \text{ for all } \mathbf{q}\in S_n \tag{15.11}$$

(b) *stability condition*

$$\text{if } \mathbf{q} \neq \mathbf{p} \text{ and } \mathbf{q}\cdot A\mathbf{p} = \mathbf{p}\cdot A\mathbf{p} \text{ , then } \mathbf{q}\cdot A\mathbf{q} < \mathbf{p}\cdot A\mathbf{q} \,. \tag{15.12}$$

(15.11) is the definition of a *Nash equilibrium*, a central notion in game theory. It means that strategy $\mathbf{p}$ is a *best reply against itself*. This property alone, however, does not guarantee non-invadability, since it permits that another strategy $\mathbf{q}$ is an *alternative best reply*. The stability condition states that in such a case, $\mathbf{p}$ fares better against $\mathbf{q}$ than $\mathbf{q}$ against itself.

It is not astonishing that human and animal games lead to similar concepts (Nash equilibrium and ESS). "Blind selection" and "rational decision" lead both to adaptive solutions. (After all, rational behaviour is itself a product of selection.)

(15.11) implies, by setting $\mathbf{q}=\mathbf{e}_i$, that

$$(A\mathbf{p})_i \le \mathbf{p}\cdot A\mathbf{p} \tag{15.13}$$

for $i = 1,\ldots,n$. By summation, one sees that for all i with $p_i>0$ (corresponding to pure strategies which are effectively played), equality must hold. The Nash equilibrium condition (15.11) means that there exists a constant c such that $(A\mathbf{p})_i \le c$ for all i, with equality if $p_i>0$. In particular, $\mathbf{p}\in\text{int } S_n$ is a Nash equilibrium iff its coordinates satisfy the linear system of equations

$$\begin{aligned} (A\mathbf{x})_1 &= \cdots = (A\mathbf{x})_n \\ x_1+\cdots+x_n &= 1 \,. \end{aligned} \tag{15.14}$$

Theorem. $\mathbf{p}\in S_n$ *is an ESS if and only if*

$$\mathbf{p}\cdot A\mathbf{x} > \mathbf{x}\cdot A\mathbf{x} \tag{15.15}$$

holds for all $\mathbf{x} \neq \mathbf{p}$ *in some neighbourhood of* $\mathbf{p}$ *in* S_n.

For the proof, let us start by assuming that $\mathbf{p}$ is evolutionarily stable. It will be enough to show that every $\mathbf{x}$ occurring in (15.15) can be written as $\epsilon\mathbf{q}+(1-\epsilon)\mathbf{p}$, for small ϵ, and then apply (15.9). Actually we can choose $\mathbf{q}$ in the compact set $C = \{\mathbf{x}\in S_n : x_i=0 \text{ for some } i \in \text{supp}(\mathbf{p})\}$, which consists of the faces which do not contain $\mathbf{p}$. For every $\mathbf{q}\in C$, (15.9) holds for all $\epsilon<\bar{\epsilon}(\mathbf{q})$. It is easy to see that $\bar{\epsilon}(\mathbf{q})$ can be chosen to be continuous. Then $\bar{\epsilon} = \min\{\bar{\epsilon}(\mathbf{q}):\mathbf{q}\in C\}$ is strictly positive, and (15.9) holds for all $\epsilon\in(0,\bar{\epsilon})$. It is enough, then, to multiply (15.9) by ϵ, and to add

$$(1-\epsilon)\mathbf{p}\cdot A((1-\epsilon)\mathbf{p}+\epsilon\mathbf{q})$$

to both sides. With $\mathbf{x} = (1-\epsilon)\mathbf{p}+\epsilon\mathbf{q}$, this yields (15.15) for all $\mathbf{x}\neq\mathbf{p}$ in an appropriate neighbourhood of $\mathbf{p}$. The converse is similar.

It follows that if $\mathbf{p}\in\text{int } S_n$ is evolutionarily stable, then there is no other ESS: indeed, (15.15) must then hold for all $\mathbf{x}\in S_n$. There are games without ESS, however, and games with several ESS (which have then to lie on bd S_n).

In the "hawk-dove" game, one easily computes that the strategy $\mathbf{p} = (\frac{G}{C},\frac{C-G}{C})$ is an ESS. Indeed

$$\mathbf{p}\cdot A\mathbf{x} - \mathbf{x}\cdot A\mathbf{x} = \frac{1}{2C}(G-Cx_1)^2 \tag{15.16}$$

is strictly positive for all $x_1\neq\frac{G}{C}$, so that (15.15) is satisfied. Since $\mathbf{p}\in\text{int } S_2$, $\mathbf{p}$ is then the unique ESS.

Exercise 1: Show that the games given by

$$\text{(a)} \begin{bmatrix} 1 & 0 & 0 \\ 0 & 1 & 0 \\ 0 & 0 & 1 \end{bmatrix} \qquad \text{(b)} \begin{bmatrix} 0 & 1 & -1 \\ -1 & 0 & 1 \\ 1 & -1 & 0 \end{bmatrix}$$

have three resp. no ESS.

Exercise 2: Let $\mathbf{p}\in\text{int } S_n$ be a Nash equilibrium. Show that $\mathbf{p}$ is an ESS if and only if

$$\xi\cdot A\xi < 0 \quad \text{for all} \quad \xi\neq\mathbf{0} \quad \text{with} \quad \sum_{i=1}^{n}\xi_i = 0\,. \tag{15.17}$$

Exercise 3: For games which do not satisfy the linearity assumption, we define $\mathbf{p}$ to be a local ESS if $\mathbf{p}\cdot\mathbf{F}(\mathbf{x})>\mathbf{x}\cdot\mathbf{F}(\mathbf{x})$ holds for all $\mathbf{x}\neq\mathbf{p}$ in some neighbourhood of $\mathbf{p}$. Show that this is the case if and only if

(a) $\mathbf{q}\cdot\mathbf{F}(\mathrm{p}) \leq \mathbf{p}\cdot\mathbf{F}(\mathbf{p})$ for all $\mathbf{q}\in S_n$, and

(b) $\mathbf{p}\cdot\mathbf{F}(\mathbf{q}) > \mathbf{q}\cdot\mathbf{F}(\mathbf{q})$ whenever equality holds in (a) and $\mathbf{q}\neq\mathbf{p}$ is sufficiently close to $\mathbf{p}$.

(Hint: Show first the analogue of (15.9).) Show by examples that several ESS can coexist in int S_2. Show that the characterization given by the previous exercise is still valid.

Exercise 4: Show that in the proof of the Theorem, one can use

$$\bar{\epsilon}(\mathbf{q}) = \frac{(\mathbf{p}-\mathbf{q})\cdot A\mathbf{p}}{(\mathbf{p}-\mathbf{q})\cdot A(\mathbf{p}-\mathbf{q})} \quad \text{if } \mathbf{q}\cdot A\mathbf{q} > \mathbf{p}\cdot A\mathbf{q}$$

$$= 1 \text{ otherwise} \quad .$$

15.6. Notes

The ritualized fighting of stags is described in Krebs and Davies (1984). The "Hawk-Dove" game and the notion of evolutionarily stable strategy were introduced in Maynard Smith (1972), Maynard Smith and Price (1973) and Maynard Smith (1974b). A good survey of the tremendous development during the first decade of "Evolutionary Game Theory" is offered by Maynard Smith (1982). The connection between Nash equilibria and ESS is discussed in Bomze (1986), Parker and Hammerstein (1985) and Riechert and Hammerstein (1983). An important example is the war of attrition: see Hammerstein and Parker (1982). The notion of local ESS (Exercise 15.5.3) was first formulated in Pohley and Thomas (1983). Exercise 15.5.2 is from Haigh (1975). For the sex ratio game in **15.4** we refer to Shaw and Mohler (1953), Maynard Smith (1982), Sigmund (1987a) and **24.6**. In Karlin and Lessard (1986) one finds many genetic models for the sex ratio. Evolutionary views of the "prisoner's dilemma" and the "tragedy of the commons" are discussed in Axelrod (1984) and Masters (1983). The "phenotypic" view of strategies is one-sided; of great biological importance is also the possibility to analyse the structure of strategies in terms of informations and actions. For a thorough discussion of this aspect we refer to Selten (1983). After the basic book by John von Neumann and Morgenstern (1953), game theory has progressed in many directions. We refer to Owen (1982) for a good introduction.

16. Game Dynamics

16.1 The evolution of phenotypes

Much of game theory is static. The notion of evolutionary stability, however, relies upon implicit dynamical considerations. In certain situations, the underlying dynamics can be modelled by a differential equation on the simplex S_n.

Thus let us assume that the population is divided into n phenotypes E_1 to E_n with frequencies x_1 to x_n. Since E_i corresponds to a (pure or mixed) strategy, its fitness f_i will be a function of the state $\mathbf{x}$ of the population. If the population is very large, and if the generations blend continuously into each other, we may assume that the state $\mathbf{x}(t)$ evolves on S_n as a differentiable function of t. The rate of increase $\dot{x}_i/x_i$ of the phenotype E_i is a measure of its evolutionary success. Following the basic tenet of Darwinism, we may express this success as the difference between the fitness $f_i(\mathbf{x})$ of E_i and the average fitness $\bar{f}(\mathbf{x}) = \sum x_i f_i(\mathbf{x})$ of the population. Thus we obtain

$$\frac{\dot{x}_i}{x_i} = \text{fitness of } E_i - \text{average fitness}$$

which yields the game dynamical equation

$$\dot{x}_i = x_i(f_i(\mathbf{x}) - \bar{f}(\mathbf{x})) \qquad i = 1,\ldots,n \ . \tag{16.1}$$

As in **12.2**, we see that the simplex S_n is invariant under (16.1): if $\mathbf{x} \in S_n$ then $\mathbf{x}(t) \in S_n$ for all $t \in \mathbb{R}$. In fact, (12.5), (12.6) and (12.7) are all special cases of (16.1). From now on, we shall only consider the restriction of (16.1) to S_n.

Let us stress right away that we derived (16.1) by assuming that the success of E_i corresponds to its rate of increase, and thus tacitly that "like begets like". We have neglected the effect of segregation and recombination, i.e. of sexual reproduction. This is characteristic for the whole "strategic", i.e. game theoretic, approach, and not just for its dynamical aspect. Evolutionary stability is primarily a phenotypic notion. We shall discuss in chapters 28 and 29 how it relates to the Mendelian machinery of inheritance.

16.2 Game dynamics and ESS

Let us check first of all that the game dynamics (16.1) agrees with our notion of evolutionary stability. Let us consider a linear game with N pure strategies and payoff matrix U. Let E_1 and E_2 be two phenotypes corresponding to the strategies $\mathbf{q}$ and $\mathbf{p}$ in S_N. If x_1 and x_2 are their respective frequencies, then their payoffs are given by

$$f_1(\mathbf{x}) = \mathbf{q} \cdot U(x_1\mathbf{q} + x_2\mathbf{p})$$

$$f_2(\mathbf{x}) = \mathbf{p} \cdot U(x_1\mathbf{q} + x_2\mathbf{p}) \quad .$$

Since $x_2 = 1 - x_1$, it is enough to describe the evolution of x_1, which we denote by x. According to (16.1)

$$\begin{aligned} \dot{x} &= x(f_1(\mathbf{x}) - \bar{f}(\mathbf{x})) \\ &= x(1-x)(f_1(\mathbf{x}) - f_2(\mathbf{x})) \qquad (16.2) \\ &= x(1-x)[x(\mathbf{p} \cdot U\mathbf{p} - \mathbf{q} \cdot U\mathbf{p} + \mathbf{q} \cdot U\mathbf{q} - \mathbf{p} \cdot U\mathbf{q}) - (\mathbf{p} \cdot U\mathbf{p} - \mathbf{q} \cdot U\mathbf{p})] \quad . \end{aligned}$$

The fixed points $x = 0$ and $x = 1$ correspond to one or the other of the two phenotypes. That the $\mathbf{p}$-phenotype cannot be invaded by the $\mathbf{q}$-phenotype means that $x = 0$ is asymptotically stable: the frequency x of $\mathbf{q}$ is decreasing whenever it is sufficiently small. But this is obviously the case if and only if either (a) $\mathbf{q} \cdot U\mathbf{p} < \mathbf{p} \cdot U\mathbf{p}$ or (b) $\mathbf{q} \cdot U\mathbf{p} = \mathbf{p} \cdot U\mathbf{p}$ and $\mathbf{q} \cdot U\mathbf{q} < \mathbf{p} \cdot U\mathbf{q}$. This corresponds precisely to the notion of ESS as given in (15.11) and (15.12).

Exercise: Give a similar interpretation for the notion of (local) ESS for nonlinear games described in Exercise 15.5.3.

16.3 Simple properties of the game dynamical equation

Let J be nontrivial subset of $\{1,...,n\}$. The "face"

$$S_n(J) = \{\mathbf{x} \in S_n : x_i = 0 \text{ for all } i \notin J\}$$

of the simplex S_n is itself an invariant simplex for (16.1), and the restriction of (16.1) to $S_n(J)$ is again an equation of a similar form. The boundary bd S_n, which is the union of all such faces, is also invariant, and so is the interior int S_n.

Exercise 1: Check the "quotient rule": for $x_j > 0$

$$\left(\frac{x_i}{x_j}\right)^{\cdot} = \frac{x_i}{x_j}(f_i(\mathbf{x}) - f_j(\mathbf{x})) \quad . \qquad (16.3)$$

The addition of a function $\Psi: S_n \to \mathbf{R}$ to all the f_i does not affect equation (16.1) on S_n. Indeed, with $g_i(\mathbf{x}) = f_i(\mathbf{x}) + \Psi(\mathbf{x})$, one has $\bar{g}(\mathbf{x}) = \sum x_i g_i(\mathbf{x}) = \bar{f}(\mathbf{x}) + \Psi(\mathbf{x})$ (on the simplex S_n) and therefore $g_i(\mathbf{x}) - \bar{g}(\mathbf{x}) = f_i(\mathbf{x}) - \bar{f}(\mathbf{x})$ for all $\mathbf{x} \in S_n$ and all i.

Of particular interest is the case of linear f_i. There exists, then, an $n \times n$-matrix $A = (a_{ij})$ such that $f_i(\mathbf{x}) = (A\mathbf{x})_i$. Equation (16.1) takes the form

$$\dot{x}_i = x_i((A\mathbf{x})_i - \mathbf{x} \cdot A\mathbf{x}) \quad . \tag{16.4}$$

The equilibria of (16.4) in $\text{int}S_n$ are the solutions of

$$(A\mathbf{x})_1 = \cdots = (A\mathbf{x})_n \tag{16.5}$$

$$x_1 + \cdots + x_n = 1$$

satisfying $x_i > 0$ for $i = 1,\ldots,n$. In general, there exists one or no such solution. In degenerate cases, the solutions form a linear manifold in $\text{int}S_n$. The equilibria on the faces $S_n(J)$ are obtained in a similar way.

Exercise 2: The addition of a constant c_j to the j-th column of A does not change (16.4) on S_n. By adding appropriate constants, one may transform A into a simpler form, having 0 in the diagonal, for example, or 0 in the last row.

Exercise 3: Show that the equation

$$\dot{x}_i = x_i(r_i + (B\mathbf{x})_i - \mathbf{x} \cdot (\mathbf{r} + B\mathbf{x}))$$

(with $\mathbf{r} \in \mathbf{R}^n$ and B an $n \times n$ matrix) reduces to an equation of the form (16.4) on S_n.

Exercise 4: Apply the "barycentric transformation"

$$y_i = \frac{x_i c_i}{\sum_j x_j c_j}$$

(with $c_j > 0$) to (16.4). A rest point $\mathbf{p} \in \text{int}\, S_n$ can thereby be moved into the central point $\mathbf{m} = \frac{1}{n}\mathbf{1} \in S_n$ (cf. **12.4**).

16.4 Evolutionarily stable states

Let us return to the game theoretical interpretation of the dynamics considered in this chapter. There is an underlying game with N pure strategies R_1 to R_N and a payoff function which we shall (for simplicity) assume to be linear and hence given by an $N \times N$ matrix U. A strategy is defined by a point in S_N: the phenotypes E_1 to E_n correspond therefore to n points $\mathbf{p}^1,\ldots,\mathbf{p}^n \in S_N$. The state of the population is defined by their frequencies x_1 to x_N, i.e. by a point $\mathbf{x} \in S_N$. With $a_{ij} = \mathbf{p}^i \cdot U\mathbf{p}^j$ (the payoff for a $\mathbf{p}^i$-strategist in a world of $\mathbf{p}^j$-opponents), we obtain as fitness $f_i(\mathbf{x})$ of the phenotype E_i the expression

$$f_i(\mathbf{x}) = \sum_j a_{ij}x_j = (A\mathbf{x})_i \tag{16.6}$$

so that the game dynamical equation (16.1) reduces to (16.4).

It will be convenient to stress the parallels between the pure strategies R_1 to R_N, points in S_N and the $N \times N$ payoff matrix U on the one hand and the phenotypes E_1 to E_n, points in S_n and the $n \times n$ fitness matrix A on the other. In particular, we shall say that a point $\mathbf{p} \in S_n$ is a *Nash equilibrium* if

$$\mathbf{q} \cdot A\mathbf{p} \leq \mathbf{p} \cdot A\mathbf{p} \tag{16.7}$$

for all $\mathbf{q} \in S_n$, and an *evolutionarily stable state* if in addition

$$\mathbf{p} \cdot A\mathbf{q} > \mathbf{q} \cdot A\mathbf{q} \tag{16.8}$$

for all $\mathbf{q} \neq \mathbf{p}$ for which equality holds in (16.7).

If $\mathbf{p} \in S_n$ *is a Nash equilibrium of the game described by the payoff matrix* A*, then* $\mathbf{p}$ *is a rest point of* (16.4).

Indeed, by **15.5**, there exists a constant c such that $(A\mathbf{p})_i = c$ for all i with $p_i > 0$. Hence $\mathbf{p}$ satisfies the equations corresponding to (16.5) for a rest point in the face $S_n(J)$, with $J = \text{supp}(\mathbf{p})$.

Exercise 1: Show that not every rest point of (16.4) is a Nash equilibrium (there may be indices i with $p_i = 0$ but $(A\mathbf{p})_i > c$).

Theorem. *If* $\mathbf{p} \in S_n$ *is evolutionarily stable, then* $\mathbf{p}$ *is an asymptotically stable rest point of* (16.4).

Proof: Let us first check that the function

$$P(\mathbf{x}) = \prod x_i^{p_i} \tag{16.9}$$

has its unique maximum (on S_n) at the point $\mathbf{p}$. This follows from

Jensen's inequality (3.12), applied to the convex function $f = -\log$, with $\mathbf{p}, \mathbf{x} \in S_n$. One sets $0 \log 0 = 0 \log \infty = 0$. Then

$$\sum_{i=1}^{n} p_i \log \frac{x_i}{p_i} = \sum_{p_i > 0} p_i \log \frac{x_i}{p_i} \leq \log\left(\sum_{p_i>0} x_i \right) \leq \log \sum_{i=1}^{n} x_i = \log 1 = 0 \,,$$

hence

$$\sum_{i=1}^{n} p_i \log x_i \leq \sum_{i=1}^{n} p_i \log p_i$$

and thus $P(\mathbf{x}) \leq P(\mathbf{p})$ with equality iff $\mathbf{x} = \mathbf{p}$.

If $P > 0$ (i.e. for all $\mathbf{x} \in S_n$ with $x_i > 0$ whenever $p_i > 0$) one has

$$\frac{\dot{P}}{P} = (\log P)^{\cdot} = \left(\sum p_i \log x_i\right)^{\cdot} = \sum_{p_i>0} p_i \frac{\dot{x}_i}{x_i} = \sum p_i((A\mathbf{x})_i - \mathbf{x} \cdot A\mathbf{x})$$

$$= \mathbf{p} \cdot A\mathbf{x} - \mathbf{x} \cdot A\mathbf{x} \quad .$$

Since $\mathbf{p}$ is evolutionarily stable, (15.15) implies $\dot{P} > 0$ for all $\mathbf{x} \neq \mathbf{p}$ in some neighbourhood of $\mathbf{p}$. The function P, then, is a *strict local Ljapunov function* for (16.4), and all orbits near $\mathbf{p}$ converge to $\mathbf{p}$.

In fact, $\mathbf{p}$ is evolutionarily stable iff P is a strict local Ljapunov function for (16.4).

If $\mathbf{p} \in \operatorname{int} S_n$ *is evolutionarily stable, then it is a globally stable rest point for* (16.4), since $\dot{P}(\mathbf{x}) > 0$ for all $\mathbf{x} \in \operatorname{int} S_n$ $(\mathbf{x} \neq \mathbf{p})$.

It is not easy to decide whether there exist real populations in an evolutionarily stable state which is mixed (i.e. contains at least two phenotypes). Mixed states are extremely common, of course, but one has to look for strategic *equilibria* consisting of several phenotypes for which the frequency dependent fitnesses are equal. This is of course very difficult , but seems to have been accomplished in a few cases.

Exercise 2: How does the game dynamical equation look like for non-linear games (cf. Ex. 15.5.3)? Show that Theorem 16.4 is still valid.

Exercise 3: If the payoff matrix A satisfies (15.17), then the game has a *unique* ESS. (Hint: You may use the fact, shown in **19.4**, that every game has a Nash equilibrium). This ESS is globally stable. Show that (15.17) is immune against adding arbitrary constants to the *rows* of A.

Exercise 4: Let C be a positive semi-definite matrix whose kernel is spanned by $\mathbf{1}$, the vector orthogonal to

$$\mathbf{R}_0^n = \{\xi \in \mathbf{R}^n : \sum_i \xi_i = 0\} \quad .$$

If A satisfies (15.17), then CA is a stable matrix, when restricted to $\mathbf{R}_0^n$. (Hint: start with the eigenvalue equation $CA\mathbf{x} = \lambda\mathbf{x}, \mathbf{x} \in \mathbf{R}_0^n, \lambda \in \mathbf{R}$, find a $\mathbf{y} \in \mathbf{R}_0^n$ with $C\mathbf{y} = \mathbf{x}$ and conclude from $\mathbf{x} \cdot A\mathbf{x} = \lambda\mathbf{y} \cdot C\mathbf{y}$ that $\lambda < 0$. Extend this idea to complex eigenvalues.)

Exercise 5: Write (16.4) as

$$\dot{\mathbf{x}} = C(\mathbf{x})A\mathbf{x} \tag{16.10}$$

with $c_{ii}(\mathbf{x}) = x_i(1 - x_i)$ and $c_{ij} = -x_i x_j$ $(i \neq j)$. Since

$$\xi \cdot C(\mathbf{x})\xi = \sum x_i \xi_i^2 - (\sum x_i \xi_i)^2 = \sum x_i(\xi_i - \bar{\xi})^2 \geq 0 \quad ,$$

the restriction of $C(\mathbf{x})$ to $\mathbf{R}_0^n$ is positive definite, if $\mathbf{x} \in \operatorname{int} S_n$. The linearization of (16.10) at a fixed point $\mathbf{p}$ is given by

$$\dot{\xi} = C(\mathbf{p})A\xi \qquad \xi \in \mathbf{R}_0^n \quad .$$

Exercise 4 then implies that an interior ESS $\mathbf{p}$ is asymptotically stable.

16.5 Examples of game dynamics

In the following, if we do not explicitly specify the phenotypes of a population, we shall assume that they correspond to the pure strategies. Thus we associate to every game a "pure strategist dynamics", without worrying about its biological relevance. The next examples are mainly intended to illustrate mathematical aspects of (16.4).

In the case $n = 2$, by setting $x = x_1$ and $1 - x = x_2$, one obtains from (16.4) a one dimensional differential equation on [0,1]:

$$\dot{x} = x(1 - x)((A\mathbf{x})_1 - (A\mathbf{x})_2) \quad . \tag{16.11}$$

In the case of the "hawk-dove" game (see **15.2**) this yields

$$\dot{x} = \frac{1}{2}x(1 - x)(G - Cx)$$

The point G/C is a global attractor in (0,1).

Exercise 1: Describe all phase portraits of (16.11). Show that the asymptotically stable rest points are just the evolutionarily stable states.

If the gain of one player is always the loss of the other, i.e. if $a_{ij} = -a_{ji}$ holds for all i and j, the game is called a *zero sum game*. In that case

$$\mathbf{x} \cdot A\mathbf{x} = -\mathbf{x} \cdot A\mathbf{x} = 0$$

and (16.4) becomes

$$\dot{x}_i = x_i(A\mathbf{x})_i \quad . \tag{16.12}$$

Exercise 2: The "rock-scissors-paper" game with payoff matrix

$$A = \begin{pmatrix} 0 & 1 & -1 \\ -1 & 0 & 1 \\ 1 & -1 & 0 \end{pmatrix}$$

is a zero sum game. Show that $x_1x_2x_3$ is a constant of motion and that the phase portrait is given by Fig. 16.1.

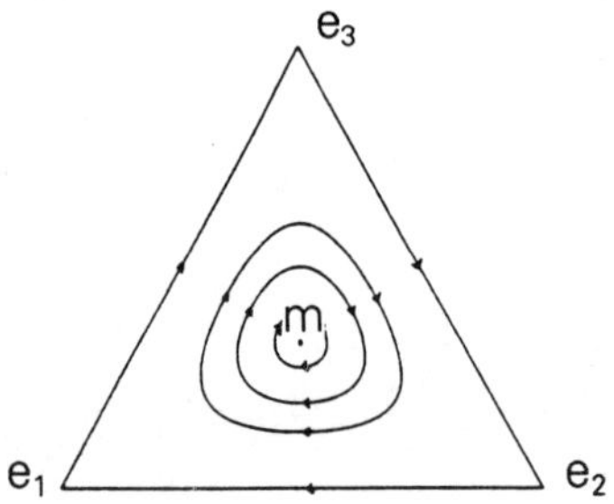

Figure 16.1

Exercise 3: Analyse the dynamics of zero sum games (16.12) (with $a_{ij} = -a_{ji}$). (Assume the existence of an interior fixed point $\mathbf{p}$).

Exercise 4: The hypercycle equation with $n = 3$ and matrix

$$A = \begin{pmatrix} 0 & 0 & k_1 \\ k_2 & 0 & 0 \\ 0 & k_3 & 0 \end{pmatrix}$$

has a globally stable rest point $\mathbf{p} \in \operatorname{int} S_3$ (see **12.6**). For which $k_i > 0$ is $\mathbf{p}$ evolutionarily stable?

Exercise 5: Analyse the general "rock-scissors-paper" game with payoff-matrix

$$A = \begin{pmatrix} 0 & a_1 & -b_3 \\ -b_1 & 0 & a_2 \\ a_3 & -b_2 & 0 \end{pmatrix} \tag{16.13}$$

where $a_i, b_i > 0$.

(a) Show that there is always an interior equilibrium $\mathbf{p}$.

(b) Show that $\mathbf{p}$ is an ESS iff $a_i \geq b_i$ and the three numbers $\sqrt{a_i - b_i}$ correspond to the lengths of the sides of a triangle.

(c) Show that $\mathbf{p}$ is asymptotically stable iff $\det A > 0$ iff $\mathbf{p} \cdot A\mathbf{p} > 0$.

(d) Show that asymptotic stability of $\mathbf{p}$ implies global stability. (Hint: After a suitable barycentric coordinate change $\mathbf{p}$ will be an ESS.)

(e) $\mathbf{p}$ is a centre iff $\det A = 0$ iff $\mathbf{p} \cdot A\mathbf{p} = 0$. (The Ljapunov function from (d) will be an invariant of motion here.)

(f) Show that (16.4) with A as above has no limit cycles. (Hint: Construct a global Ljapunov function or use Dulac's criterion - see **18.1**).

(g) What is the ω-limit of the orbits if $\mathbf{p}$ is unstable?

Exercise 6: The hypercycle equation with $n = 4$ and matrix

$$\begin{pmatrix} 0 & 0 & 0 & k_1 \\ k_2 & 0 & 0 & 0 \\ 0 & k_3 & 0 & 0 \\ 0 & 0 & k_4 & 0 \end{pmatrix}$$

has a globally stable rest point $\mathbf{p} \in \operatorname{int} S_n$ (see **12.6**). Show that it is not evolutionarily stable (for any choice of rate constants $k_i > 0$.)

A further example for an equilibrium in int S_3 which is asymptotically (but not globally) stable without being evolutionarily stable is obtained with the matrix

$$A = \begin{pmatrix} 0 & 6 & -4 \\ -3 & 0 & 5 \\ -1 & 3 & 0 \end{pmatrix} . \tag{16.14}$$

The central point $\mathbf{m} = (\frac{1}{3}, \frac{1}{3}, \frac{1}{3})$ is a rest point and asymptotically stable, since its eigenvalues $\frac{1}{3}(-1 \pm i\sqrt{2})$ have negative real parts. But $\mathbf{e}_1 = (1,0,0)$ is an evolutionarily stable point on bd S_3 (see Fig. 16.2). It follows that $\mathbf{m}$ is not evolutionarily stable (being in the interior, it would have to be the unique stable state).

Exercise 7: Check all this and find a global Ljapunov function.

Let us now return to the "hawk-dove" game and assume that there occurs, in addition to the phenotypes E_1 (hawk) and E_2 (dove), a third phenotype E_3 playing "hawk" and "dove" with probabilities $p_1 = \frac{G}{C}$

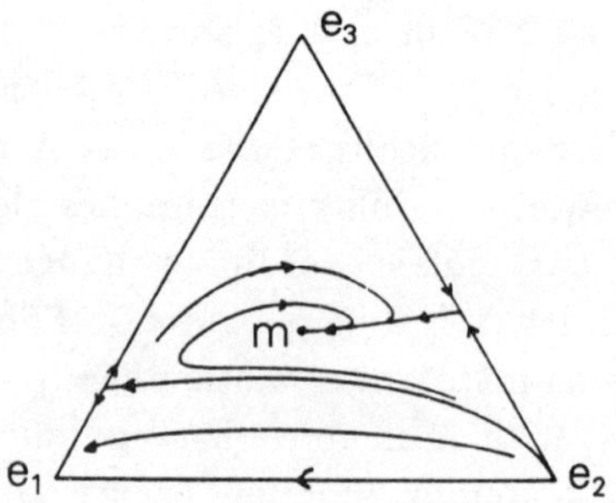

Figure 16.2

and $p_2 = \dfrac{C-G}{C}$ given by the equilibrium frequencies (see **15.2**). The payoff matrix then is

$$A = \begin{pmatrix} \dfrac{G-C}{2} & G & \dfrac{G(C-G)}{2\,C} \\ 0 & \dfrac{G}{2} & \dfrac{G(C-G)}{2\,C} \\ \dfrac{G(G-C)}{2\,C} & \dfrac{G(G+C)}{2\,C} & \dfrac{G(C-G)}{2\,C} \end{pmatrix} . \tag{16.15}$$

The phenotype E_3 cannot be invaded by E_1 or E_2, since the corresponding *strategy* is evolutionarily stable. The corresponding *state* $\mathbf{e}_3$ is not evolutionarily stable, however, and neither is the mixture $\mathbf{p} = p_1\mathbf{e}_1 + p_2\mathbf{e}_2$ in this extended game. Indeed, one has the following result:

Exercise 8: There exists a line F of rest points through $\mathbf{e}_3$ and $\mathbf{p}$ (see Fig. 16.3). All orbits converge to F, remaining on the constant level curves of the function

$$Q(\mathbf{x}) = x_1^{p_1} x_2^{p_2} x_3^{-1} . \tag{16.16}$$

Exercise 9: Analyse (16.4) with the matrix

$$A = \begin{pmatrix} 0 & 10 & 1 \\ 10 & 0 & 1 \\ 1 & 1 & 1 \end{pmatrix} . \tag{16.17}$$

Show that the phase portrait is given by Fig. 16.4. The phenotype E_3 is

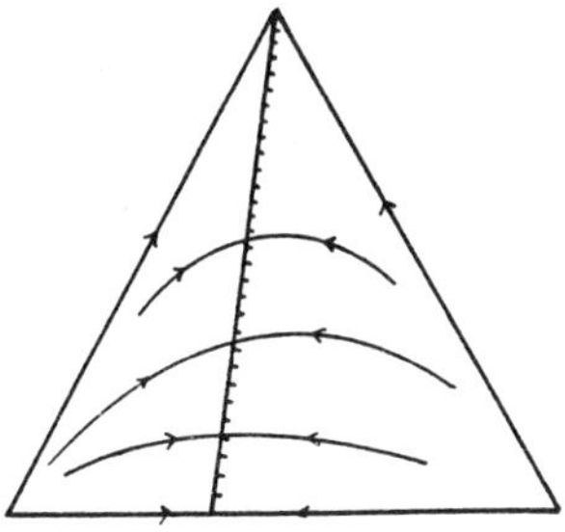

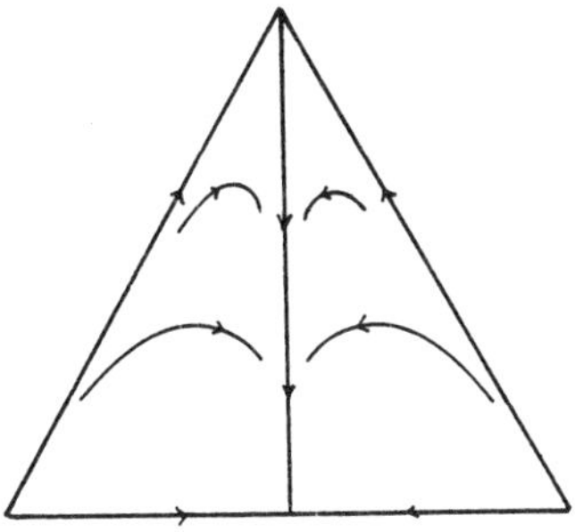

Figures 16.3 and 16.4

stable against invasion by E_1 alone, or by E_2 alone, but not if both phenotypes invade simultaneously.

Exercise 10: For discrete generations, the difference equation $\mathbf{x} \to T\mathbf{x}$ with

$$(T\mathbf{x})_i = x_i \frac{(A\mathbf{x})_i + C}{\mathbf{x} \cdot A\mathbf{x} + C} \tag{16.18}$$

is a candidate for game dynamics. Here, C is a positive constant corresponding to the fitness in the absence of interaction, such that $(A\mathbf{x})_i + C$ is always positive (recall that the payoff a_{ij} is the increment of fitness). Prove the following:

(a) Nash equilibria are fixed points of (16.18).

(b) An evolutionarily stable state need not be asymptotically stable for (16.18). (Hint: use a "rock-scissors-paper" type game.)

(c) Analyse (16.18) for zero sum games. Show that all orbits (up to a possible interior equilibrium) converge to the boundary. (Hint: if there is an interior equilibrium, then $\prod x_i^{p_i}$ is strictly decreasing. If there is no interior equilibrium then proceed as in Theorem 16.7.1).

(d) Analyse the discrete dynamics (16.18) for the hypercycle. Prove permanence for $C > 0$. Analyse the local stability of the interior fixed point $\mathbf{p}$. Prove that $\mathbf{p}$ is globally stable if $n \leq 3$ and $C > 0$. Analyse the asymptotic behaviour for $C = 0$.

(e) Analyse a discrete "rock-scissors-paper" game with

$$A = \begin{bmatrix} a & b & c \\ c & a & b \\ b & c & a \end{bmatrix}$$

($b > a > c \geq 0$). Show that $\dfrac{\mathbf{x}\cdot A\mathbf{x}}{x_1 x_2 x_3}$ is a global Ljapunov function, and the centre point $\mathbf{m}$ is asymptotically stable iff $bc > a^2$.

16.6 Game dynamics and the Lotka-Volterra equation

The game dynamical equation (16.4) is a cubic equation on the compact set S_n, while the Lotka-Volterra equation (9.1) is quadratic on $\mathbf{R}^n_+$. We shall presently see, however, that the game dynamical equation in n variables $x_1,\ldots,x_n$ is equivalent to the Lotka-Volterra equation in $n-1$ variables $y_1,\ldots,y_{n-1}$.

Indeed, let us define $y_n \equiv 1$ and consider the transformation $\mathbf{y} \to \mathbf{x}$ given by

$$x_i = \frac{y_i}{\sum\limits_{j=1}^{n} y_j} \qquad i = 1,\ldots,n \tag{16.19}$$

which maps

$$\{\mathbf{y} \in \mathbf{R}^n\colon y_n = 1,\ y_i \geq 0 \ \text{ for } \ i = 1,\ldots,n-1\}$$

onto

$$\hat{S}_n = \{\mathbf{x} \in S_n\colon x_n > 0\} \quad .$$

The inverse $\mathbf{x} \to \mathbf{y}$ is given by

$$y_i = \frac{y_i}{y_n} = \frac{x_i}{x_n} \qquad i = 1,\ldots,n \quad . \tag{16.20}$$

Both maps are differentiable.

Now let us consider the game dynamical equation in n variables

$$\dot{x}_i = x_i((A\mathbf{x})_i - \mathbf{x}\cdot A\mathbf{x}) \quad . \tag{16.21}$$

We shall assume that the last row of the $n \times n$ matrix $A = (a_{ij})$ consists of zeros: by exercise 16.3.2, this is no restriction of generality. By the quotient rule (16.3)

$$\dot{y}_i = \left(\frac{x_i}{x_n}\right)^{\cdot} = \left(\frac{x_i}{x_n}\right)[(A\mathbf{x})_i - (A\mathbf{x})_n] \quad . \tag{16.22}$$

Since $(A\mathbf{x})_n = 0$, this implies

$$\dot{y}_i = y_i\left(\sum_{j=1}^{n} a_{ij}x_j\right) = y_i\left(\sum_{j=1}^{n} a_{ij}y_j\right)x_n \quad . \tag{16.23}$$

By a change in velocity (cf. **12.4**) we can remove the term x_n. Since $y_n = 1$, this yields

$$\dot{y}_i = y_i\left(a_{in} + \sum_{j=1}^{n-1} a_{ij}y_j\right)$$

or (with $r_i = a_{in}$) the Lotka-Volterra equation

$$\dot{y}_i = y_i\left(r_i + \sum_{j=1}^{n-1} a_{ij}y_j\right) \quad . \tag{16.24}$$

The converse direction from (16.24) to (16.21) is analogous. Hence:

Theorem. *There exists a differentiable, invertible map from $\hat{S}_n$ onto $\mathbf{R}_+^{n-1}$ mapping the orbits of the game dynamical equation* (16.21) *onto the orbits of the Lotka-Volterra equation*

$$\dot{y}_i = y_i\left(r_i + \sum_{j=1}^{n-1} a'_{ij}\, y_j\right) \qquad i = 1,\ldots,n-1$$

where $r_i = a_{in} - a_{nn}$ and $a'_{ij} = a_{ij} - a_{nj}$.

Results about Lotka-Volterra equations can therefore be carried over to the game dynamical equation and vice versa. However, some properties are simpler to prove (or more natural to formulate) for one equation and some for the other.

Exercise: How do equations (12.5), (12.6), (12.7) and (16.13) look in Lotka-Volterra form?

16.7 Time averages and an exclusion principle

We have seen in **9.2** that a Lotka-Volterra equation admits an ω-limit point in int $\mathbf{R}_+^n$ iff it has a rest point in int $\mathbf{R}_+^n$. From this and the equivalence theorem of last section, we can infer the following *exclusion principle*:

Theorem 1. *If the game dynamical equation* (16.4) *has no rest point in* int S_n, *then every orbit converges to* bdS_n.

Exercise 1: Prove this exclusion principle directly: (a) show that there exists a $\mathbf{c} \in \mathbf{R}^n$ such that $\mathbf{c}\cdot\mathbf{z} > \mathbf{c}\cdot\mathbf{y}$ for all $\mathbf{z} \in W$ and all $\mathbf{y} \in D$, where

$W = A(\text{int}\, S_n)$ and $D = \{\mathbf{y} \in \mathbf{R}^n : y_1 = \cdots = y_n\}$ (W and D are convex); (b) show that $\sum c_i = 0$; (c) show that $V(\mathbf{x}) = \sum c_i \log x_i$ is strictly increasing along the orbits in int S_n.

Exercise 2: Show that the game with payoff matrix A admits a Nash equilibrium in int S_n iff there is no strategy $\mathbf{u}$ dominating a strategy $\mathbf{v}$ in the sense that $\mathbf{u} \cdot A\mathbf{x} > \mathbf{v} \cdot A\mathbf{x}$ for all $\mathbf{x} \in \text{int}\, S_n$. (Hint: the vector $\mathbf{c}$ from the previous exercise can be written as difference $\mathbf{u} - \mathbf{v}$ of two strategies.)

Theorem 2. *If* (16.4) *admits a unique rest point* $\mathbf{p} \in \text{int}\, S_n$, *and if the ω-limit of the orbit of* $\mathbf{x}(t)$ *is in* int S_n, *then*

$$\lim_{t \to \infty} \frac{1}{T} \int_0^T x_i(t)\, dt = p_i \qquad i = 1,\ldots,n \quad . \tag{16.25}$$

This does not follow immediately from the corresponding Theorem 9.2.2. for Lotka-Volterra equations, since a coordinate transformation or a change in velocity could affect the time average. Hence, we have the following:

Exercise 3: Prove the previous theorem. (Hint: proceed along the same lines as in the Lotka-Volterra case.)

This is of particular interest for the hypercycle equation (12.7). Since it is permanent (see **13.3**), the ω-limit $\omega(\mathbf{x})$ lies in int S_n for every $\mathbf{x} \in \text{int}\, S_n$. The unique interior equilibrium $\mathbf{p}$ is unstable, for $n \geq 5$ (see **12.5**). The point $\mathbf{p}$ is still physically relevant, however, as time average of the concentrations.

16.8 Notes

The game dynamical equation (16.4) was introduced by Taylor and Jonker (1978). In Zeeman (1980) and Hofbauer et al. (1979) the relation between ESS and asymptotically stable states is analyzed. For relations to Boltzmann's H-theorem see Alberti and Crell (1984). The approach indicated in exercises 16.4.4-5 is due to Hines (1980b). Examples (16.13) and (16.14) are due to Zeeman (1980), where a classification of the stable phase portraits for $n = 3$ can be found. In Zeeman (1981) and Schuster et al. (1981a) some biological games are analysed globally. Difference equations are investigated in Bishop and Cannings (1978) (for the "war of attrition"), Hofbauer (1984) (for the hypercycle), and Akin and Losert (1984) (for zero sum games). The equivalence between game dynamics and the Lotka-Volterra equation was shown by Hofbauer (1981a). Examples (16.15) and (16.17) are from Maynard Smith (1982). The result described in exercise 16.7.2 is due to Akin (1980). Theorem 16.7.1 is from Hofbauer (1981b) and Theorem 16.7.2 is from Schuster et al. (1980). There are other dynamic approaches to game theory, for example the Brown-Robinson dynamics or the theory of differential games (see Owen (1982) for an introduction).

17. Asymmetric Conflicts

17.1 Bimatrix games

In the game theoretic models considered so far, we have always assumed that the conflict are *symmetric* in the sense that all "players" are in the same situation: equally strong, equally hungry, etc. ... However, many conflicts are asymmetric. Thus food is more important for a starving animal than for a replete one, while the risk of injury is smaller for a stronger contestant. In fact, asymmetries are not only incidental, but quite often essential features of the game: for example, in conflicts between males and females, between parents and offspring, between owner of a habitat and intruder, or between different species. If we restrict ourselves again to conflicts settled in pairwise encounters, and finite numbers of pure strategies, we are led to *bimatrix* games.

Thus let $E_1,...,E_n$ denote the phenotypes of population X, and $F_1,...,F_m$ those of population Y. By x_i we denote the frequency of E_i, and by y_j that of F_j. Hence the state of population X is given by a point $\mathbf{x}$ in S_n, and that of population Y by a point $\mathbf{y}$ in S_m. If an E_i-individual is matched against an F_j-individual, the payoff will be a_{ij} for the former and b_{ji} for the latter. Thus the game is described by two payoff matrices $A = (a_{ij})$ and $B = (b_{ji})$. The expressions $\mathbf{x} \cdot A\mathbf{y}$ and $\mathbf{y} \cdot B\mathbf{x}$ are the average payoffs for population X and population Y, respectively. In an alternative interpretation, E_1 to E_n would correspond to the pure strategies of player X and F_1 to F_m to the pure strategies for player Y. Points $\mathbf{x} \in S_n$ and $\mathbf{y} \in S_m$ correspond to mixed strategies and $\mathbf{x} \cdot A\mathbf{y}$ would be the expected payoff for the $\mathbf{x}$-strategy against the $\mathbf{y}$-strategy.

The notion of evolutionary stability is considerably more restricted for asymmetric games than for symmetric ones. Indeed, let us assume that the phenotype E_i is stable in contests against the population Y. It must do at least as well, then, as all competing phenotypes, i.e.

$$(A\mathbf{y})_i \geq (A\mathbf{y})_k \text{ for all } k \neq i \quad .$$

But what if equality holds? There is nothing, then, to prevent E_k from invading, there is no reasonable condition analogous to (15.12). Thus, in contrast to the symmetric case, we cannot allow several "best replies". This leads to the following definition.

A pair of states (or strategies) $(\mathbf{p},\mathbf{q})$ with $\mathbf{p} \in S_n$ and $\mathbf{q} \in S_m$ is said to be ***evolutionarily stable*** if

$$\mathbf{p} \cdot A\mathbf{q} > \mathbf{x} \cdot A\mathbf{q} \quad \text{for all } \mathbf{x} \in S_n, \mathbf{x} \neq \mathbf{p} \tag{17.1}$$

and

$$\mathbf{q} \cdot B\mathbf{p} > \mathbf{y} \cdot B\mathbf{p} \quad \text{for all } \mathbf{y} \in S_m, \mathbf{y} \neq \mathbf{q} \quad . \tag{17.2}$$

Exercise: Show that such a pair must consist of pure strategies. i.e. $\mathbf{p} = \mathbf{e}_i$ and $\mathbf{q} = \mathbf{f}_j$ for some corners $\mathbf{e}_i$ of S_n and $\mathbf{f}_j$ of S_m.

Thus in asymmetric contests a mixed strategy can never be evolutionarily stable. This also reflects on symmetric contests since it is often quite possible that a small, seemingly irrelevant difference can break the symmetry between the opponents and transform the originally symmetric contest into an asymmetric one. This could explain why mixed strategies and states are rather rare in pairwise contests (cf. **15.5**).

Relaxing inequalities (17.1) and (17.2) leads to a central notion of classical game theory. A pair of strategies $(\mathbf{p},\mathbf{q}) \in S_n \times S_m$ is called a *Nash equilibrium pair* if

$$\mathbf{p} \cdot A\mathbf{q} \geq \mathbf{x} \cdot A\mathbf{q} \quad \text{for all } \mathbf{x} \in S_n \tag{17.3}$$

$$\mathbf{q} \cdot B\mathbf{p} \geq \mathbf{y} \cdot B\mathbf{p} \quad \text{for all } \mathbf{y} \in S_m \quad . \tag{17.4}$$

It can be shown that at least one such pair exists for every bimatrix game (cf. exercises 19.4.5 and 27.1.5.) In contrast to evolutionarily stable pairs, $\mathbf{p}$ and $\mathbf{q}$ can be mixed.

In non-cooperative games between rational opponents, a Nash equilibrium pair $(\mathbf{p},\mathbf{q})$ may be considered as a "solution". As long as the opponent sticks to $\mathbf{q}$, there is no reason to deviate from $\mathbf{p}$. However, such solutions are not stable, in general. Indeed, (17.3) leads to

$$\mathbf{p} \cdot A\mathbf{q} \geq (A\mathbf{q})_i \qquad i = 1,\dots,n$$

with equality for all i with $p_i > 0$. If in particular $\mathbf{p}$ is completely mixed (i.e. $\mathbf{p} \in \operatorname{int} S_n$), then

$$(A\mathbf{q})_1 = \cdots = (A\mathbf{q})_n \tag{17.5}$$

and hence $\mathbf{p} \cdot A\mathbf{q} = \mathbf{x} \cdot A\mathbf{q}$ for all $\mathbf{x} \in S_n$. Thus there is no reason *not* to deviate from $\mathbf{p}$ either. A random fluctuation altering strategy $\mathbf{p}$ will cause no harm. In the biological context, this means that the fluctuation will not be corrected by natural selection, and hence that $\mathbf{p}$ is not invasion proof.

17.2 The Battle of the Sexes

As an example of an asymmetric game, we shall discuss a conflict between males and females concerning their respective shares in ***parental investment***. In many species, raising an offspring requires a considerable amount of time and energy. Each parent might attempt to reduce its own share at the expense of the other. The outcome might depend on which sex is in a position to desert first. Whenever fertilisation is internal, for example, females risk being deserted even before giving birth to the offspring. The game is still further "rigged" against females by the fact that they produce relatively few, large gametes, and males many small ones. Females are thereby much more committed and can less afford to lose a child. Thus males are in many cases in a better position to desert. They can invest the corresponding gain in time and energy into increasing their offspring with the help of new mates.

The female counterstrategy is "coyness", i.e. the insistence upon a long engagement period before copulation. Rather than undergoing a second costly engagement (for which it might be too late in the mating season), males would do better to stay faithfully home and help raise their offspring. Roughly speaking, in a population of coy females, males would have to be faithful. Among faithful males, it would not pay a female to be coy, however: the long engagement period is an unnecessary cost. Thus the proportion of "fast" females would grow. But then "philandering" males will have their chance and spread. Females, therefore, would do well to be coy. The argument thus runs full circle.

In order to model this game-theoretically, let us assume that there are two phenotypes in the male population X, namely E_1 ("philandering") and E_2 ("faithful") with frequencies x_1 and x_2, and two phenotypes in the female population Y, namely F_1 ("coy") and F_2 ("fast") with frequencies y_1 and y_2. Let us suppose that the successful raising of an offspring increases the fitness of both parents by G. The parental investment $-C$ will be entirely borne by the female if the male deserts. Otherwise, it is shared equally by both parents. A long engagement period represents a cost $-E$ to both partners.

If a "faithful" male mates with a "coy" female, the payoff is $G - \frac{C}{2} - E$ for both. A "faithful" male and a "fast" female skip the engagement cost and their payoff is $G - \frac{C}{2}$. But a "philandering" male meeting a "fast" female makes off with G, while her payoff is $G - C$. Finally if a "philandering" male encounters a "coy" female, nothing much happens and the payoff for both is 0.

The payoff matrices therefore are

$$A = \begin{bmatrix} 0 & G \\ G - \frac{C}{2} - E & G - \frac{C}{2} \end{bmatrix} \quad B = \begin{bmatrix} 0 & G - \frac{C}{2} - E \\ G - C & G - \frac{C}{2} \end{bmatrix} . \tag{17.6}$$

We shall assume $0 < E < G < C < 2(G - E)$. No pair of phenotypes is evolutionarily stable; neither is any pair of population states (we have only to check the pure states). There exists a unique pair of mixed strategies $\mathbf{p}$ and $\mathbf{q}$ in Nash equilibrium. It is given by the solution of

$$a_{11}q_1 + a_{12}q_2 = a_{21}q_1 + a_{22}q_2 \qquad (q_2 = 1 - q_1)$$

$$b_{11}p_1 + b_{12}p_2 = b_{21}p_1 + b_{22}p_2 \qquad (p_2 = 1 - p_1)$$

i.e. by

$$p_1 = \frac{E}{C - G + E} \qquad q_1 = \frac{C}{2(G - E)} \quad . \tag{17.7}$$

This equilibrium is not stable, however. If a fluctuation decreases, say, the amount of philandering males, then the payoff for the males will not change: each phenotype has the same payoff, which depends only on the state of the female population. One cannot expect the frequency of philanderers to return to p_1. As to the female population, their payoff will even increase: but "fast" females gain more than "coy" ones, since their risk of being deserted decreases. It is only when the amount of "fast" females increases that the male payoffs change. Again, they increase: but philanderers gain more than the faithful males; hence more philanderers, hence more coy females, hence fewer philanderers, and so on. This looks like an oscillating system. The static approach cannot deal with this situation.

17.3 A differential equation for asymmetric games

As in **16.1**, we may associate a differential equation to a bimatrix game by assuming that the rate of increase $\frac{\dot{x}_i}{x_i}$ of phenotype E_i is equal to the difference between its payoff $(A\mathbf{y})_i$ and the average payoff $\mathbf{x} \cdot A\mathbf{y}$. This, and the corresponding assumption concerning phenotype F_j, leads to the differential equation

$$\begin{aligned} \dot{x}_i &= x_i((A\mathbf{y})_i - \mathbf{x} \cdot A\mathbf{y}) \quad i = 1,\ldots,n \\ \dot{y}_j &= y_j((B\mathbf{x})_j - \mathbf{y} \cdot B\mathbf{x}) \quad j = 1,\ldots,m \end{aligned} \tag{17.8}$$

on the (invariant) space $S_n \times S_m$ of all states of X and Y.

Again, one checks easily that the boundary faces of $S_n \times S_m$ (i.e. the products of a face of S_n with S_m or with a face of S_m) are invariant under (17.8), and that the restriction of (17.8) to such a boundary face yields an equation of similar form. One obtains the boundary faces by setting some x_i or y_j equal to 0. Each such face in turn can be decomposed into boundary and interior, the boundary consisting of faces again. It is enough, therefore, to consider the restriction of (17.8) to subsets of the following form:

(a) at least one of the population consists of only one phenotype;
(b) both populations consist of several phenotypes.

This means:

(a) $x_i \equiv 1$ or $y_j \equiv 1$ for some i or some j;
(b) $x_i > 0$ for several i, and $y_j > 0$ for several j.

There is no loss of generality in studying only the restrictions of (17.8) to sets of the following type:

(a') $S_n \times \{\mathbf{f}_1\}$ with $\mathbf{f}_1 = \{1,0,\ldots,0\} \in S_m$;
(b') int $S_n \times S_m$.

All other restrictions of type (a) or (b) are of the same form as (a') or (b'). It remains to consider the case (b'). Indeed one has:

Exercise: (a') leads to equation (12.5).

We stress that (17.8) not only assumes asexual reproduction, but also that the phenotypes of one population are unlinked with that of the other. For games between different species, this is fine; but for the Battle of the Sexes, or conflicts between owners and intruders belonging to one species, it does not look convincing. Again, we refer to chapter 28 and 29 for related models taking genetics into account.

17.4 Rest points

The rest points of (17.8) in int $S_n \times S_m$ are the strictly positive solutions of the equations

$$(A\mathbf{y})_1 = \cdots = (A\mathbf{y})_n \qquad \sum_{j=1}^{m} y_j = 1 \tag{17.9}$$

$$(B\mathbf{x})_1 = \cdots = (B\mathbf{x})_m \qquad \sum_{i=1}^{n} x_i = 1\ . \tag{17.10}$$

For $n > m$, (17.9) has solutions only if the matrix A is degenerate, while the solutions of (17.10) form a linear manifold of dimension at least $n - m$. The set of rest points in int $S_n \times S_m$ is thus either empty - this is the generic case - or it contains an $(n - m)$-dimensional subset.

An isolated rest point can thus only exist for $n = m$. If it exists, it is unique. We shall presently see that the rest points of (17.8) in int $S_n \times S_m$ cannot be sources or sinks. Isolated rest points are saddles or centres.

Indeed (17.9) implies $(A\mathbf{y})_i = \mathbf{x} \cdot A\mathbf{y}$ and hence

$$\frac{\partial \dot{x}_i}{\partial x_j} = \frac{\partial}{\partial x_j} x_i((A\mathbf{y})_i - \mathbf{x} \cdot A\mathbf{y}) = x_i(-\frac{\partial}{\partial x_j}(\mathbf{x} \cdot A\mathbf{y}))\ . \tag{17.11}$$

But for $1 \le j \le n-1$ one has

$$\frac{\partial}{\partial x_j}(\mathbf{x} \cdot A\mathbf{y}) =$$

$$= \frac{\partial}{\partial x_j}(x_1(A\mathbf{y})_1 + \cdots + x_{n-1}(A\mathbf{y})_{n-1} + (1 - x_1 - \cdots - x_{n-1})(A\mathbf{y})_n)$$

$$= (A\mathbf{y})_j - (A\mathbf{y})_n = 0 \tag{17.12}$$

and hence $\dfrac{\partial \dot{x}_i}{\partial x_j} = 0$ for $1 \le i, j \le n-1$. A similar relation holds for the y_j, so that the Jacobian of (17.8) at a rest point in int $S_n \times S_m$ takes (after elimination of x_n and y_n) the form

$$J = \begin{bmatrix} O & C \\ D & O \end{bmatrix}$$

where the two blocks of zeros in the diagonal are an $(n-1) \times (n-1)$ and an $(m-1) \times (m-1)$ matrix respectively. For the characteristic polynomial $p(\lambda) = \det(J - \lambda I)$ (where I is the identity matrix) one obtains

$$p(\lambda) = (-1)^{n+m} p(-\lambda) \tag{17.13}$$

as can be seen by changing the signs of the first $n-1$ columns and the last $m-1$ rows of $J - \lambda I$. Thus if λ is an eigenvalue of J, then $-\lambda$ is also an eigenvalue. Sinks and sources are therefore excluded. In particular, (17.8) admits sinks only at the corners of $S_n \times S_m$. This corresponds to the fact that mixed evolutionarily stable states are excluded for asymmetric games. It may be conjectured that if an isolated rest point in $S_n \times S_m$ is a centre then it is *totally stable*, i.e. stable for both $t \to \pm\infty$. It would then behave like the linearized equation. We will see in **27.4** that it cannot be asymptotically stable.

Exercise 1: If the ω-limit of an orbit of (17.8) is in int $S_n \times S_m$ then the time average exists and corresponds to a rest point in int $S_n \times S_m$.

Exercise 2: If there is no rest point in int $S_n \times S_m$, then all orbits converge to bd $S_n \times S_m$. (Hint: see **16.7**).

Exercise 3: If there is a manifold of fixed points in int $S_n \times S_m$, then there exists a corresponding decomposition of the state space into invariant manifolds. (Hint: construct constants of motion).

Exercise 4: Show that the characteristic polynomial of J is given by

$$p(\lambda) = (-\lambda)^{n-m} \det(\lambda^2 I - DC) \quad . \tag{17.14}$$

Exercise 5: How much of this is valid if the payoffs are nonlinear?

17.5 The 2 × 2-case

Let us now take a closer look at the case $n = m = 2$, which occurs in the Battle of the Sexes. Since we may add, as in the symmetric case, a constant to every column of A and B (cf. exercise 16.3.2), we may assume without restricting generality that the diagonal terms are zero. Hence the matrices are

$$A = \begin{pmatrix} 0 & a_{12} \\ a_{21} & 0 \end{pmatrix} \qquad B = \begin{pmatrix} 0 & b_{12} \\ b_{21} & 0 \end{pmatrix} \quad . \tag{17.15}$$

Since $x_2 = 1 - x_1$ and $y_2 = 1 - y_1$, it is enough to consider the variables x_1 and y_1, which we shall denote by x and y. (17.8) now becomes

$$\dot{x} = x(1-x)(a_{12} - (a_{12} + a_{21})y)$$

$$\dot{y} = y(1-y)(b_{12}-(b_{12}+b_{21})y) \tag{17.16}$$

on the square $Q = \{(x,y): 0 \leq x,\ y \leq 1\} \cong S_2 \times S_2$.

If $a_{12}a_{21} \leq 0$, then $\dot{x}$ does not change its sign in Q. In this case x is either constant, or converges monotonically to 0 or 1. A similar result holds for $b_{12}b_{21} \leq 0$. Thus it only remains to investigate the case when $a_{12}a_{21} > 0$ and $b_{12}b_{21} > 0$. In this case (17.16) admits a unique rest point in $\operatorname{int} Q$, namely

$$\mathbf{F} = \left(\frac{b_{12}}{b_{12}+b_{21}}, \frac{a_{12}}{a_{12}+a_{21}} \right) .$$

The Jacobian at $\mathbf{F}$ is

$$J = \begin{pmatrix} 0 & -(a_{12}+a_{21})\dfrac{b_{12}b_{21}}{(b_{12}+b_{21})^2} \\ -(b_{12}+b_{21})\dfrac{a_{12}a_{21}}{(a_{12}+a_{21})^2} & 0 \end{pmatrix} \tag{17.17}$$

and the eigenvalues are $\pm\lambda$ with

$$\lambda^2 = \frac{a_{12}a_{21}b_{12}b_{21}}{(a_{12}+a_{21})(b_{12}+b_{21})} . \tag{17.18}$$

If $a_{12}b_{12} > 0$, then $\mathbf{F}$ is a saddle, and almost all orbits in int Q converge to one or the other of two opposite corners of Q (see Fig. 17.1). If $a_{12}b_{12} < 0$, then $\mathbf{F}$ is a centre. In this case all orbits in $\operatorname{int} Q$ are periodic orbits surrounding $\mathbf{F}$ (see Fig. 17.2).

This can best be seen by dividing the right hand side of (17.16) by the function $xy(1-x)(1-y)$, which is positive in $\operatorname{int} Q$. This corresponds to a change in velocity and does not affect the orbits (see **12.4**). In $\operatorname{int} Q$, we obtain

$$\begin{aligned} \dot{x} &= \frac{a_{12}-(a_{12}+a_{21})y}{y(1-y)} = \frac{a_{12}}{y} - \frac{a_{21}}{1-y} \\ \dot{y} &= \frac{b_{12}}{x} - \frac{b_{21}}{1-x} . \end{aligned} \tag{17.19}$$

This system is *Hamiltonian*, i.e. it can be written as

$$\dot{x} = \frac{\partial H}{\partial y} \qquad \dot{y} = -\frac{\partial H}{\partial x} \tag{17.20}$$

with

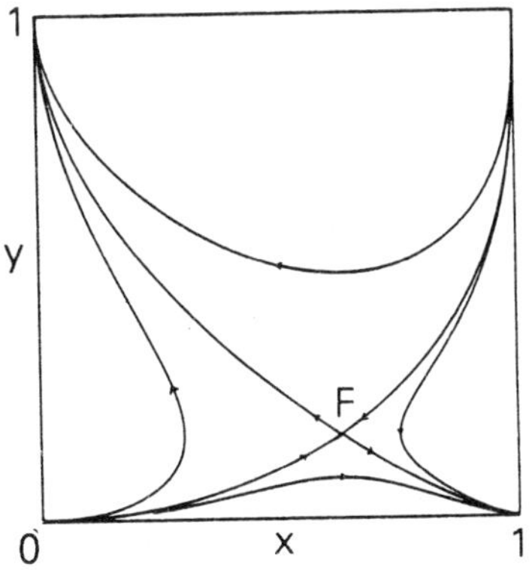

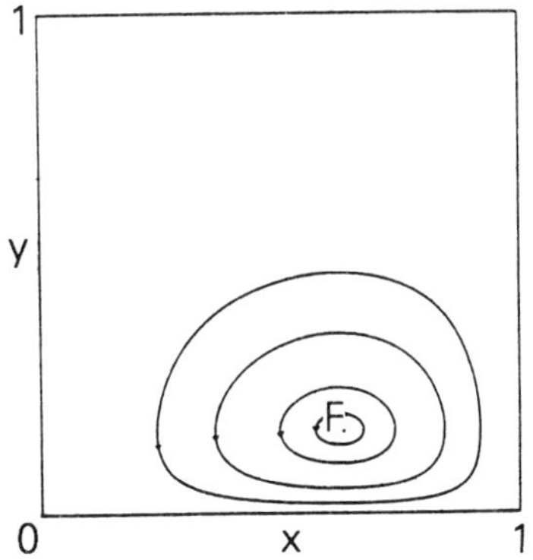

Figures 17.1, 17.2

$$H(x,y) = a_{12}\log y + a_{21}\log(1-y) - b_{12}\log x - b_{21}\log(1-x) \quad (17.21)$$

H is a constant of motion for (17.19), since

$$\dot{H} = \frac{\partial H}{\partial x}\dot{x} + \frac{\partial H}{\partial y}\dot{y} \equiv 0 \quad . \quad (17.22)$$

Now if (17.16) admits an equilibrium $\mathbf{F}$ in int$Q$ which is a centre, then the diverse conditions on the signs of the a_{ij} and b_{ji} imply that H attains its unique extremum in int$Q$ at the point $\mathbf{F}$. All other orbits are periodic and surround $\mathbf{F}$.

Exercise 1: For the time average along a periodic orbit with period T, one has

$$\left(\frac{1}{T}\int_0^T x(t)\,dt,\ \frac{1}{T}\int_0^T y(t)\,dt\right) = \mathbf{F} \quad .$$

Let us return now to the Battle of the Sexes described in **17.2**. After the addition of appropriate constants, the matrices A and B from (17.6) turn into

$$A = \begin{pmatrix} 0 & \frac{C}{2} \\ G - \frac{C}{2} - E & 0 \end{pmatrix} \qquad B = \begin{pmatrix} 0 & -E \\ G - C & 0 \end{pmatrix} \quad . \quad (17.23)$$

The Nash equilibrium given by (17.7) is a centre surrounded by periodic orbits. Equation (17.16) takes the form

$$\dot{x} = x(1-x)\left(\frac{C}{2} - (G-E)y\right)$$
$$\dot{y} = y(1-y)(-E + (C+E-G)x) \quad . \tag{17.24}$$

Up to the factors $(1-x)$ and $(1-y)$, this looks like the Lotka-Volterra equation (7.1). We conclude that the relations between the sexes resemble those between predators and prey and are subject to perpetual oscillations.

Exercise 2: Discuss the game dynamics for the prisoners' dilemma.

17.6 Notes

The Battle of the Sexes is due to Dawkins (1976). Asymmetric biological games were introduced in Maynard Smith and Parker (1976). Among the many follow-up papers we mention Taylor (1979), Hammerstein (1979), Hammerstein and Parker (1982) (see also Maynard Smith (1982) for further references). Selten (1980) proved in a very general framework that evolutionarily stable strategies for asymmetric games are never mixed. Equation (17.8) is analyzed in Schuster et al. (1981b). The special case of the Battle of the Sexes was studied in Schuster and Sigmund (1981) and - under a more general dynamics - in Eshel and Akin (1983). Game dynamics for asymmetric conflicts with innerspecific interaction are studied in Taylor (1979) and Schuster et al. (1981c). We refer to Chapter 27 for more on asymmetric games.

Interlude: The Replicator Equation

Halfway through the book, it may be fitting to look back at the four parts lying behind us. Like a red thread, the equation

$$\dot{x}_i = x_i\left(\sum_j a_{ij}x_j - \sum_{rs} a_{rs}x_r x_s\right)$$

runs through all of them. In sociobiology it occurs as the game dynamical equation modelling the evolution of behavioural phenotypes. In macromolecular evolution, it describes networks of autocatalytic reactions and in particular hypercyclic feedback. Disguised as the Lotka-Volterra equation, it holds centre stage in mathematical ecology. The restriction to the symmetric case $a_{ij} = a_{ji}$, finally, yields the continuous counterpart of the discrete selection equation in population genetics. The x_i are in the first case the probabilities of certain strategies, in the second case concentrations of polynucleotides, in the third case relative densities of interacting populations and in the last case frequencies of alleles in a gene pool. The binary coefficients a_{ij} correspond to payoffs, biological or chemical interactions or survival rates.

The common biological denominator behind the four mathematical models can best be described by the notion of "replicator". In the definition of Dawkins, who coined the term, these are any entities which can get copied and which satisfy the following two conditions:

(a) their properties can affect their probability of being copied;

(b) the line of descendent copies must be - at least in principle - unlimited.

A gene, in this sense, is a replicator as long as it remains in the germ line. A gene in a body cell, on the other hand, can get copied a few times only and hence is no replicator.

Segments of RNA or DNA are replicators. Sexually reproducing organisms are not, since they do not yield identical copies; but for the simple ecological models which we discussed so far, we took no heed of such individual differences, and hence viewed organisms as replicators. Similarly, phenotypes are not replicators in the strict sense, but it may sometimes be advantageous to treat them as such. Groups and species can also be viewed as replicators. Their copies are the populations which descend from them.

Replicators are the abstract "units of selection". Just as in the physical world forces act on different ranges - as gravitational, electromagnetic, strong or weak interactions - so selection takes place on many levels.

We shall design equations of the type

$$\dot{x}_i = x_i(f_i(\mathbf{x}) - \bar{f}(\mathbf{x})) \quad \text{with} \quad \bar{f}(\mathbf{x}) = \sum_{i=1}^{n} x_i f_i(\mathbf{x})$$

on the state space S_n as replicator equations, and in the special case of linear rate functions $f_i(\mathbf{x}) = \sum a_{ij} x_j$ as first order replicator systems. They come up in many biological models and lead to lots of mathematical problems: the next half of this book is devoted to some of them.

PART V: COMPETE, PREDATE OR COOPERATE: THE STRUGGLE FOR PERMANENCE

18. Ecological Equations for Two Species

18.1 The theorem of Poincaré-Bendixson

Two dimensional differential equations cannot behave very wildly. The reason for this lies in the well known *Jordan curve theorem.* Any periodic orbit γ in the plane divides the plane into two disjoint connected sets, the interior and the exterior. Two points in the interior (or two points in the exterior) can be connected by a path which never intersects γ, but every path between the interior and the exterior has to cross γ.

This theorem, which is intuitively obvious, requires a long and difficult proof. One of its deep consequences is the *theorem of Poincaré-Bendixson*:

Theorem. *Let $\dot{\mathbf{x}} = \mathbf{f}(\mathbf{x})$ be an ODE defined on an open set $G \subseteq \mathbf{R}^2$. Let $\omega(\mathbf{x})$ be a nonempty compact ω-limit set. Then, if $\omega(\mathbf{x})$ contains no rest point, it must be a periodic orbit.*

It is, incidentally, quite possible that $\omega(\mathbf{x})$ is empty, or unbounded. It may also happen that $\omega(\mathbf{x})$ consists neither of one equilibrium nor of one periodic orbit.

As an immediate consequence of the theorem of Poincaré-Bendixson, one obtains that if $K \subseteq G$ is nonempty, compact and forward invariant, then K must contain a fixed point or a periodic orbit. If γ is a periodic orbit which, together with its interior Γ, is contained in G, then Γ contains an equilibrium.

An important technique for excluding periodic orbits is the *method of Bendixson-Dulac*: A differential equation $\dot{\mathbf{x}} = \mathbf{f}(\mathbf{x})$ defined on a simply connected subset G of $\mathbf{R}^2$ (i.e. a set G without 'holes') and satisfying $\operatorname{div} \mathbf{f}(\mathbf{x}) > 0$ for all $\mathbf{x} \in G$ admits no periodic orbit. (Here,

$$\operatorname{div} \mathbf{f}(\mathbf{x}) = \frac{\partial f_1}{\partial x_1}(\mathbf{x}) + \frac{\partial f_2}{\partial x_2}(\mathbf{x}) \tag{18.1}$$

is the *divergence* of $\mathbf{f}$. It is the trace of the Jacobian.)

Indeed, if γ is such a periodic orbit and Γ its interior, then Γ is an invariant set and has its area preserved under the flow: but this cannot be, since $\operatorname{div} \mathbf{f}(\mathbf{x}) > 0$ means that the flow is area-expanding (cf. **19.5**). More precisely, the *theorem of Green* implies

$$\int_\Gamma \operatorname{div}\mathbf{f}(\mathbf{x})\, d(x_1,x_2) = \pm \int_0^T [f_2(\mathbf{x}(t))\dot{x}_1(t) - f_1(\mathbf{x}(t))\dot{x}_2(t)]\, dt$$

where T is the period of γ. The left hand side is positive, while the right hand side is 0 since $f_1 = \dot{x}_1$ and $f_2 = \dot{x}_2$. This is a contradiction.

As a corollary, one obtains: if there exists a positive function B on G such that the vector field $B\mathbf{f}$ has positive divergence at every point, then $\dot{\mathbf{x}} = \mathbf{f}(\mathbf{x})$ admits no periodic orbit. (Such a function B is said to be a *Dulac function*). Indeed, $\mathbf{f}$ differs from $B\mathbf{f}$ only by a change in velocity, which does not affect the orbits.

Of course the same conclusion holds if $\operatorname{div} B\mathbf{f}(\mathbf{x}) < 0$ everywhere.

Exercise: If there exists a positive function B such that $\operatorname{div} B\mathbf{f}(\mathbf{x}) = 0$ for all $\mathbf{x} \in G$, then there exists a constant of motion for $\dot{\mathbf{x}} = \mathbf{f}(\mathbf{x})$.

18.2 Periodic orbits for two dimensional Lotka-Volterra equations

Theorem. *The two dimensional Lotka-Volterra equation*

$$\begin{aligned} \dot{x} &= x(a + bx + cy) \\ \dot{y} &= y(d + ex + fy) \end{aligned} \tag{18.2}$$

admits no isolated periodic orbits.

Indeed, let γ be a periodic solution of (18.2). Then there must be a rest point in the interior of γ (and hence of $\mathbf{R}^2_+$). The two lines

$$a + bx + cy = 0$$

$$d + ex + fy = 0$$

must therefore intersect in a unique point of int $\mathbf{R}^2_+$. In particular

$$\Delta = bf - ce \neq 0 \quad . \tag{18.3}$$

We shall now apply the theorem of Bendixson-Dulac, using as Dulac function

$$B(x,y) = x^{\alpha-1}y^{\beta-1} \tag{18.4}$$

with coefficients α and β which will be specified later on. Denoting the right hand side of (18.2) by $P(x,y)$ and $Q(x,y)$, we compute as divergence of the vector field (BP,BQ):

$$\frac{\partial}{\partial x}(BP) + \frac{\partial}{\partial y}(BQ) =$$

$$= \frac{\partial}{\partial x}[x^{\alpha}y^{\beta-1}(a+bx+cy)] + \frac{\partial}{\partial y}[x^{\alpha-1}y^{\beta}(d+ex+fy)]$$

$$= \alpha x^{\alpha-1}y^{\beta-1}(a+bx+cy) + x^{\alpha}y^{\beta-1}b + \beta x^{\alpha-1}y^{\beta-1}(d+ex+fy) + x^{\alpha-1}y^{\beta}f$$

$$= B[\alpha(a+bx+cy) + bx + \beta(d+ex+fy) + fy] \quad .$$

We choose α and β such that

$$\begin{aligned} \alpha b + \beta e &= -b \\ \alpha c + \beta f &= -f \quad . \end{aligned} \tag{18.5}$$

This is possible by (18.3) and leads to

$$\frac{\partial}{\partial x}(BP) + \frac{\partial}{\partial y}(BQ) = \delta B \tag{18.6}$$

with

$$\delta = a\alpha + d\beta \quad . \tag{18.7}$$

Since we have assumed that a periodic orbit γ exists, the theorem of Bendixson-Dulac implies $\delta = 0$. But then, (18.6) becomes

$$-\frac{\partial}{\partial x}(BP) = \frac{\partial}{\partial y}(BQ) \quad . \tag{18.8}$$

This is just the *integrability condition* for the two dimensional vector field $(BQ,-BP)$. Hence there exists a function $V = V(x,y)$ on int $\mathbf{R}^2_+$ such that

$$\frac{\partial V}{\partial x} = BQ \qquad \frac{\partial V}{\partial y} = -BP \quad . \tag{18.9}$$

The derivative of $t \to V(x(t), y(t))$ then satisfies

$$\dot{V} = \frac{\partial V}{\partial x}\dot{x} + \frac{\partial V}{\partial y}\dot{y} = PQ(B-B) \equiv 0 \tag{18.10}$$

and hence V is a *constant of motion*, just as in **7.3**. If a periodic orbit exists, it is surrounded by other periodic orbits: no periodic orbit is isolated.

Exercise 1: Show that the constant δ from (18.7) is the trace of the Jacobian at the interior fixed point.

Exercise 2: Show that (18.2) has a continuum of periodic orbits iff the eigenvalues at the interior equilibrium are purely imaginary (i.e. $\delta = 0$ and $\Delta > 0$). (Hint: Compute the Hessian of V). Show that three phase portraits are possible.

Exercise 3: Compute the constant of motion V from (18.9) explicitly. Show that it is in general of the form

$$V(x,y) = x^p y^q (A + Bx + Cy) \quad .$$

Exercise 4: Give an alternative proof of Theorem 18.2 by showing that every two dimensional Lotka-Volterra equation (18.2) admits a global Ljapunov function of the above form.

18.3 Limit cycles

There is a fair amount of empirical evidence for periodic oscillations in predator-prey systems. The most impressive data is furnished by the centuries old records of the Hudson's Bay Company on the population densities of hares and lynxes. The regular outbreaks of budworms in large forests also point at the existence of stable cycles in host-parasite systems.

What one is looking for are *periodic attractors* or at least *limit cycles.* A periodic orbit γ is an attractor if $\omega(\mathbf{x}) = \gamma$ for all initial conditions $\mathbf{x}$ in some neighbourhood of γ; it is a limit cycle if $\omega(\mathbf{x}) = \gamma$ for at least one $\mathbf{x} \notin \gamma$.

Such limit cycles do not occur for linear systems: as we have seen in the last section, they don't exist for two dimensional Lotka-Volterra systems either. The periodic orbits in the classical predator prey equation (7.1) are not stable: random fluctuations may send the state to another orbit with different frequency and amplitude, and the slightest intraspecific competition destroys the periodicity altogether. To get a more robust cycling, we must look for nonlinear interactions.

Exercise: Show that the unit circle is a periodic attractor for

$$\begin{aligned} \dot{x} &= x - y - x(x^2 + y^2) \\ \dot{y} &= x + y - y(x^2 + y^2) \quad . \end{aligned} \tag{18.11}$$

(Hint: use $V(x,y) = (1 - x^2 - y^2)^2$ as Ljapunov function.)

18.4 The predator-prey model of Gause

Let x and y denote the densities of prey and predator, respectively. In the absence of predators, the prey population should converge towards a limit $K > 0$. Thus $\dot{x} = xg(x)$ with

(a) $g(x) > 0$ for $x < K$, $g(x) < 0$ for $x > K$ and $g(K) = 0$.

The predators will reduce the rate of increase $\dot{x}$ of the prey by $yp(x)$, where $p(x)$ is the amount of prey killed by one predator. Thus

(b) $p(0) = 0$ and $p(x) > 0$ for $x > 0$.

The rate of increase $\dot{y}/y$ of the predators, finally, shall be given by $-d + q(x)$. The constant $d > 0$ corresponds to the mortality of predators in the absence of prey, and $q(x)$ is a positive, monotonically increasing function. Thus

(c) $q(0) = 0$ and $\frac{dq}{dx}(x) > 0$ for $x > 0$.

This yields the equation

$$\begin{aligned} \dot{x} &= xg(x) - yp(x) \\ \dot{y} &= y(-d + q(x)) \ . \end{aligned} \tag{18.12}$$

Once more, $x \equiv 0$ and $y \equiv 0$ are solutions of (18.12). Hence $\mathbf{R}^2_+$ is invariant. Since q increases monotonically, there is at most one $\bar{x} > 0$ with $q(\bar{x}) = d$. If $\bar{x}$ does not exist, the predators vanish. We shall exclude this case and assume that $\bar{x}$ exists. The y-isocline, then, is the vertical axis $x = \bar{x}$. The x-isocline is given by the equation

$$y = \frac{xg(x)}{p(x)} \tag{18.13}$$

i.e. by the graph of a function of x defined on the interval $(0,K)$. Obviously, the isoclines intersect at most once. If there is no intersection (i.e. if $\bar{x} \geq K$), the predators will vanish, as may be seen from the signs of $\dot{x}$ and $\dot{y}$ (cf. Fig. 18.1). But if the intersection $\mathbf{F} = (\bar{x},\bar{y})$ exists (i.e. in the case $\bar{x} < K$) then $\mathbf{F}$ is the unique equilibrium in int $\mathbf{R}^2_+$. This is the case which we shall consider from now on.

Exercise: Compute the Jacobian at $\mathbf{F}$. Show that $\mathbf{F}$ is a sink iff

$$\frac{d}{dx}\left(\frac{xg(x)}{p(x)}\right) \tag{18.14}$$

is negative at $x = \bar{x}$, i.e. if the slope of the x-isocline at $\mathbf{F}$ is negative. If the slope if positive, $\mathbf{F}$ is unstable.

On bd $\mathbf{R}^2_+$, there are two fixed points, namely (0,0) and $\mathbf{P} = (K,0)$ (see Fig. 18.2-3). We check immediately that both are saddles. The stable manifold of $\mathbf{P}$ is given by the positive x-axis. The unstable manifold consists of two orbits of which only one lies in $\mathbf{R}^2_+$.

Let $\mathbf{x}$ be a point on this orbit. Clearly $\omega(\mathbf{x})$ is nonempty and compact. Using the theorem of Poincaré-Bendixson, we obtain the following two alternatives:

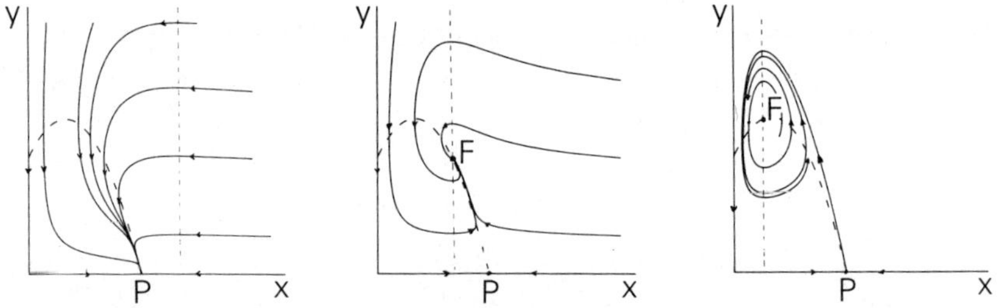

Figures 18.1,2,3

(a) if $\omega(\mathbf{x})$ contains no equilibrium, then $\omega(\mathbf{x})$ is a periodic orbit γ (see Fig. 18.3). This orbit must surround a rest point, i.e. the point $\mathbf{F}$. Obviously γ is a limit cycle: in fact every orbit in the exterior spirals towards γ;

(b) if $\omega(\mathbf{x})$ contains an equilibrium, this must be the point $\mathbf{F}$, since (0,0) and $(K,0)$ are out of question. With the help of the signs of $\dot{x}$ and $\dot{y}$, one checks that every orbit in int $\mathbf{R}^2_+$ converges to $\mathbf{F}$: such an orbit, indeed, cannot cross the unstable manifold of $\mathbf{P}$ (see Fig. 18.2). In this case $\mathbf{F}$ is globally stable.

18.5 Holling-type interaction

The response functions $p(x)$ and $q(x)$ in (18.12) are often assumed to be proportional to $\frac{x}{a+x}$ (with $a > 0$): this displays a saturation effect for large x which is very common in ecology and chemical kinetics: the corresponding type of interaction is associated with the names of Holling, in the former field, and Michaelis-Menten in the latter. If $g(x)$, furthermore, is linear, then (18.12) becomes

$$\begin{aligned} \dot{x} &= rx(1 - \frac{x}{K}) - y\frac{cx}{a+x} \\ \dot{y} &= y(-d + \frac{bx}{a+x}) \end{aligned} \tag{18.15}$$

where all parameters are positive.

Exercise 1: If either $b \leq d$ or $K \leq \frac{ad}{b-d}$, then all orbits of (18.15) in int $\mathbf{R}^2_+$ converge to the fixed point $(K,0)$.

Let us assume for the rest of the section that $b > d$ and $K > \frac{ad}{b-d}$. Then (18.15) admits a unique interior fixed point $\mathbf{F} = (\bar{x},\bar{y})$ with

$$\bar{x} = \frac{ad}{b-d} \quad . \tag{18.16}$$

Exercise 2: Using Exercise 18.4 show that $\mathbf{F}$ is a sink iff

$$K < a + 2\bar{x} \quad . \tag{18.17}$$

Theorem. *The equilibrium* $\mathbf{F}$ *is globally stable for* (18.15) *iff* $K \leq a + 2\bar{x}$.

For the proof we consider the Dulac function

$$B(x,y) = \frac{a+x}{x}y^{\alpha-1} \tag{18.18}$$

where α will be chosen later. Denoting the right hand side of (18.15) by P and Q, we obtain

$$\frac{\partial}{\partial x}(BP) + \frac{\partial}{\partial y}(BQ) = \tag{18.19}$$

$$= y^{\alpha-1}x^{-1}[rx(1-\frac{a}{K} - \frac{2x}{K}) + \alpha(-ad + (b-d)x)].$$

At $x = \bar{x}$, (18.19) is negative in view of our assumption $K < a + 2\bar{x}$. Hence we can find a constant α such that the parabola $rx(1 - \frac{a}{K} - \frac{2x}{K})$

lies below the line $\alpha((b-d)x-ad)$. If we choose, for example, $\alpha = \frac{r}{b-d}(1-\frac{a}{K})$, then the bracket in (18.19) can be rewritten as

$$-\frac{2r}{K}(x-\frac{K-a}{2})^2 + r(1-\frac{a}{K})(-\bar{x}+\frac{K-a}{2}) \leq 0 \quad . \tag{18.20}$$

Hence periodic orbits are excluded.

If $K > a+2\bar{x}$ then by the criterion in exercise 18.4 the point $\mathbf{F}$ is a source and **18.4** shows that there must be a limit cycle around $\mathbf{F}$.

Let us take a look at what happens if we let the parameter K increase. For $K \leq a+2\bar{x}$ all orbits in $\operatorname{int}\mathbf{R}^2_+$ spiral towards $\mathbf{F}$. At $K = a+2\bar{x}$, $\mathbf{F}$ stops being a sink but is still asymptotically stable. For $K > a+2\bar{x}$, $\mathbf{F}$ is a source, and in a neighbourhood of $\mathbf{F}$ everything flows away. However, it is as if the differential equation did not know this at points which are further off: there, the orbits are still approaching $\mathbf{F}$. It is intuitively obvious that this must lead to at least one periodic orbit in the intermediate zone.

Behind this example there lurks a general principle.

18.6 Hopf bifurcations

Let G be an open subset of $\mathbf{R}^n$, and

$$\dot{\mathbf{x}} = \mathbf{f}_\mu(\mathbf{x}) \tag{18.21}$$

a family of differential equations depending on some parameter $\mu \in (-\epsilon,\epsilon)$. Let $\mathbf{P}_\mu$ be a rest point of (18.21). Let us assume that all eigenvalues of the Jacobian J_μ have negative real parts, with the exception of one pair of complex conjugate eigenvalues

$$\alpha(\mu) \pm i\beta(\mu)$$

(with $\alpha(\mu), \beta(\mu) \in \mathbf{R}$) for which the sign of the real part $\alpha(\mu)$ is that of μ, and which satisfies $\beta(0) \neq 0$. In particular, then, $\mathbf{P}_\mu$ is a sink (and hence asymptotically stable) for $\mu < 0$, but unstable for $\mu > 0$.

Let us add to this three "technical conditions":

(a) the components of $\mathbf{f}_\mu(\mathbf{x})$ are analytic (i.e. given by power series);

(b) $\frac{d\alpha}{d\mu}(0) > 0$;

(c) $\mathbf{P}_0$ is asymptotically stable.

The *Theorem of Hopf* asserts that under these conditions, and for sufficiently small positive values of the parameter μ, the unstable equilibrium $\mathbf{P}_\mu$ is surrounded by a periodic attractor.

This theorem states that under the given conditions (which are quite often satisfied, but sometimes rather difficult to check), a periodic solution splits off from equilibrium if the parameter value μ crosses a critical threshold. The rest point, formerly asymptotically stable, becomes unstable: in its place, the periodic orbit is now the attractor. A stable equilibrium turns into a stable oscillation (see Fig. 18.4). The example given in the last section was characteristic.

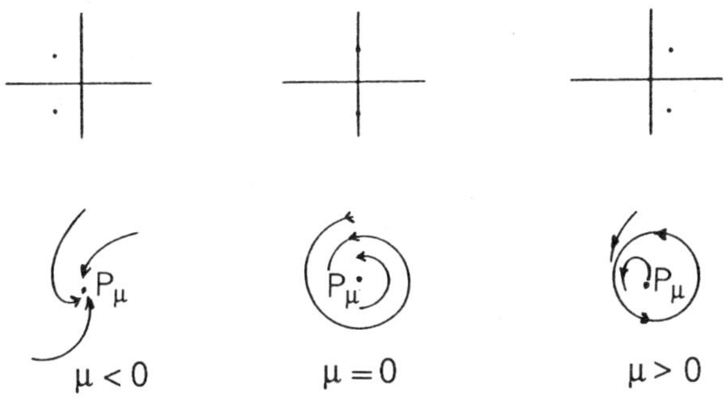

Figure 18.4

In order to understand the "technical conditions" better, let us return to the planar case. By a smooth change in coordinates, the Taylor series of degree 3 of $\mathbf{f}_\mu(\mathbf{x})$ can be brought into the *normal form*

$$\begin{aligned}\dot{x}_1 &= (d\mu + a(x_1^2 + x_2^2))x_1 - (\omega + c\mu + b(x_1^2 + x_2^2))x_2 \\ \dot{x}_2 &= (\omega + c\mu + b(x_1^2 + x_2^2))x_1 + (d\mu + a(x_1^2 + x_2^2))x_2 \ .\end{aligned} \tag{18.22}$$

In polar coordinates, this becomes

$$\begin{aligned}\dot{r} &= (d\mu + ar^2)r \\ \dot{\theta} &= \omega + c\mu + br^2 \ .\end{aligned} \tag{18.23}$$

The first of these equations does not depend on θ. If its right hand side is 0 for an $r > 0$, this corresponds to a periodic orbit of this radius. If $d \neq 0$ and $a \neq 0$, there are periodic orbits of radius

$$r = \sqrt{\frac{-d\mu}{a}} \ .$$

Depending on the signs of d and a, they occur for $\mu > 0$ or $\mu < 0$. If $a < 0$ they are attracting, and if $a > 0$ repelling; one speaks of *supercritical* and *subcritical* Hopf bifurcations, respectively. If $d \neq 0$ and $a = 0$, there are no periodic orbits except for $\mu = 0$ (for which parameter value the plane is filled with them). The Hopf bifurcation, in this case, is *degenerate.*

Note that from (18.23) $d = \frac{d\alpha}{d\mu}(0)$. The second technical condition means $d > 0$, therefore, and the third technical condition $a < 0$.

In higher dimensions, the same phenomena occur (roughly speaking) on the two dimensional "centre manifold" corresponding to the eigenvalues $\pm i\omega$. Since the other eigenvalues have negative real part, all nearby orbits converge to this centre manifold.

18.7 Cooperative systems

A differential equation $\dot{\mathbf{x}} = \mathbf{f}(\mathbf{x})$ defined on $G \subseteq \mathbf{R}^n$ is said to be *cooperative* if $\frac{\partial f_i}{\partial x_j}(\mathbf{x}) \geq 0$ for all $\mathbf{x} \in G$ and all $i \neq j$ (the growth of every component is enhanced by an increase in any other component).

Theorem. *The orbits of a two dimensional cooperative system converge either to an equilibrium or to infinity.*

Proof: Let us denote the orthants of $\mathbf{R}^2$ by $C_1 = \mathbf{R}^2_+$, $C_2 = \{(x_1, x_2) : x_1 \leq 0, x_2 \geq 0\}$, $C_3 = -\mathbf{R}^2_+$ and $C_4 = -C_2$. If $\dot{\mathbf{x}}(t_0) \in C_1$ for some $t_0 \in \mathbf{R}$, then $\dot{\mathbf{x}}(t) \in C_1$ for all $t \geq t_0$: if for example $\dot{x}_1(t) = 0$ but $\dot{x}_2(t) \geq 0$ then

$$\ddot{x}_1 = \dot{f}_1(\mathbf{x}(t)) = \frac{\partial f_1}{\partial x_1}\dot{x}_1 + \frac{\partial f_1}{\partial x_2}\dot{x}_2 \geq 0$$

and hence $\dot{x}_1$ cannot become negative. The same argument shows that $\dot{\mathbf{x}}(t_0) \in C_3$ implies $\dot{\mathbf{x}}(t) \in C_3$ for all $t \geq t_0$. If $\dot{\mathbf{x}}(t_0) \in C_2$, then either $\dot{\mathbf{x}}(t)$ enters C_1 or C_3 for some subsequent time, or it remains in C_2 for all $t \geq t_0$; similarly for $\dot{\mathbf{x}}(t_0) \in C_4$. In every case, the components $x_1(t)$ and $x_2(t)$ are ultimately monotone and hence converge either to a limit or to infinity.

The same result holds for *competitive systems*, defined by $\frac{\partial f_i}{\partial x_j}(\mathbf{x}) \leq 0$ for all $i \neq j$. The orbits of a two dimensional competitive system converge either to equilibrium or to infinity. One has only to replace C_1 and C_3 by C_2 and C_4.

Let us assume that two species (with densities x and y) are competing with each other. Rather than assume linear interactions as in the Lotka-Volterra equation (8.8), we shall describe the interaction by the more general differential equation

$$\begin{aligned} \dot{x} &= xS(x,y) \\ \dot{y} &= yW(x,y) \end{aligned} \tag{18.24}$$

where we only assume that the growth of each species affects the other one adversely, i.e. that

$$\frac{\partial S}{\partial y} \leq 0 \quad \text{and} \quad \frac{\partial W}{\partial x} \leq 0 \tag{18.25}$$

and that growth to infinity is impossible. Equation (18.24), then, is competitive and has no orbits converging to infinity: it follows that every orbit must converge to an equilibrium.

The same is true for symbiosis (where (18.25), of course, is replaced by

$$\frac{\partial S}{\partial y} \geq 0 \quad \text{and} \quad \frac{\partial W}{\partial x} \geq 0) \quad . \tag{18.26}$$

Exercise 1: Prove directly that (18.25) implies that all orbits of (18.24) converge to equilibria, using the fact that the isoclines are given by graphs of functions of y (resp. x). These graphs divide $\mathbf{R}^2_+$ into subsets which are either forward or backward invariant.

Exercise 2: Show that in $\mathbf{R}^2$, the change of coordinates $(y_1,y_2) = (x_1,-x_2)$ transforms a cooperative system into a competitive system.

Exercise 3: Show that in $\mathbf{R}^n$, the time reversal of a cooperative system yields a competitive system.

18.8 Notes

The classical monograph on two-dimensional ODEs is Andronov et al. (1973). It contains a proof of the nonexistence of limit cycles for two dimensional Lotka-Volterra equations (a result which is due to Bautin), see also Coppel (1966). For the Poincaré-Bendixson theory, we also refer to Hirsch and Smale (1974), and for Hopf bifurcations to Arnold (1983), Guckenheimer and Holmes (1983), Marsden and McCracken (1976) and Hassard et al (1981). A classification of all two dimensional phase portraits of Lotka-Volterra equations can be found in Bomze

(1983). The model (18.12) was proposed by Gause (1934). Another general predator-prey model was considered by Kolmogoroff (1936). We refer to Freedman (1980) and Bazykin (1985) for a mathematical analysis of these systems, and to Holling (1973) and May (1973) for an ecological discussion. The graphical method in Exercise 18.4 is due to Rosenzweig and MacArthur (1963) and Theorem 18.5 to Hsu et al. (1978). More general global stability results were shown in Cheng et al. (1981), and Cheng (1981) proved the uniqueness of the limit cycle of (18.15). We refer to Conway and Smoller (1986) for related equations and to Waltman (1983) for a survey of generalisations of (18.15) to several predators, which display interesting phenomena. The results in **18.7** are due to Hadeler and Glas (1983) and to Hirsch (1982): we refer to these papers and to Smale (1976) for more on higher dimensional cases.

19. Criteria for Permanence

19.1 Permanence and persistence for replicator equations

For the replicator equation

$$\dot{x}_i = x_i(f_i(\mathbf{x}) - \bar{f}) \tag{19.1}$$

on S_n and for ecological equations of the type

$$\dot{x}_i = x_i f_i(\mathbf{x}) \tag{19.2}$$

on $\mathbf{R}^n_+$, we can rarely expect to obtain a full description of the attractors. For many purposes, however, the precise asymptotic behaviour is less important than the question of *extinction*. There are several mathematical concepts dealing with this aspect. In particular, a system of type (19.1) or (19.2) is said to be *permanent* if there exists a compact set K in the interior of the state space such that all orbits in the interior end up in K. This means that the boundary is a *repellor* (for (19.2), we have to consider the points at infinity as part of the boundary of $\mathbf{R}^n_+$).

Equivalently, permanence means for (19.1) that there exists a $\delta > 0$ such that

$$\delta < \liminf_{t\to+\infty} x_i(t) \tag{19.3}$$

for all i, whenever $x_i(0) > 0$ for all i. (Thus we recover the definition from **13.1**.) For (19.2), permanence requires in addition that there exists a D such that

$$\limsup_{t\to+\infty} x_i(t) \le D \tag{19.4}$$

for all i, whenever $\mathbf{x} \in \operatorname{int} \mathbf{R}^n_+$. If (19.4) holds even for all $\mathbf{x} \in \mathbf{R}^n_+$, we shall say that the orbits of (19.2) are *uniformly bounded* (obviously a minimal concession to reality).

Condition (19.3) means that if all types are initially present, selection will not lead to extinction. Conversely, if some originally missing component is introduced through mutation, it will spread. The "threshold" δ is a uniform one, independent of the initial condition. Thus permanence is a more stringent property than *strong persistence* (which requires (19.3) with $\delta = 0$) and *persistence* (which requires

$$\limsup_{t \to +\infty} x_i(t) > 0 \tag{19.5}$$

for all orbits in the interior of the state space).

We have met so far one sufficient condition for permanence: the existence of an *average Ljapunov function.* For (19.1) this was proved in **13.2**. The same proof applies for (19.2) with uniformly bounded orbits.

Exercise 1: Check this. Why does one need the boundedness condition to prove (19.3)?

A useful necessary condition for permanence is an *index theorem* implying the existence of an interior rest point. We shall derive this result in **19.3** after some preliminary theory on index and degree.

For most examples, the terms f_i in (19.1) and (19.2) are linear. This yields the first order replicator equation

$$\dot{x}_i = x_i((A\mathbf{x})_i - \mathbf{x} \cdot A\mathbf{x}) \tag{19.6}$$

on S_n, resp. the Lotka-Volterra equation

$$\dot{x}_i = x_i(r_i + (A\mathbf{x})_i) \tag{19.7}$$

on $\mathbf{R}^n_+$. In **19.5** and **19.6**, we shall present useful necessary and sufficient conditions for the permanence of such systems.

Exercise 2: Show that the modified "rock-scissors-paper" game (19.6) with

$$A = \begin{bmatrix} 0 & 1+\epsilon & -1 \\ -1 & 0 & 1+\epsilon \\ 1+\epsilon & -1 & 0 \end{bmatrix} \tag{19.8}$$

is persistent, but not strongly persistent for $\epsilon < 0$. It is strongly persistent but not permanent for $\epsilon = 0$.

19.2 Brouwer's degree and Poincaré's index

Let U be a bounded open subset of $\mathbf{R}^n$ and $\mathbf{f}$ a vector field defined on a neighbourhood of its closure $\bar{U}$. A point $\mathbf{x} \in U$ is said to be *regular* if $\det D_{\mathbf{x}}\mathbf{f} \neq 0$. A point $\mathbf{y} \in \mathbf{R}^n$ is said to be a *regular value* if all $\mathbf{x} \in \bar{U}$ with $\mathbf{f}(\mathbf{x}) = \mathbf{y}$ are regular.

As a consequence of the *implicit function theorem*, $\mathbf{f}$ is locally invertible around every such $\mathbf{x}$, and hence the roots of $\mathbf{f}(\mathbf{x}) = \mathbf{y}$ are isolated. The *theorem of Sard* states that almost every $\mathbf{y} \in \mathbf{R}^n$ is a regular value, and hence that regular values are dense in $\mathbf{R}^n$.

Now suppose $\mathbf{y} \notin \mathbf{f}(\mathrm{bd}\, U)$. Then the *Brouwer degree* of $\mathbf{f}$ at the value $\mathbf{y} \in \mathbf{R}^n$ is defined by

$$\deg(\mathbf{f},\mathbf{y}) = \sum_{\mathbf{f}(\mathbf{x})=\mathbf{y}} \operatorname{sgn} \det D_{\mathbf{x}}\mathbf{f} \tag{19.9}$$

if $\mathbf{y}$ is a regular value, and by

$$\deg(\mathbf{f},\mathbf{y}) = \lim_{n\to\infty} \deg(\mathbf{f},\mathbf{y}_n) \tag{19.10}$$

if $\mathbf{y}$ is not, where $\mathbf{y}_n$ is a sequence of regular values converging to $\mathbf{y}$ (it can be shown that this limit is well defined). For a regular value $\mathbf{y}$, the degree counts the difference between the number of orientation preserving and orientation reserving roots of $\mathbf{f}(\mathbf{x}) = \mathbf{y}$.

The basic property of the degree is its *homotopy invariance:* if $\mathbf{f}_t$ (with $t \in [0,1]$) is a continuous family of mappings from $\bar{U}$ to $\mathbf{R}^n$, and if $\mathbf{y}$ does not belong to any $\mathbf{f}_t(\mathrm{bd}\, U)$, then $t \to \deg(\mathbf{f}_t,\mathbf{y})$ is constant and hence $\deg(\mathbf{f}_1,\mathbf{y}) = \deg(\mathbf{f}_0,\mathbf{y})$. It follows that if $\mathbf{f}$ and $\mathbf{g}$ are defined on $\bar{U}$ and coincide on $\mathrm{bd}\, U$, then $\deg(\mathbf{f},\mathbf{y}) = \deg(\mathbf{g},\mathbf{y})$ for all $\mathbf{y} \notin \mathbf{f}(\mathrm{bd}\, U)$. One just has to apply the previous result to the family $\mathbf{f}_t = t\mathbf{g} + (1-t)\mathbf{f}$, noting that $\mathbf{f}_0 = \mathbf{f}$ and $\mathbf{f}_1 = \mathbf{g}$.

Furthermore, if $\mathbf{y}_0$ and $\mathbf{y}_1$ belong to the same component of $\mathbf{R}^n \setminus \mathbf{f}(\mathrm{bd}\, U)$, i.e. if there is a continuous path $t \to \mathbf{y}_t$ not intersecting $\mathbf{f}(\mathrm{bd}\, U)$ for $t \in [0,1]$, then $\deg(\mathbf{f},\mathbf{y}_0) = \deg(\mathbf{f},\mathbf{y}_1)$. The degree $\deg(\mathbf{f},\mathbf{y})$ depends only on the values of $\mathbf{f}$ on $\mathrm{bd}\, U$ and on the component of $\mathbf{R}^n \setminus \mathbf{f}(\mathrm{bd}\, U)$ containing $\mathbf{y}$.

Of particular interest, of course, are the zeros of the vector field $\mathbf{f}$, i.e. the equilibria of the differential equation $\dot{\mathbf{x}} = \mathbf{f}(\mathbf{x})$. If the value $\mathbf{y}$ is not otherwise specified, the degree of $\mathbf{f}$ is understood to be $\deg(\mathbf{f},\mathbf{0})$. If $\mathbf{f}$ and $\mathbf{g}$ are two vector fields on $\bar{U}$ such that the degree is defined (which means that they do not vanish on $\mathrm{bd}\, U$) and if they never point in opposite directions on $\mathrm{bd}\, U$ (i.e. if there is no $\mathbf{x} \in \mathrm{bd}\, U$ and no $\lambda > 0$ such that $\mathbf{f}(\mathbf{x}) + \lambda\mathbf{g}(\mathbf{x}) = \mathbf{0}$) then $\deg(\mathbf{f},\mathbf{0}) = \deg(\mathbf{g},\mathbf{0})$. This can be seen by looking at the family $\mathbf{f}_t = t\mathbf{g} + (1-t)\mathbf{f}$, which has no zero on bd U.

As a consequence, we obtain the celebrated *fixed point theorem of Brouwer: For any map* $\mathbf{h}$ *of the unit ball* $D = \{\mathbf{x} \in \mathbf{R}^n : \|\mathbf{x}\| \leq 1\}$ *into itself there exists an* $\mathbf{x} \in D$ *such that* $\mathbf{h}(\mathbf{x}) = \mathbf{x}$.

Indeed, if $\mathbf{h}(\mathbf{x}) \neq \mathbf{x}$ for all $\mathbf{x} \in D$, then the vector field $\mathbf{f}(\mathbf{x}) = \mathbf{h}(\mathbf{x}) - \mathbf{x}$ points inward for all $\mathbf{x} \in$ bd D. On bd D, therefore, it never points into the opposite direction of $\mathbf{g}(\mathbf{x}) = -\mathbf{x}$. But then $\deg(\mathbf{f},\mathbf{0}) = \deg(\mathbf{g},\mathbf{0}) = (-1)^n \neq 0$, and so there is some $\mathbf{x} \in D$ with $\mathbf{f}(\mathbf{x}) = \mathbf{0}$, which is a contradiction. (We have assumed here that the map $\mathbf{h}$ is differentiable. The theorem of Brouwer holds, in fact, for any continuous $\mathbf{h}$. This can be proved by approximating $\mathbf{h}$ uniformly by a sequence of differentiable vector fields $\mathbf{h}_n$.)

Exercise 1: If $\dot{\mathbf{x}} = \mathbf{f}(\mathbf{x})$ is a differential equation for which the ball D (or any homeomorphic set) is forward invariant, then D contains an equilibrium.

Now let $\hat{\mathbf{x}}$ be an isolated equilibrium of the differential equation $\dot{\mathbf{x}} = \mathbf{f}(\mathbf{x})$ defined on the open set $U \subseteq \mathbf{R}^n$. The *Poincaré index* of $\hat{\mathbf{x}}$ with respect to the vector field $\mathbf{f}$ is defined as

$$i(\hat{\mathbf{x}}) = \deg(\mathbf{f},\mathbf{0}) \tag{19.11}$$

(where $\mathbf{f}$, here, denotes the restriction to a closed ball $\bar{B} \subseteq U$ containing $\hat{\mathbf{x}}$ but no other rest point. It can easily be shown that this degree does not depend on the particular choice of B.)

If $\hat{\mathbf{x}}$ is regular, then $i(\hat{\mathbf{x}})$ is just the sign of $\det D_{\hat{\mathbf{x}}}\mathbf{f}$. Hence

$$i(\hat{\mathbf{x}}) = (-1)^\sigma \tag{19.12}$$

where σ is the number of real negative eigenvalues of the Jacobian $D_{\hat{\mathbf{x}}}\mathbf{f}$. For $n = 2$, for example, the index of a centre, a sink or a source is $+1$, while that of a saddle is -1.

There exists an important *topological characterization* of $i(\hat{\mathbf{x}})$, valid also for non-regular $\hat{\mathbf{x}}$. Let us consider the map

$$\mathbf{x} \to \mathbf{h}(\mathbf{x}) = \frac{\mathbf{f}(\mathbf{x})}{\|\mathbf{f}(\mathbf{x})\|} \tag{19.13}$$

from bd B to the unit sphere in $\mathbf{R}^n$. The degree of such a map between two $(n-1)$-dimensional spheres can be defined just as before. Then

$$i(\hat{\mathbf{x}}) = \deg(\mathbf{h},\mathbf{y}) \tag{19.14}$$

where $\mathbf{y}$ is any regular value of $\mathbf{h}$ (it does not matter which one, since bd B has empty boundary). In Fig. 19.1, we sketch a few examples for

$n = 2$. If $\hat{\mathbf{x}}$ is non-regular, the index $i(\hat{\mathbf{x}})$ may take other values than ± 1.

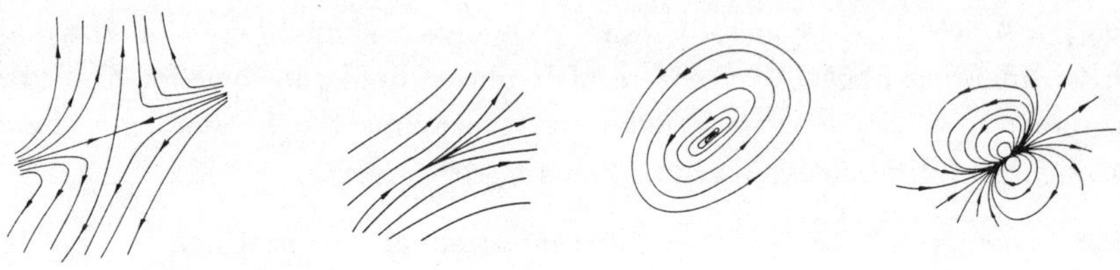

Figure 19.1

Exercise 2: Show that for the differential equation $\dot{z} = z^n$ on $\mathbf{R}^2$ (written in complex notation), the origin has index n.

The amazing *theorem of Poincaré-Hopf* states that if M is a compact manifold with boundary and if $\mathbf{f}$ is a vector field on M pointing outward at the boundary points and having only finitely many fixed points, then

$$\sum i = \chi(M) \tag{19.15}$$

where the sum is over all equilibria in M and the right hand side, the so called *Euler characteristic*, is a topological invariant depending only on M. The sum of the indices, thus, does not depend on the vector field.

We shall illustrate this with some two dimensional examples. If B is the unit disk in $\mathbf{R}^2$, then $\chi(B) = 1$. Let us consider a planar vector field admitting a closed orbit γ and defined in its interior Γ. Along the boundary of Γ, i.e. along γ, it will point outward (at least after a continuous transformation). It follows from the index theorem that Γ must contain a rest point. If all equilibria in Γ are regular, there must be an odd number $2n + 1$ of them, of which exactly n are saddles. In particular, if there is a unique rest point in Γ, it cannot be a saddle.

Next let us look at a vector field on the two-dimensional sphere M, for which $\chi(M) = 2$. They must have at least one fixed point. If all fixed points are regular, there must be at least two of them. If M is the two dimensional surface of the solid torus in $\mathbf{R}^3$, on the other hand, there exist vector fields without rest points: $\chi(M) = 0$. If there are only regular equilibria, they must be even in number, half of them being saddles.

Exercise 3: Draw such vector fields.

19.3 An index theorem for permanent systems

Theorem. *If (19.2) is permanent, then the degree of the vector field is $(-1)^n$ with respect to any bounded set U (with $\bar{U} \subseteq \operatorname{int} \mathbf{R}^n_+$) containing all interior ω-limits. In particular, there exists a rest point in* $\operatorname{int} \mathbf{R}^n_+$.

Proof: Let $K \subseteq \operatorname{int} \mathbf{R}^n_+$ be a compact set containing all ω-limits in its interior. Let $\tau(\mathbf{x})$ be the time of first entrance into $\operatorname{int} K$:

$$\tau(\mathbf{x}) = \inf\{t \geq 0 : \mathbf{x}(t) \in \operatorname{int} K\} \quad .$$

It is easy to see that τ is defined on $\operatorname{int} \mathbf{R}^n_+$, upper semicontinuous and therefore locally bounded. Let

$$T = \max_{\mathbf{x} \in K} \tau(\mathbf{x})$$

be the maximum time for orbits leaving K to return to K. The set

$$K^+ = \{\mathbf{x}(t) : \mathbf{x} \in K, 0 \leq t \leq T\}$$

is compact and forward invariant.

Now let $B \subseteq \operatorname{int} \mathbf{R}^n_+$ be homeomorphic to a ball and contain K^+. The entrance time $\tau(\mathbf{x})$ will again attain an upper bound T^1 on B. Let $\mathbf{h}(\mathbf{x})$ denote the right hand side of (19.2), and consider the following homotopy:

$$\begin{aligned} \mathbf{h}_t(\mathbf{x}) &= \mathbf{h}(\mathbf{x}) && \text{for } t = 0 \\ &= \frac{\mathbf{x}(t) - \mathbf{x}(0)}{t} && \text{for } t > 0 \ . \end{aligned} \tag{19.16}$$

Clearly $\mathbf{h}_t(\mathbf{x}) \neq \mathbf{0}$ for all $\mathbf{x} \in \operatorname{bd} B$ and $t \in \mathbf{R}$, since there are no fixed or periodic points on $\operatorname{bd} B$. Thus the degree of the vector field $\mathbf{x} \to \mathbf{h}_t(\mathbf{x})$ with respect to B is defined for all $t \in [0, T^1]$ and independent of t. For $t = T^1$, the vector field $\mathbf{h}_t(\mathbf{x})$ points inward along $\operatorname{bd} B$, and so its degree is $(-1)^n$. Thus the degree of $\mathbf{h}_0(\mathbf{x}) = \mathbf{h}(\mathbf{x})$ with respect to B (or any other open bounded set containing K) is $(-1)^n$.

The same result holds obviously for (19.1), except that the degree is $(-1)^{n-1}$, since the dimension is reduced by 1.

19.4 Saturated rest points and a general index theorem

Before proceeding further, it will be useful to define a rest point $\mathbf{p}$ of (19.2) (resp. (19.1)) as *saturated* if $f_i(\mathbf{p}) \leq 0$ (resp. $f_i(\mathbf{p}) \leq \bar{f}(\mathbf{p})$) whenever $p_i = 0$ (for $p_i > 0$, of course, the equality sign must hold). Every rest point in the interior of the state space is trivially saturated. For a rest point on the boundary, the condition means that selection doesn't "call for" the missing species.

To be saturated is an eigenvalue condition. Indeed, the Jacobian of (19.2) at $\mathbf{p}$ is of the form

$$\frac{\partial \dot{x}_i}{\partial x_j} = \delta_{ij} f_i(\mathbf{p}) + p_i \frac{\partial f_i}{\partial x_j}(\mathbf{p}) \quad .$$

(where δ_{ij} is the Kronecker delta: $\delta_{ij} = 1$ if $i = j$, $\delta_{ij} = 0$ otherwise). If $p_i = 0$, this reduces to $f_i(\mathbf{p})$ for $i = j$ and to 0 for $i \neq j$, so $f_i(\mathbf{p})$ is an eigenvalue for the (left) eigenvector $\mathbf{e}_i$. Hence we call it a *transversal eigenvalue*. It measures the rate of approach to the face $x_i = 0$ near the rest point $\mathbf{p}$. Similarly the transversal eigenvalues of (19.1) are of the form $f_i(\mathbf{p}) - \bar{f}(\mathbf{p})$. The rest point $\mathbf{p}$ is saturated iff all its transversal eigenvalues are nonpositive. With this concept at hand we can state a general *index theorem for replicator equations*.

Theorem 1. *There exists at least one saturated rest point for* (19.1). *If all saturated rest points are regular, the sum of their indices is* $(-1)^{n-1}$ (*and hence their number is odd.*)

Proof: Let us consider

$$\dot{x}_i = x_i[f_i(\mathbf{x}) - \bar{f}(\mathbf{x}) - n\epsilon] + \epsilon \tag{19.17}$$

as a perturbation of (19.1). The small $\epsilon > 0$ represents biologically an immigration. Clearly (19.17) maintains the relation $\sum \dot{x}_i = 0$ on S_n. On the boundary, the flow points now into int S_n. The sum of the indices of the rest points of (19.17) in int S_n is therefore $(-1)^{n-1}$.

For every $\epsilon > 0$, there must consequently exist at least one rest point $\mathbf{p}(\epsilon)$ of (19.17) in int S_n. Then

$$f_i(\mathbf{p}(\epsilon)) - \bar{f}(\mathbf{p}(\epsilon)) - n\epsilon = -\frac{\epsilon}{p_i(\epsilon)} < 0 \quad . \tag{19.18}$$

Any accumulation point $\mathbf{p}$ of $\mathbf{p}(\epsilon)$ for $\epsilon \to 0$ satisfies $f_i(\mathbf{p}) - \bar{f}(\mathbf{p}) \leq 0$ by continuity and is therefore a saturated rest point of (19.1). This shows the first claim.

If every saturated rest point $\mathbf{p}$ of (19.1) is regular, there are only finitely many of them (since S_n is compact), and each one continues, by the implicit function theorem, to a unique family $\mathbf{p}(\epsilon)$ of fixed points of (19.17), as long as ϵ is sufficiently small. $\mathbf{p}(\epsilon) \in \operatorname{int} S_n$ for $\epsilon > 0$ since $f_i(\mathbf{p}) < \bar{f}(\mathbf{p})$ implies $p_i(\epsilon) > 0$ by (19.18). But as we have seen, fixed points of (19.17) accumulate for $\epsilon \to 0$ only near saturated rest points of (19.1). Hence every rest point of (19.17) lies on one of these families $\mathbf{p}(\epsilon)$, for ϵ sufficiently small. As the index does not change when ϵ reaches 0 (by regularity), the above index relation for $\epsilon > 0$ holds also for $\epsilon = 0$.

Exercise 1: Give a heuristic argument (by drawing a picture) that regular saturated rest points move to the interior and nonsaturated ones to the exterior of S_n, if immigration terms $\epsilon > 0$ are introduced.

Exercise 2: Show that a "degenerate" saturated rest point (with at least one transversal eigenvalue being 0) can move inwards, outwards, disappear, or split up into several rest points, if immigration is introduced. (As a typical example, consider a one- or two-dimensional Lotka-Volterra equation with $\mathbf{0}$ as degenerate saturated rest point, i.e. $r_i = 0$.)

Exercise 3: Prove the *index theorem for ecological equations:* If (19.2) has uniformly bounded orbits, then it has a saturated fixed point, and if all saturated rest points are regular, then the sum of their indices is $(-1)^n$.

Exercise 4: Let $\hat{\mathbf{x}} \in \mathbf{R}^n_+$ be an isolated fixed point of (19.2) and U the intersection of an isolating neighbourhood of $\hat{\mathbf{x}}$ with $\operatorname{int} \mathbf{R}^n_+$. Let $-\epsilon$ be the vector with components $-\epsilon_i < 0$ and $\mathbf{h}$ the vector field with components $x_i f_i(\mathbf{x})$ on U. Show that $\text{bd-ind}(\hat{\mathbf{x}}) := \lim_{\epsilon \to 0} \deg(\mathbf{h}, -\epsilon)$ is well defined. Show that $\text{bd-ind}(\hat{\mathbf{x}}) = i(\hat{\mathbf{x}})$ if $\hat{\mathbf{x}}$ is saturated and regular, but that $\text{bd-ind}(\hat{\mathbf{x}}) = 0$ if $\hat{\mathbf{x}}$ is not saturated, and that Theorem 19.4.1 extends to isolated fixed points for this "boundary-index". If bd-ind $(\hat{\mathbf{x}}) \neq 0$, then there exist points $\mathbf{x} \in \mathbf{R}^n_+$ arbitrarily close to $\hat{\mathbf{x}}$ such that $f_i(\mathbf{x}) < 0$ for all i with $\hat{x}_i = 0$. Prove the equivalent statements for (19.1).

For equations (19.6) and (19.7), the rest point $\mathbf{p}$ is saturated if $(A\mathbf{p})_i \le \mathbf{p}\cdot A\mathbf{p}$ resp. $r_i + (A\mathbf{p})_i \le 0$ whenever $p_i = 0$. The saturated rest points $\mathbf{p}$ of (19.6) are the Nash equilibria for the (symmetric) game with payoff matrix A:

$$\mathbf{x}\cdot A\mathbf{p} \le \mathbf{p}\cdot A\mathbf{p} \quad \text{for all } \mathbf{x} \in S_n$$

(the strategy $\mathbf{p}$ is a best reply to itself, cf. **15.5**). Thus one obtains, as immediate corollaries of the previous theorem, the existence of a Nash equilibrium and the odd number theorem for regular Nash equilibria, both classical results in game theory.

Exercise 5: Show that the same "dynamical" proof works also for bimatrix games (cf. chapter 17).

Exercise 6: Show that a persistent system (19.1) always has a rest point $\mathbf{x} \in S_n$ satisfying $f_i(\mathbf{x}) = \bar{f}(\mathbf{x})$ for all i (but $\mathbf{x}$ might lie on the boundary).

We shall call (19.1) ***robustly persistent*** if it remains persistent under small perturbations of the f_i. (Here "small" means that the values of the $f_i(\mathbf{x})$ and of the partial derivatives $\frac{\partial f_i}{\partial x_j}$ change only a little.) Similarly (19.6) will be called robustly persistent if it remains persistent under small perturbations of the a_{ij}.

We can now extend the index theorem 19.3 to robustly persistent systems.

Theorem 2. *If (19.1) is robustly persistent then it has an interior fixed point. Moreover the degree of (19.1) is $(-1)^{n-1}$ with respect to any open set U with $\bar{U} \subseteq \operatorname{int} S_n$ which contains all interior fixed points. In particular for (19.6) the interior fixed point is unique and has index $(-1)^{n-1}$.*

Proof: We note first that there are no saturated rest points on the boundary. Indeed, if $\mathbf{p}$ were such a rest point with $f_i(\mathbf{p}) \le \bar{f}(\mathbf{p})$ for all i with $p_i = 0$, then a suitable perturbation would turn $\mathbf{p}$ into a hyperbolic fixed point with all transversal eigenvalues negative. But then the stable manifold of $\mathbf{p}$ meets $\operatorname{int} S_n$. Thus there are interior orbits converging to $\mathbf{p}$, contradicting persistence.

Therefore the set S of saturated fixed points (which is always closed) is a compact subset of $\operatorname{int} S_n$. Let U be any open neighbourhood of S with $\bar{U} \subseteq \operatorname{int} S_n$ which is homeomorphic to a ball. For every $\mathbf{x} \in S_n \setminus U$ we have $\max_i(f_i(\mathbf{x}) - \bar{f}(\mathbf{x})) > 0$ since there is no saturated fixed point outside U. By compactness we get a $c > 0$ such that

$$\max_i [f_i(\mathbf{x}) - \bar{f}(\mathbf{x})] \geq c > 0 \quad \text{for} \quad \mathbf{x} \in S_n \setminus U \quad .$$

Inserting this into (19.18) we observe that the perturbed equation (19.17) cannot have a fixed point in $S_n \setminus U$ either, as long as $\epsilon < \frac{c}{n}$. By homotopy the degree of (19.1) with respect to U coincides with the degree of (19.17) with respect to U (or with respect to $\operatorname{int} S_n$), which is $(-1)^{n-1}$. For (19.6) the above implies that the interior fixed point exists and is unique and therefore regular. So its index is $(-1)^{n-1}$.

Exercise 7: Prove a similar result for robustly persistent systems (19.2) with uniformly bounded orbits.

19.5 Necessary conditions for permanence

Theorem 1. *If (19.6) or (19.7) is permanent, then there exists a unique interior rest point $\hat{\mathbf{x}}$ and for each $\mathbf{x}$ in the interior of the state space,*

$$\lim_{T \to +\infty} \frac{1}{T} \int_0^T \mathbf{x}(t)\, dt = \hat{\mathbf{x}} \quad .$$

Proof: Theorem 19.3 implies that there exists at least one interior rest point. If there were two of them, the line l joining them would consist only of rest points, and intersect the boundary of the state space. But since the boundary is a repellor, there cannot be fixed points arbitrarily near-by. The convergence of the time averages, finally, follows from the Theorems 9.2.2 and 16.7.2, and the fact that by permanence, if $\mathbf{x}$ is in the interior of the state space, then so is $\omega(\mathbf{x})$.

Theorem 2. *Let (19.7) be permanent and denote the Jacobian at the unique interior rest point $\hat{\mathbf{x}}$ by D. Then*

$$(-1)^n \det D > 0 \quad , \tag{19.19}$$

$$tr\, D < 0 \quad , \tag{19.20}$$

$$(-1)^n \det A > 0 \quad . \tag{19.21}$$

Proof: The conditions on the signs of the determinants are simple consequences of the index theorem 19.3. Indeed, since $\hat{\mathbf{x}}$ is the unique solution of $-\mathbf{r} = A\mathbf{x}$, the matrix A is nonsingular. Besides, Theorem 19.3 implies that $i(\hat{\mathbf{x}}) = (-1)^n$. Now

$$d_{ij} = \hat{x}_i a_{ij} \tag{19.22}$$

and so D is also nonsingular. Thus $i(\hat{\mathbf{x}})$ is just the sign of $\det D$. This establishes (19.19) and (via (19.22)) also (19.21).

In order to prove the trace condition (19.20), we shall have to use the *Liouville formula*: if $\dot{\mathbf{x}} = \mathbf{f}(\mathbf{x})$ is defined on the open set $U \subseteq \mathbf{R}^n$ and if $G \subseteq U$ has volume V, then the volume $V(t)$ of $G(t) = \{\mathbf{y} = \mathbf{x}(t) : \mathbf{x} \in G\}$ satisfies

$$\dot{V}(t) = \int_{G(t)} \operatorname{div} \mathbf{f}(\mathbf{x}) d(x_1,...,x_n) \quad . \tag{19.23}$$

Indeed, the substitution rule for integrals implies that if $\mathbf{g} : G \to \mathbf{g}(G)$ is invertible and $\det D_{\mathbf{x}}\mathbf{g} > 0$ for all $\mathbf{x} \in G$, then the volume of $\mathbf{g}(G)$ is given by

$$\int_G \det D_{\mathbf{x}}\mathbf{g}\, d(x_1,...,x_n) \quad .$$

If one sets $\mathbf{g} : \mathbf{x} \to \mathbf{x}(T)$, lets T converge to 0 and uses the fact that the derivative of $\det D_{\mathbf{x}}\mathbf{g}$ is the trace of $D_{\mathbf{x}}\mathbf{f}$, this yields (19.23).

Now let us assume that (19.7) is permanent, and multiply its right hand side with the positive function

$$B(\mathbf{x}) = \prod x_i^{s_i - 1} \tag{19.24}$$

where the s_i will be specified later. The resulting equation $\dot{x}_i = h_i(\mathbf{x})$ with

$$h_i(\mathbf{x}) = x_i(r_i + \sum a_{ij} x_j) B(\mathbf{x}) \tag{19.25}$$

differs from (19.7) just by a change in velocity and is therefore also permanent. Since

$$\frac{\partial h_i}{\partial x_i} = B(\mathbf{x}) \Big[s_i(r_i + (A\mathbf{x})_i) + x_i a_{ii} \Big] \quad ,$$

we obtain

$$\operatorname{div} \mathbf{h}(\mathbf{x}) = B(\mathbf{x}) \Big[\sum s_i(r_i + (A\mathbf{x})_i) + \sum x_i a_{ii} \Big] \tag{19.26}$$

which at the rest point $\hat{\mathbf{x}}$ reduces to

$$\operatorname{div} \mathbf{h}(\hat{\mathbf{x}}) = B(\hat{\mathbf{x}}) \sum \hat{x}_i a_{ii} = B(\hat{\mathbf{x}}) \operatorname{tr} D \quad . \tag{19.27}$$

Hence

$$\operatorname{div} \mathbf{h}(\mathbf{x}) = B(\mathbf{x}) \{ \sum_i s_i \Big[\sum_j a_{ij}(x_j - \hat{x}_j) \Big] + \sum a_{ii}(x_i - \hat{x}_i) + \operatorname{tr} D \}$$

$$= B(\mathbf{x})\left[\sum_j (x_j - \hat{x}_j)(\sum_i s_i a_{ij} + a_{jj}) + \operatorname{tr} D\right] \quad . \tag{19.28}$$

Since A is nonsingular, we may choose s_i such that for all j

$$\sum_i s_i a_{ij} + a_{jj} = 0 \quad . \tag{19.29}$$

Then

$$\operatorname{div}\mathbf{h}(\mathbf{x}) = B(\mathbf{x}) \operatorname{tr} D \quad . \tag{19.30}$$

Since there exists a ball in int $\mathbf{R}^n_+$ which, in time T, gets shrunk (see the proof of the index theorem 19.3), Liouville's formula (19.23) and (19.30) imply (19.20).

The latter part of this proof shows that if $\operatorname{tr} D$ is positive, negative or zero, then the flow of the Lotka-Volterra equation is (essentially) volume expanding, contracting, or preserving, respectively. Thus the Lotka-Volterra equation (19.7) and its linearization affect volumes in the same way.

Theorem 3. *Let* (19.6) *with* $a_{ii} = 0$ *be permanent and* $\hat{\mathbf{x}}$ *be the interior fixed point. Then*

$$\hat{\mathbf{x}} \cdot A\hat{\mathbf{x}} > 0 \tag{19.31}$$

and

$$(-1)^{n-1} \det A > 0 \quad . \tag{19.32}$$

Proof: We transform (19.6) into

$$\dot{y}_i = y_i(r_i + (A'\mathbf{y})_i) \qquad i = 1,\dots,n-1 \quad , \tag{19.33}$$

as described in **16.5**, with $a'_{ij} = a_{ij} - a_{nj}$, and apply Theorem 19.5.2.

Computing the Jacobian of (19.6),

$$\frac{\partial \dot{x}_i}{\partial x_j} = \delta_{ij}((A\mathbf{x})_i - \mathbf{x} \cdot A\mathbf{x}) + x_i(a_{ij} - (A\mathbf{x})_j - (\mathbf{x}A)_j)$$

we see that its trace at $\hat{\mathbf{x}}$ is given by

$$\sum_{i=1}^n \left.\frac{\partial \dot{x}_i}{\partial x_i}\right|_{\hat{\mathbf{x}}} = \sum_i \hat{x}_i a_{ii} - 2\hat{\mathbf{x}} \cdot A\hat{\mathbf{x}} \quad .$$

Since we restrict (19.6) to the simplex we have to omit the one eigenvalue corresponding to the left eigenvector $\mathbf{1}$, which is $-\hat{\mathbf{x}} \cdot A\hat{\mathbf{x}}$ (see for instance exercise 12.2.1). Thus the divergence of (19.6) *within* S_n at $\hat{\mathbf{x}}$ is given by

$$\sum \hat{x}_i a_{ii} - \hat{\mathbf{x}} \cdot A\hat{\mathbf{x}} \quad . \tag{19.34}$$

Since we assume $a_{ii} = 0$, (19.20) implies (19.31).

Furthermore, applying Cramer's rule to the system $(A\hat{\mathbf{x}})_i = \hat{\mathbf{x}} \cdot A\hat{\mathbf{x}}$ yields

$$\hat{x}_n \det A = (\hat{\mathbf{x}} \cdot A\hat{\mathbf{x}}) \det A_n \tag{19.35}$$

with

$$A_n = \begin{bmatrix} a_{11} & \cdots & a_{1,n-1} & 1 \\ \cdot & & \cdot & \cdot \\ \cdot & & \cdot & \cdot \\ \cdot & & \cdot & \cdot \\ a_{n1} & \cdots & a_{n,n-1} & 1 \end{bmatrix} .$$

Clearly

$$\det A_n = \begin{bmatrix} a'_{11} & \cdots & a'_{1,n-1} & 0 \\ \cdot & & \cdot & \cdot \\ \cdot & & \cdot & \cdot \\ \cdot & & \cdot & \cdot \\ a'_{n-1,1} & \cdots & a'_{n-1,n-1} & 0 \\ a_{n1} & \cdots & a_{n,n-1} & 1 \end{bmatrix} = \det A' \quad .$$

Since by (19.21) $(-1)^{n-1} \det A' > 0$, (19.31) and (19.35) imply (19.32).

Exercise 1: Subzero sum games are games with $a_{ii} = 0$ and $a_{ij} + a_{ji} \leq 0$. An example is (19.8) with $\epsilon \leq 0$. Show that a subzero sum game cannot have attractors in $\operatorname{int} S_n$ or in the interior of a subface.

Exercise 2: If the replicator equation (19.6) admits a unique interior fixed point $\hat{\mathbf{x}}$, then its index (within S_n) is given by

$$i(\hat{\mathbf{x}}) = \operatorname{sgn} \frac{\det A}{\hat{\mathbf{x}} \cdot A\hat{\mathbf{x}}} \quad .$$

Hint: (19.35) and $\operatorname{sgn} \det A_n = i(\hat{\mathbf{x}})$. Note that $a_{ii} = 0$ is not required. As a consequence, $\hat{\mathbf{x}} \cdot A\hat{\mathbf{x}} = 0$ iff $\det A = 0$.

Exercise 3: Show that (19.32) holds for permanent equations (19.6) with all $a_{ij} \geq 0$.

We conclude with another necessary condition for permanence of Lotka-Volterra systems.

Theorem 4. *Assume that* (19.7) *is permanent or robustly persistent and admits a unique rest point* $\mathbf{y}$ *in the interior of the face* $x_k = 0$. *Then the minor* $A^{(k)}$ *obtained by deleting the k-th row and column of A satisfies*

$$(-1)^{n-1}\det A^{(k)} > 0 \quad .$$

Proof: It is enough to show that the transversal eigenvalue at $\mathbf{y}$ is given by

$$-\hat{x}_k \frac{\det A}{\det A^{(k)}} \quad . \tag{19.36}$$

Since $A\hat{\mathbf{x}} + \mathbf{r} = \mathbf{0}$, Cramer's rule implies that this expression is equal to

$$\frac{1}{\det A^{(k)}} \det\left(a_{i1},\ldots,r_i,\ldots,a_{in}\right)$$

with $\mathbf{r} = (r_i)$ in the k-th column. Developing this along the k-th row yields

$$\frac{1}{\det A^{(k)}} \sum_j a'_{kj}(-1)^{k+j} \det{}_j^k(a_{i1},\ldots,r_i,\ldots,a_{in})$$

where $a'_{kj} = a_{kj}$ if $k \neq j$, $a'_{kk} = r_k$ and $\det_j^k$ indicates that the k-th row and the j-th column of det $(a_{i1},\ldots,r_i,\ldots,a_{in})$ are deleted. But we also have $A^{(k)}\mathbf{y} + \mathbf{r}^{(k)} = \mathbf{0}$ (where $\mathbf{r}^{(k)}$ is obtained from $\mathbf{r}$ by removing the k-th component), so that the previous expression coincides with

$$r_k + \sum_{j \neq k} a_{kj} y_j \quad .$$

Now this is just the transversal eigenvalue of $\mathbf{y}$, which has to be positive.

Exercise 4: Let (19.6) admit an interior fixed point $\hat{\mathbf{x}}$ and a fixed point $\mathbf{y}$ in the interior of the face $x_k = 0$. Show that the transversal eigenvalue at $\mathbf{y}$ has the same sign as

$$\frac{-(\mathbf{y} \cdot A\mathbf{y}) \det A}{(\hat{\mathbf{x}} \cdot A\hat{\mathbf{x}}) \det A^{(k)}} \quad .$$

19.6 Sufficient conditions for permanence

The following theorem is a very useful strengthening of Theorem 13.2.2 on average Ljapunov functions.

Theorem 1. *The replicator system* (19.6) *is permanent if there exists a* $\mathbf{p} \in \operatorname{int} S_n$ *such that*

$$\mathbf{p} \cdot A\mathbf{x} > \mathbf{x} \cdot A\mathbf{x} \tag{19.37}$$

holds for all rest points $\mathbf{x} \in \operatorname{bd} S_n$.

It is remarkable that only rest points are involved: the ω-limit sets on the boundary may be considerably more complicated. The similarity with the ESS condition (15.15) is also intriguing.

Proof: It is enough to show that

$$P(\mathbf{x}) = \prod x_i^{p_i} \tag{19.38}$$

is an average Ljapunov function. Clearly $P(\mathbf{x}) = 0$ for $\mathbf{x} \in \operatorname{bd} S_n$ and $P(\mathbf{x}) > 0$ for $\mathbf{x} \in \operatorname{int} S_n$. The function

$$\Psi(\mathbf{x}) = \mathbf{p} \cdot A\mathbf{x} - \mathbf{x} \cdot A\mathbf{x} \tag{19.39}$$

satisfies $\dot{P} = P\Psi$ in $\operatorname{int} S_n$. It remains to show that for every $\mathbf{y} \in \operatorname{bd} S_n$, there is a $T > 0$ such that

$$\int_0^T \Psi(\mathbf{y}(t))\, dt > 0 \quad . \tag{19.40}$$

The proof proceeds by induction on the number r of positive components of $\mathbf{y}$. For $r = 1$, (19.40) is an obvious consequence of (19.37), since $\mathbf{y}$ then is a corner of S_n and hence a boundary rest point. Assume now that (19.40) is valid for $r = 1,\ldots,m-1$ and let $I = \operatorname{supp}(\mathbf{y})$ have cardinality m. We have to distinguish two cases:

(a) $\mathbf{y}(t)$ converges to the boundary of the simplex

$$S_n(I) = \{\mathbf{x} \in S_n : x_i = 0 \text{ for all } i \notin I\} \quad .$$

Then $\omega(\mathbf{y})$ is contained in a union of faces of dimension $\leq m-1$. By our assumption, (19.40) holds for all $\mathbf{z} \in \omega(\mathbf{y})$. The argument in Theorem 13.2.2 shows then that (19.40) holds for $\mathbf{y}$ as well.

(b) $\mathbf{y}(t)$ does not converge to $\operatorname{bd} S_n(I)$.

In that case, there exists an $\epsilon > 0$ and a sequence $T_k \to +\infty$ such that $y_i(T_k) > \epsilon$ for all $i \in I$ and $k = 1,2,\ldots$. Let us write

$$\overline{y}_i(T) = \frac{1}{T}\int_0^T y_i(t)\,dt \quad \text{and} \quad a(T) = \frac{1}{T}\int_0^T \mathbf{y}(t) \cdot A\mathbf{y}(t)\,dt \quad .$$

Since the sequences $\bar{y}_i(T_k)$ and $a(T_k)$ are obviously bounded, we may obtain a subsequence - which we shall again denote by T_k - such that $\bar{y}_i(T_k)$ and $a(T_k)$ converge: the limits will be denoted by $\bar{x}_i$ and $\bar{a}$. For $i \in I$, we have

$$(\log y_i)^{\cdot} = (A\mathbf{y})_i - \mathbf{y} \cdot A\mathbf{y} \quad .$$

Integrating this from 0 to T_k and dividing by T_k, we obtain

$$\frac{1}{T_k}(\log y_i(T_k) - \log y_i(0)) = (A\bar{\mathbf{y}}(T_k))_i - a(T_k) \quad . \tag{19.41}$$

Since $\log y_i(T_k)$ is bounded, the left hand side converges to 0. Hence

$$(A\bar{\mathbf{x}})_i = \bar{a} \text{ for all } i \in I \quad . \tag{19.42}$$

Furthermore $\sum \bar{x}_i = 1$, $\bar{x}_i \geq 0$ and $\bar{x}_i = 0$ for $i \notin I$. Hence $\bar{a} = \bar{\mathbf{x}} \cdot A\bar{\mathbf{x}}$ holds and $\bar{\mathbf{x}}$ is an equilibrium in $S_n(I)$. Now

$$\frac{1}{T_k}\int_0^{T_k} \Psi(\mathbf{y}(t))\,dt = \sum_{i=1}^n p_i \frac{1}{T_k}\int_0^{T_k} [(A\mathbf{y})_i - \mathbf{y} \cdot A\mathbf{y}]\,dt$$

converges to

$$\sum_{i=1}^n p_i[(A\bar{\mathbf{x}})_i - \bar{\mathbf{x}} \cdot A\bar{\mathbf{x}}] = \mathbf{p} \cdot A\bar{\mathbf{x}} - \bar{\mathbf{x}} \cdot A\bar{\mathbf{x}}$$

which is positive by (19.37). This implies (19.40) and hence permanence.

Thus in order to prove permanence, we have to find out whether there exists a positive solution $\mathbf{p}$ for the linear inequalities

$$\sum_{i:x_i=0} p_i[(A\mathbf{x})_i - \mathbf{x} \cdot A\mathbf{x}] > 0 \tag{19.43}$$

(where $\mathbf{x}$ runs through the boundary rest points). The coefficients $(A\mathbf{x})_i - \mathbf{x} \cdot A\mathbf{x}$ are the transversal eigenvalues of $\mathbf{x}$. The number of inequalities may be quite large, but the following exercise shows that the problem is finite. Applications of Theorem 19.6.1 will be given in **20.4** and **20.5**.

Exercise 1: If there is a continuum of rest points on the boundary, then it is sufficient to test the inequalities (19.43) on the extremal points of the continuum.

Exercise 2: Show that similarly a Lotka-Volterra equation (19.7) with uniformly bounded orbits is permanent if there exists a $\mathbf{p} \in \operatorname{int} \mathbf{R}^n_+$ such that

$$\mathbf{p} \cdot (\mathbf{r} + A\mathbf{x}) > 0 \tag{19.44}$$

holds for all fixed points $\mathbf{x} \in \operatorname{bd} \mathbf{R}^n_+$, i.e. if there exists a positive solution $\mathbf{p}$ for

$$\sum_{i:x_i=0} p_i[r_i + (A\mathbf{x})_i] > 0 \tag{19.45}$$

(where $\mathbf{x}$ runs through the boundary rest points).

Exercise 3: Show that the set of solutions $\mathbf{p}$ of (19.43) is an open convex subset of $\operatorname{int} S_n$. For the hypercycle, every $\mathbf{p} \in \operatorname{int} S_n$ works. Show that in general, this subset need not contain the interior fixed point.

Exercise 4: Investigate the inhomogeneous hypercycle

$$\dot{x}_i = x_i(b_i + k_i x_{i-1} - \bar{f})$$

with $b_i, k_i > 0$. Show that it is permanent iff it admits an interior rest point. (Hint: Use either Theorem 19.6.1 with $p_i = \dfrac{1}{k_i}$ or imitate the argument in **13.3** .)

A geometric condition for permanence is given by the following:

Theorem 2. *Consider a Lotka-Volterra equation* (19.7) *with uniformly bounded orbits. If the set*

$$D = \{\mathbf{x} \in \mathbf{R}^n_+ : \mathbf{r} + A\mathbf{x} \le \mathbf{0}\} \tag{19.46}$$

(where no species increases) is disjoint from the convex hull C of all boundary fixed points, then (19.7) *is permanent.*

Proof: A rest point $\mathbf{z}$ of (19.7) lies in D iff it is saturated. The assumption $C \cap D = \emptyset$ implies that there are no saturated rest points on the boundary. Theorem 19.4.1 shows then the existence of an interior rest point $\hat{\mathbf{x}}$; this point is unique, since otherwise we would have a line of fixed points intersecting $\operatorname{bd} \mathbf{R}^n_+$, i.e., a nonempty intersection of C and D. It follows that A is nonsingular. The convex set C can be separated from the convex set $\hat{D} = \{\mathbf{x} \in \mathbf{R}^n : \mathbf{r} + A\mathbf{x} \le \mathbf{0}\}$ by a hyperplane. Since A is nonsingular we can write the separating functional in the form $\mathbf{p}A$, with $\mathbf{p} \in \mathbf{R}^n$. Then

$$\mathbf{p} \cdot A\mathbf{z} > \mathbf{p} \cdot A\mathbf{x} \tag{19.47}$$

for all $\mathbf{x} \in \hat{D}$ and all fixed points $\mathbf{z} \in \operatorname{bd} \mathbf{R}^n_+$. Since the interior fixed point $\hat{\mathbf{x}}$ lies in $\hat{D}$ we obtain in particular

$$\mathbf{p} \cdot A\mathbf{z} > \mathbf{p} \cdot A\hat{\mathbf{x}} = -\mathbf{p} \cdot \mathbf{r} \quad . \tag{19.48}$$

Thus $\mathbf{p}$ is a solution of (19.44).

We are left to show that $\mathbf{p}$ is positive. Let $\mathbf{v}$ be any vector in $\mathbf{R}^n$ with $A\mathbf{v} \leq \mathbf{0}$. Then $\hat{\mathbf{x}} + c\mathbf{v}$ belongs to $\hat{D}$ for every $c > 0$ since

$$\mathbf{r} + A(\hat{\mathbf{x}} + c\mathbf{v}) = cA\mathbf{v} \leq \mathbf{0} \quad .$$

Thus

$$\mathbf{p} \cdot A\mathbf{z} > \mathbf{p} \cdot A\hat{\mathbf{x}} + c\mathbf{p} \cdot A\mathbf{v} \quad . \tag{19.49}$$

This can hold for arbitrarily large c only if $\mathbf{p} \cdot A\mathbf{v} \leq 0$ for all such $\mathbf{v}$. Therefore $\mathbf{p} \geq \mathbf{0}$. Since the set of solutions $\mathbf{p}$ of (19.48) is open, as $\mathbf{z}$ varies in a compact set, we can find a solution $\mathbf{p} > \mathbf{0}$. Hence (19.7) is permanent.

As a simple special case, let us consider two species competition (cf. **8.3**). In the presence of an interior rest point there are two possible types of behaviour.
(a) *global stability* (and hence permanence) if the interior fixed point is above the line joining the two one-species equilibria;
(b) *bistability* (and thus competitive exclusion) if it is below.

Exercise 5: Show that the geometric condition of Theorem 2 is actually equivalent to the one given in exercise 2.

Exercise 6: Check the three species system (9.18) from the point of view of Theorem 2.

These examples might suggest that the geometric condition $C \cap D = \emptyset$ is not only sufficient, but also necessary for (at least a robust type of) permanence. This will be shown for $n = 3$ in **22.2**. For $n \geq 4$ this is not true, however. See exercise 22.4.5 for a counterexample.

19.7 Notes

This chapter and the following one contain mostly material from Hofbauer and Sigmund (1987). Persistence (now usually called weak persistence) was introduced by Freedman and Waltman (1977). A completely different approach yielding sufficient conditions for persistence is due to Butler and Waltman (1986). Butler et al. (1986) give conditions under which persistence implies permanence. We also refer to Svirezhev (1986) for a discussion of Lagrange type stability for ecological models. The first to show that permanence implies the existence of an interior equilibrium were Hutson and Moran (1982) and Sieveking (1983). The notion of saturated rest points and applications of the index theorem 19.4.1 to ecological equations were discussed in Hofbauer (1988c), a generalization to semiflows on $\mathbf{R}^n_+$ or S_n is given in Hofbauer (1988b). Theorem 19.6.1 is due to Jansen (1987), Theorem 19.5.2 and 19.5.3 to Amann and Hofbauer (1985). For more on the notions of degree and index, we refer to Milnor (1965), and for the Liouville formula to Amann (1983). Inhomogeneous hypercycles (ex. 19.6.4) were introduced in Eigen and Schuster (1979).

20. Replicator Networks

20.1 A periodic attractor for $n = 4$

We have seen in **18.2** that the two dimensional Lotka-Volterra equation admits no limit cycle. The three dimensional one does, however. This follows from the fact that *the replicator equation*

$$\dot{x}_i = x_i((A\mathbf{x})_i - \mathbf{x} \cdot A\mathbf{x}) \tag{20.1}$$

on S_n admits a Hopf bifurcation for $n = 4$.

We shall use the matrix

$$A = \begin{bmatrix} 0 & 0 & -\mu & 1 \\ 1 & 0 & 0 & -\mu \\ -\mu & 1 & 0 & 0 \\ 0 & -\mu & 1 & 0 \end{bmatrix} . \tag{20.2}$$

It is obvious that $\mathbf{m} = (\frac{1}{4}, \frac{1}{4}, \frac{1}{4}, \frac{1}{4})$, i.e. the barycentre of S_4, is a rest point for (20.1). Since A is circulant, the Jacobian D of (20.1) at $\mathbf{m}$ will also be circulant. A simple computation shows that the first row of D is given by

$$\frac{-1+\mu}{8}, \quad \frac{-1+\mu}{8}, \quad \frac{-1-\mu}{8}, \quad \frac{1+\mu}{8} . \tag{20.3}$$

The eigenvalues can be easily computed by (9.20). The eigenvalue

$$\gamma_0 = \frac{1}{4}(-1+\mu)$$

belongs to the eigenvector $\mathbf{1}$ which is orthogonal to S_4. Since we are only interested in the restriction of (20.1) to S_4, we may exclude it from further considerations. The other eigenvalues are

$$\gamma_1 = \frac{1}{4}(\mu - i)$$

$$\gamma_2 = \frac{1}{4}(-1-\mu) \tag{20.4}$$

$$\gamma_3 = \frac{1}{4}(\mu+i) \quad .$$

If μ varies from $-\frac{1}{2}$ to $+\frac{1}{2}$, then γ_2 will be negative, while the complex conjugate pair γ_1 and γ_3 crosses the imaginary axis, from left to right, for $\mu = 0$.

For $\mu < 0$, therefore , $\mathbf{m}$ is a sink; for $\mu > 0$, it is unstable; for $\mu = 0$, finally, the matrix A describes the hypercycle with $n = 4$: as we know from **12.6**, $\mathbf{m}$ is asymptotically stable in this case.

All conditions for the occurrence of a Hopf bifurcation are therefore satisfied (see **18.6**). For sufficiently small values $\mu > 0$, there will be a periodic attractor in the neighbourhood of $\mathbf{m}$ (cf. Fig. 20.1).

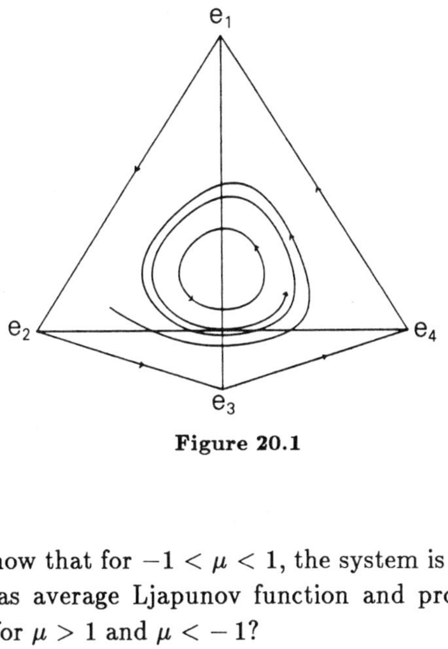

Figure 20.1

Exercise 1: Show that for $-1 < \mu < 1$, the system is permanent. (Hint: take $P = \prod x_i$ as average Ljapunov function and proceed as in **19.6**.) What happens for $\mu > 1$ and $\mu < -1$?

Exercise 2: Analyse the general 4-species system (20.1) with cyclic symmetry

$$A = \begin{bmatrix} 0 & a_1 & a_2 & a_3 \\ a_3 & 0 & a_1 & a_2 \\ a_2 & a_3 & 0 & a_1 \\ a_1 & a_2 & a_3 & 0 \end{bmatrix} \quad .$$

(a) If $a_1 + a_3 \geq a_2 > 0$, then $P = x_1x_2x_3x_4$ is a global Ljapunov function.
(b) Show that a Hopf bifurcation occurs at $a_2 = 0$ if $a_1 \neq a_3$, and leads to periodic attractors for small $a_2 < 0$ if $a_1 + a_3 > 0$.
(c) The system is permanent iff

$$a_1a_3 \leq 0 \quad \text{and} \quad -(a_1+a_3) < a_2 < a_1 + a_3$$

or

$$a_1a_3 > 0 \quad \text{and} \quad -\frac{(a_1-a_3)^2}{a_1+a_3} < a_2 < a_1 + a_3 \quad .$$

(d) If $a_2 > a_1 + a_3 > 0$ then $V = x_1x_3 + x_2x_4$ is a global Ljapunov function.

20.2 Cyclic symmetry

We have seen in **9.5** and **20.1** that the assumption of *cyclic symmetry*, biologically unrealistic as it may be, helps to display interesting properties of replicator equations. We shall investigate now why it leads to a Hopf bifurcation for $n \geq 4$, but not for $n = 3$.

We shall count the indices modulo n, denote the coefficients $a_{i,i+k}$ (which do not depend on i) by a_k and assume without restriction of generality that $a_0 = 0$. The replicator equation becomes

$$\dot{x}_i = x_i\Big(\sum_{j=1}^{n} a_j x_{i+j} - \bar{f}\Big) \quad . \tag{20.5}$$

The central point $\mathbf{m} = \frac{1}{n}\mathbf{1}$ is an inner equilibrium for (20.5). The Jacobian at $\mathbf{m}$ is again circulant, its first row being $c_0,\dots,c_{n-1}$ with

$$c_i = \frac{1}{n}(a_i - 2\bar{a}) \quad \text{and} \quad \bar{a} = \frac{1}{n}\sum_{j=1}^{n-1} a_j \quad . \tag{20.6}$$

According to (9.20), its eigenvalues are $\gamma_0 = -\bar{a}$ (with eigenvector $\mathbf{1}$ orthogonal to S_n, and henceforth omitted) and

$$\gamma_j = \sum_{s=0}^{n-1} c_s\, e^{2\pi ijs/n} = \frac{1}{n}\sum_{s=1}^{n-1} a_s\, e^{2\pi ijs/n} \tag{20.7}$$

for $j = 1,\dots,n-1$.

The exercises at the end of this section yield a proof of the following:

Theorem. *If* m *is a sink, then it is globally stable. If* m *is a source, all orbits in* int S_n *(except* m*) converge to* bd S_n.

In the case $n = 3$, which corresponds to the "rock - scissors - paper" game from exercise 16.5.2, the eigenvalues are

$$\gamma_{1,2} = -\frac{1}{6}[(a_1 + a_2) \pm i\sqrt{3}(a_2 - a_1)] \quad . \tag{20.8}$$

The function $P = x_1 x_2 x_3$ satisfies

$$\dot{P} = \frac{3P}{2}(a_1 + a_2) \sum_{j=1}^{3} (x_j - \frac{1}{3})^2 \quad . \tag{20.9}$$

Since there are no limit cycles for $n = 3$, we know that Hopf bifurcations leading to the emergence of periodic attractors cannot occur. The reason is that when $\mathrm{Re}\,\gamma_{1,2} = -\frac{1}{6}(a_1 + a_2) = 0$, the centre m is not asymptotically stable. P is a constant of motion and all orbits circle around m (see Fig. 16.1). The Hopf bifurcation is *degenerate.*

For $n \geq 4$, however, m is asymptotically stable if one pair of complex eigenvalues lies on the imaginary axis and all other eigenvalues to its left (as shown in the next exercise). Hence supercritical Hopf bifurcations leading to stable limit cycles occur.

Exercise 1: Introduce complex coordinates

$$y_p = \sum_{j=1}^{n} x_j \exp(2\pi i j p/n) \quad (p = 0,\dots,n-1) \tag{20.10}$$

and check that $y_0 = 1$, $\bar{y}_p = y_{n-p} (p = 1,\dots,n-1)$ and

$$x_j = \frac{1}{n} \sum_{p=0}^{n-1} y_p \exp(-2\pi i j p/n) \quad (j = 1,\dots,n) \quad . \tag{20.11}$$

Exercise 2: (20.5) leads to

$$\dot{y}_p = \sum_m w_m \bar{y}_m y_{p+m} - y_p \bar{f} \tag{20.12}$$

and

$$\bar{f} = \sum_{m=0}^{n-1} \mathrm{Re}\, w_m |y_m|^2 \tag{20.13}$$

(where $w_0 = -\gamma_0$ and $w_m = \gamma_m$ for $m \neq 0$).

Exercise 3: Check that with $P = x_1 x_2 \cdots x_n$ one has

$$\dot{P} = -nP \sum_{m=1}^{n-1} \mathrm{Re}\, w_m |y_m|^2 \quad . \tag{20.14}$$

The above theorem follows from (20.14).

Exercise 4: If $\mathrm{Re}\, w_1 = \mathrm{Re}\, w_{n-1} = 0$ and $\mathrm{Re}\, w_i < 0$ for $i = 2,\ldots,n-2$ $(n \geq 4)$, then $\dot{P} \geq 0$, so that every orbit in $\mathrm{int}\, S_n$ converges to the maximal invariant subset M of $\{\dot{P} = 0\} = \{y_m = 0$ for $m = 2,\ldots,n-2\}$. Show that $M = \{\mathbf{y} = (1,0,\ldots,0)\}$ and hence that $\mathbf{m}$ is asymptotically stable.

Exercise 5: Show that (9.18) is equivalent to (20.5) with $n = 3$.

20.3 Permanence and irreducibility

To a replicator equation (20.1) with $a_{ij} \geq 0$ for all i and j, we associate a *directed graph* whose vertices correspond to the replicating species $M_1,\ldots,M_n$: an arrow from M_j to M_i denotes that $a_{ij} > 0$, i.e. that M_j enhances the reproduction of M_i.

A directed graph is said to be *irreducible* if for any two species M_i and M_j there exists an orientated path leading from M_i to M_j. Each species, then, favors directly or indirectly the reproduction of every other.

Theorem. *If (20.1) with $a_{ij} \geq 0$ is permanent, then its graph is irreducible.*

Proof: Let us assume that the system is not irreducible. There exists, then, a proper subset $D \subseteq \{1,\ldots,n\}$ which is *closed* in the sense that no arrow leads out of it, i.e. that $a_{ji} = 0$ for all $i \in D$, $j \notin D$. Let us define

$$M = \max_{1 \leq j \leq n} \sum_{s=1}^{n} a_{js} \qquad m = \min_{i \in D} \sum_{t \in D} a_{it} \quad . \tag{20.15}$$

We may assume that some a_{js} are positive, since otherwise the system would have only fixed points, which does not agree with permanence. This implies $M > 0$. We shall show that we also have $m > 0$.

This can be seen indirectly. If $m = 0$, there exists an $i \in D$ such that $a_{it} = 0$ for all $t \in D$. In this case we replace D by the closed set $D \backslash \{i\}$. If we still have $m = 0$, we repeat the procedure. We end up either with $m > 0$ or with a set containing a single element k, which satisfies $a_{kk} = 0$. Clearly, $a_{jk} = 0$ for all j. In that case, x_k does not occur in the equations (16.5) for the unique interior rest point guaranteed by permanence: this is a contradiction.

Now we may define

$$G = \{\mathbf{x} \in \operatorname{int} S_n \colon Mx_j < mx_i \text{ for all } i \in D, j \notin D\} \quad .$$

The set G is forward invariant. Indeed, the quotient rule implies for $\mathbf{x} \in G, i \in D$ and $j \notin D$:

$$\begin{aligned}(\frac{x_j}{x_i})^{\cdot} &= (\frac{x_j}{x_i})(\sum_s a_{js}x_s - \sum_t a_{it}x_t) \le (\frac{x_j}{x_i})(\sum_{s\notin D} a_{js}x_s - \sum_{t\in D} a_{it}x_t) \\ &\le (\frac{x_j}{x_i})(\max_{s\notin D} x_s \sum_{s\notin D} a_{js} - \min_{t\in D} x_t \sum_{t\in D} a_{it}) \\ &\le (\frac{x_j}{x_i})(M \max_{s\notin D} x_s - m \min_{t\in D} x_t) < 0 \quad . \end{aligned} \tag{20.16}$$

Hence $\mathbf{x}(t)$ remains in G and converges to $\operatorname{bd} S_n$ for $t \to +\infty$, a contradiction to permanence.

20.4 Permanence of catalytic networks

By a *catalytic network*, we shall understand a replicator equation (20.1) with $a_{ij} \ge 0$ and $a_{ii} = 0$. Hypercycles are simple examples of catalytic networks: they guarantee permanence, but of course many other types of networks can also do that job.

We have seen that the directed graph of a permanent catalytic network is irreducible. It is tempting to conjecture that it must even be *Hamiltonian*, i.e. contain a closed path visiting every vertex exactly once. This would mean that such a network has to contain a full hypercycle. It turns out, however, that this is valid only in low dimensions.

Theorem 1. *If $n \le 5$, then the graph of a permanent catalytic network is Hamiltonian.*

Proof: We know from (19.32) that the sign of $\det A$ is $(-1)^{n-1}$. Now

$$\det A = \sum_\sigma \operatorname{sgn} \sigma \, a_{1\sigma(1)} a_{2\sigma(2)} \cdots a_{n\sigma(n)}$$

where the summation extends over all permutations σ of $\{1,...,n\}$. Since $a_{ii} = 0$ we need only consider permutations without fixed elements. Every permutation σ can be split into some k elementary cycles, and $\operatorname{sgn} \sigma = (-1)^{n-k}$. If $n \le 5$, k can be 1 or 2. Any permutation which is the product of two smaller cycles has sign $(-1)^{n-2}$. In order that

$\operatorname{sgn}\det A = (-1)^{n-1}$, in accordance with (19.32), there must be at least one permutation σ consisting of a single cycle such that

$$a_{1\sigma(1)}a_{2\sigma(2)}\cdots a_{n\sigma(n)} > 0 \quad .$$

This corresponds to a closed feedback loop visiting every vertex precisely once.

Exercise 1: For $n = 6$ and

$$A = \begin{bmatrix} 0 & 1 & 0 & 0 & 0 & 3 \\ 2 & 0 & 0 & 0 & 0 & 0 \\ 1 & 0 & 0 & 2 & 0 & 0 \\ 0 & 3 & 1 & 0 & 0 & 0 \\ 0 & 0 & 0 & 3 & 0 & 1 \\ 0 & 0 & 0 & 0 & 1 & 0 \end{bmatrix}$$

the replicator equation is permanent although the graph is not Hamiltonian (see Fig. 20.2). (Hint: Apply Theorem 19.6.1 with $\mathbf{p} = \frac{1}{49}(9,1,2,13,23,1)$. The inequalities corresponding to boundary fixed points $\mathbf{x}$ with $\mathbf{x}\cdot A\mathbf{x} = 0$ are trivially satisfied, since A is irreducible. The only remaining fixed points are $(\frac{1}{2},\frac{2}{3},0,0,0,0)$, $(0,0,\frac{1}{3},\frac{2}{3},0,0)$, $(0,0,0,0,\frac{1}{2},\frac{1}{2})$, $\frac{1}{28}(5,0,9,2,9,3)$ and $\frac{1}{41}(9,6,0,5,18,3)$.)

Exercise 2: Show that for $n = 3$ a catalytic network is permanent iff there exists a unique interior fixed point. List the graphs of all such networks and check that $\det A > 0$.

Theorem 2. *For a catalytic network with $n = 4$, a necessary and sufficient condition for permanence is that there exists an interior fixed point $\hat{\mathbf{x}}$ and* $\det A < 0$.

Proof: Necessity follows from Theorems 19.3 and 19.5.3. In order to prove sufficiency, we shall use Theorem 19.6.1 and look for a positive solution $\mathbf{p}$ of (19.37).

As a first step we shall show that one only has to take into account those boundary equilibria $\mathbf{x}$ which lie on an edge and are isolated within that edge:

(a) For the corners of S_n, (19.37) is satisfied for every choice $\mathbf{p} > \mathbf{0}$, since they must have some positive eigenvalue: each column of A contains a positive element.

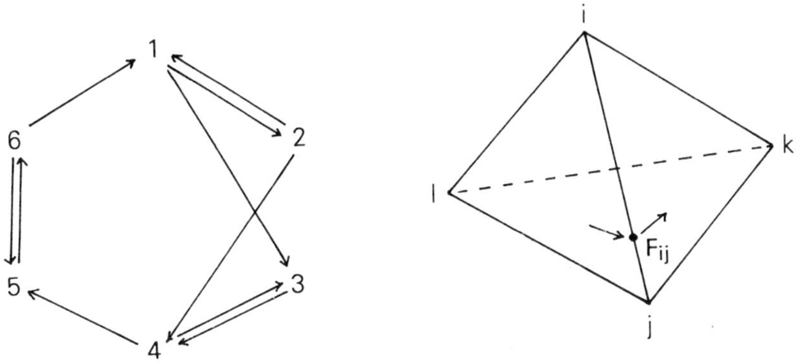

Figures 20.2 and 20.3

(b) Next we show that (19.37) is satisfied for every choice of $\mathbf{p} > \mathbf{0}$ if $\mathbf{x}$ is a rest point in the interior of the 3-face $x_k = 0$. We need only to consider the case where $\mathbf{x}$ is isolated. (Indeed, (19.37) needs only to be checked at the extremal points of continua of boundary equilibria). Then $A^{(k)}$, the 3×3-minor of A obtained by deleting its k-th row and column, is nonsingular. As shown in exercise 19.5.4, the transversal eigenvalue of $\mathbf{x}$, i.e. $(A\mathbf{x})_k - \mathbf{x} \cdot A\mathbf{x}$, has the same sign as

$$\frac{-(\hat{\mathbf{x}} \cdot A\hat{\mathbf{x}}) \det A}{(\mathbf{x} \cdot A\mathbf{x}) \det A^{(k)}} \quad .$$

Since the restriction of the catalytic network to the face $x_k = 0$ is permanent by exercise 20.4.2, we have $\det A^{(k)} > 0$. Hence $(A\mathbf{x})_k > \mathbf{x} \cdot A\mathbf{x}$ and (19.37) is trivially satisfied at $\mathbf{x}$.

(c) Thus we have only to check (19.37) for the rest points on edges. Let F_{ij} denote such a point on the edge from $\mathbf{e}_i$ to $\mathbf{e}_j$, and let Γ^k_{ij} be its eigenvalue in the direction $\mathbf{e}_k$. Then

$$\Gamma^k_{ij} = \frac{a_{ki}a_{ij} + a_{kj}a_{ji} - a_{ij}a_{ji}}{a_{ij} + a_{ji}} \quad . \tag{20.17}$$

We note first that F_{ij} cannot be saturated. Indeed, suppose that $\Gamma^k_{ij} \le 0$ and $\Gamma^l_{ij} \le 0$ for the two indices k and l different from i and j. Both the conditions on the interior rest point $\hat{\mathbf{x}}$ and the sign of $\det A$ are open.

Thus we may, without affecting them, slightly perturb the coefficients such that every fixed point is regular and $\Gamma^k_{ij}, \Gamma^l_{ij} < 0$. But the index of every saturated regular rest point on an edge is -1. Since $\det A < 0$ implies that the index of $\hat{\mathbf{x}}$ is also -1, this yields a contradiction to the index theorem 19.4.1.

For (19.37) we are left with (up to 6) inequalities of the form

$$p_k \Gamma^k_{ij} + p_l \Gamma^l_{ij} > 0 \quad . \tag{20.18}$$

Now

$$\Gamma^k_{ij} < 0 \quad \text{implies} \quad \Gamma^j_{ik} > 0 \quad \text{and} \quad \Gamma^i_{jk} > 0 \quad . \tag{20.19}$$

Indeed, $\Gamma^k_{ij} < 0$ implies $a_{ki} < a_{ji}$ and $a_{kj} < a_{ij}$, and hence $\Gamma^j_{ik} > 0$ and $\Gamma^i_{jk} > 0$. Similarly

$$\Gamma^i_{lk} < 0 \quad \text{and} \quad \Gamma^k_{jl} < 0 \quad \text{implies} \quad \Gamma^j_{il} > 0 \quad . \tag{20.20}$$

Indeed, $\Gamma^i_{lk} < 0$ implies $a_{il} < a_{kl}$, and $\Gamma^k_{jl} < 0$ implies $a_{kl} < a_{jl}$, so that $\Gamma^j_{il} > 0$ follows. By (20.19), at least two of the six inequalities (20.18) are trivially satisfied. We shall write $l \to k$ if $\Gamma^l_{ij} < 0$ and $\Gamma^k_{ij} > 0$ (see Fig. 20.3). If this relation has no cycles, (20.18) is easy to solve: whenever $l \to k$, one simply chooses p_k much larger than p_l. (20.20) implies that there can be no cycles of length 3. So we are left with the case of a cycle of length 4. After some suitable rearranging of the indices, this means

$$\Gamma^1_{23}, \Gamma^2_{34}, \Gamma^3_{41}, \Gamma^4_{12} < 0 \tag{20.21}$$

while all other Γ^i_{jk} are positive by (20.19). It is easy to see that the system (20.18) admits a solution $\mathbf{p} > 0$ iff

$$\Gamma^1_{23}\Gamma^2_{34}\Gamma^3_{41}\Gamma^4_{12} < \Gamma^1_{34}\Gamma^2_{41}\Gamma^3_{12}\Gamma^4_{23} \quad . \tag{20.22}$$

It remains to show that (20.22) holds. Now

$$
\begin{aligned}
0 &< -\Gamma^1_{23}\frac{a_{23}+a_{32}}{a_{32}-a_{12}} \qquad (a_{32} > a_{12} \text{ since } \Gamma^1_{23} < 0) \\
&= a_{23} - \frac{a_{13}a_{32}}{a_{32}-a_{12}} < a_{23} < a_{43} \qquad (\text{because } \Gamma^2_{34} < 0) \\
&< a_{43} + \frac{a_{13}a_{34}}{a_{14}-a_{34}} \qquad (a_{14} > a_{34} \text{ since } \Gamma^3_{41} < 0) \\
&= \Gamma^1_{34}\frac{a_{34}+a_{43}}{a_{14}-a_{34}} \quad .
\end{aligned}
$$

Multiplying

$$0 < -\Gamma^1_{23}\frac{a_{23}+a_{32}}{a_{32}-a_{12}} < \Gamma^1_{34}\frac{a_{34}+a_{43}}{a_{14}-a_{34}}$$

with its circular permutations yields (20.22) and concludes the proof.

20.5 Essentially hypercyclic networks

A large class of interaction matrices displays an *essentially hypercyclic structure:* $a_{ij} > 0$ if $i = j+1 \pmod n$, $a_{ii} = 0$ and $a_{ij} \le 0$, or (in a selfexplanatory notation)

$$A = \begin{bmatrix} 0 & - & - & \cdot & \cdot & \cdot & - & + \\ + & 0 & - & \cdot & \cdot & \cdot & - & - \\ - & + & 0 & \cdot & \cdot & \cdot & - & - \\ \cdot & \cdot & \cdot & \cdot & \cdot & \cdot & \cdot & \cdot \\ \cdot & \cdot & \cdot & \cdot & \cdot & \cdot & \cdot & \cdot \\ - & - & - & \cdot & \cdot & \cdot & + & 0 \end{bmatrix}. \tag{20.23}$$

As we shall presently see, permanence of such networks can be characterized in terms of M-matrices. An $n \times n$ matrix C with off-diagonal terms $c_{ij} \le 0$ $(i \ne j)$ is said to be a (nonsingular) *M-matrix* if one of the following equivalent conditions is satisfied (the equivalence of these conditions will be shown in **21.1**):

(i) There is a $\mathbf{p} > 0$ such that $C\mathbf{p} > \mathbf{0}$;
(ii) all leading principal minors of C are positive;
(iii) all real eigenvalues of C are positive;
(iv) for all $\mathbf{x} > \mathbf{0}$ there is an i with $(C\mathbf{x})_i > 0$.

The equivalence of (i) and (iv) can be reformulated as

either there is a $\mathbf{p} > \mathbf{0}$ such that $C\mathbf{p} > \mathbf{0}$
or there is a $\mathbf{p} > \mathbf{0}$ such that $C\mathbf{p} \le \mathbf{0}$.

We recall that a *principal submatrix* A_I of an $n \times n$ matrix A is a matrix (a_{ij}) with $i,j \in I \subseteq \{1,\dots,n\}$. If I is of the form $\{1,2,\dots,k\}$, $k = 1,\dots,n$, then A_I is said to be a *leading principal submatrix.* The *minors* are the corresponding determinants. We also recall that $\mathbf{p} > 0$ means that $p_i > 0$ for all i.

Theorem. *For an essentially hypercyclic matrix A, the following conditions are equivalent:*

(1) *The replicator equation* (20.1) *is permanent;*
(2) *there is an inner equilibrium* $\hat{\mathbf{x}}$ *of* (20.1) *where the Jacobian* D *has negative trace;*
(3) *there is an inner equilibrium* $\hat{\mathbf{x}}$ *of* (20.1) *with* $\hat{\mathbf{x}} \cdot A\hat{\mathbf{x}} > 0$;
(4) *the matrix* C *obtained by moving the top row of* A *to the bottom is an M-matrix;*
(5) *there is a* $\mathbf{p} > \mathbf{0}$ *such that* $\mathbf{p}A > \mathbf{0}$.

Proof: The implication from (1) to (2) has been shown in Theorem 19.5.2, the equivalence of (2) and (3) follows from (19.34). (3) implies (4) since $(A\hat{\mathbf{x}})_i = \hat{\mathbf{x}} \cdot A\hat{\mathbf{x}} > 0$, hence $A\hat{\mathbf{x}} > \mathbf{0}$ and thus $C\hat{\mathbf{x}} > \mathbf{0}$. Condition (5) follows from (4) because there is a vector $\mathbf{p} > \mathbf{0}$ such that $(p_2,\ldots,p_n,p_1)C > \mathbf{0}$. The implication from (5) to (1), finally, is due to the fact that

$$P(\mathbf{x}) = \prod_{i=1}^{n} x_i^{p_i}$$

is an average Ljapunov function. We have only to note that $\Psi(\mathbf{x}) = \mathbf{p} \cdot A\mathbf{x} - \mathbf{x} \cdot A\mathbf{x}$ satisfies (13.3). We proceed indirectly. If there exists an $\mathbf{x} \in \operatorname{bd} S_n$ such that for all $T > 0$

$$\int_0^T \Psi(\mathbf{x}(t))\,dt \leq 0$$

then $\bar{f}(\mathbf{x}) = \mathbf{x} \cdot A\mathbf{x}$ satisfies

$$\frac{1}{T}\int_0^T \bar{f}(\mathbf{x}(t))\,dt \geq \frac{1}{T}\int_0^T \mathbf{p} \cdot A\mathbf{x}(t)\,dt > \epsilon \tag{20.24}$$

for some suitable $\epsilon > 0$, since $\mathbf{p}A > \mathbf{0}$. This implies $x_i(t) \to 0$ for all i. Indeed, $x_i(t) \to 0$ is satisfied for at least one i (since $\mathbf{x} \in \operatorname{bd} S_n$). Now if $x_{i+1} > 0$ then

$$\frac{\dot{x}_{i+1}}{x_{i+1}} = a_{i+1,i}x_i + \sum_{j \neq i} a_{i+1,j}x_j - \mathbf{x} \cdot A\mathbf{x} \quad .$$

Integrating from 0 to T and using the fact that the sum of the right hand side is nonpositive, one gets

$$\frac{1}{T}(\log x_{i+1}(T) - \log x_{i+1}(0)) \leq a_{i+1,i}\frac{1}{T}\int_0^T x_i(t)\,dt - \frac{1}{T}\int_0^T \bar{f}(\mathbf{x}(t))\,dt \quad .$$

The first integral on the right hand side converges to 0, while the second one is larger than ϵ by (20.24). Hence

$$x_{i+1}(T) \leq e^{-\epsilon T}$$

and so $x_{i+1}(T) \to 0$ for $T \to +\infty$. All components of $\mathbf{x}(t)$ vanish, therefore, for $t \to +\infty$, which is a contradiction.

Exercise 1: Check that each of the conditions (2), (3), (4) and (5) of the theorem is satisfied for the hypercycle equation (12.7).

Exercise 2: For $n = 3$, essentially hypercyclic matrices are given by

$$A = \begin{bmatrix} 0 & - & + \\ + & 0 & - \\ - & + & 0 \end{bmatrix} .$$

Show that the corresponding replicator equation is permanent iff it admits a globally stable interior equilibrium, and that this holds iff $\det A > 0$. (Compare with exercise 16.5.5.)

Exercise 3: Consider a hypercycle of autocatalysts

$$\dot{x}_i = x_i(a_i x_i + b_{i-1} x_{i-1} - \bar{f}) \tag{20.25}$$

with $a_i, b_i > 0$. Show that $a_i < b_i$ for all i is a necessary condition for permanence. Show that (20.25) is permanent iff the matrix

$$C = \begin{bmatrix} \alpha_1 & -1 & -1 & \cdots & -1 & 0 \\ 0 & \alpha_2 & -1 & \cdots & -1 & -1 \\ -1 & 0 & \alpha_3 & \cdots & -1 & -1 \\ \cdot & \cdot & \cdot & & \cdot & \cdot \\ \cdot & \cdot & \cdot & & \cdot & \cdot \\ \cdot & \cdot & \cdot & & \cdot & \cdot \\ -1 & -1 & -1 & \cdots & 0 & \alpha_n \end{bmatrix} \tag{20.26}$$

with

$$\alpha_k = \frac{b_k}{a_k} - 1 \tag{20.27}$$

is an M-matrix. Show that this is equivalent to $\alpha_1\alpha_2\alpha_3 > 1$, i.e. $\prod_{i=1}^{3} (b_i - a_i) > \prod_{i=1}^{3} a_i$ for $n = 3$; to

$$\alpha_1\alpha_2\alpha_3\alpha_4 > \alpha_1 + \alpha_2 + \alpha_3 + \alpha_4 + \alpha_1\alpha_3 + \alpha_2\alpha_4$$

for $n = 4$; and to $b > a(n-1)$ if $a_i = a$ and $b_i = b$ for all i (n arbitrary).

Exercise 4: Consider two counterrotating hypercycles:

$$\dot{x}_i = x_i(a_{i-1}x_{i-1} + b_{i+1}x_{i+1} - \bar{f}) \tag{20.28}$$

with $a_i, b_i > 0$. Show that permanence implies either $a_i < b_i$ or $b_i < a_i$ for all i. If $b_i < a_i$ for all i, then (20.28) is permanent if

$$\begin{bmatrix} \beta_1 & -1 & -1 & \cdots & \alpha_n & 0 & 0 \\ 0 & \beta_2 & -1 & \cdots & -1 & \alpha_1 & 0 \\ 0 & 0 & \beta_3 & \cdots & -1 & -1 & \alpha_2 \\ \vdots & \vdots & \vdots & & \vdots & \vdots & \vdots \end{bmatrix}$$

is an M-matrix (with α_k as in (20.27) and $\beta_k = \frac{a_k}{b_k} - 1$). This is always the case if $n \leq 4$. Show that in the symmetric case $a_i = a$, $b_i = b$ this reduces to

$$\frac{b}{a} + 2 + \frac{a}{b} > n \quad . \tag{20.29}$$

20.6 Notes

The Hopf bifurcation in **20.1** and the analysis of cyclic symmetry in **20.2** are from Hofbauer et al. (1980). Limit cycles for Lotka-Volterra equations were displayed in Coste et al. (1979). In Hofbauer et al. (1980) and Schuster et al. (1980) graphs were used to classify small catalytic networks. Theorem 20.3 is from Sigmund and Schuster (1984), **20.4** follows Amann (1988) (see also Hofbauer and Sigmund (1987)). Essentially hypercyclic networks and in particular the examples of exercises 20.5.3 and 20.5.4 were studied in Amann and Hofbauer (1985), while the results in exercise 20.5.2 were first obtained by Zeeman (1980).

21. Stability of n-Species Communities

21.1 Mutualism and M-matrices

This chapter deals with the relationship between stability properties of the Lotka-Volterra equation

$$\dot{x}_i = x_i(r_i + \sum_{j=1}^{n} a_{ij}x_j) \tag{21.1}$$

and algebraic properties of the interaction matrix $A = (a_{ij})$. For illustration, we shall start with a section on *mutualism* where this relationship is particularly clear. As we saw in exercise 8.3.6, a mutualistic two-species system has unbounded solutions if $a_{21}a_{12} > a_{11}a_{22}$, and a globally stable rest point if $a_{12}a_{21} < a_{11}a_{22}$. The following theorem generalizes this simple dichotomy to higher dimensions.

Theorem. *Consider (21.1) with $a_{ij} \geq 0$ for all $i \neq j$, and assume that it admits an interior rest point $\hat{\mathbf{x}}$. Then the following statements are equivalent:*

(M1) *All orbits in $\mathbf{R}^n_+$ are uniformly bounded as $t \to +\infty$;*

(M2) *The matrix A is stable (i.e. all eigenvalues of A have negative real part);*

(M3) *The leading principal minors of A alternate in sign:*

$$(-1)^k \det(a_{ij})_{1 \leq i,j \leq k} > 0 \quad ;$$

(M4) *For all $\mathbf{c} > \mathbf{0}$ there exists an $\mathbf{x} > \mathbf{0}$ such that*

$$A\mathbf{x} + \mathbf{c} = \mathbf{0} \quad ; \tag{21.2}$$

(M5) *There exists an $\mathbf{x} > \mathbf{0}$ with $A\mathbf{x} < \mathbf{0}$;*

(M6) *The rest point $\hat{\mathbf{x}}$ is globally asymptotically stable and all (boundary) orbits are uniformly bounded as $t \to +\infty$.*

For the proof, we shall need a result valid for general (not necessarily mutualistic) interactions.

Lemma: *If the matrix A has a left eigenvector $\mathbf{v} \geq \mathbf{0}$ with eigenvalue $\lambda > 0$, then (21.1) has interior solutions which are unbounded as $t \to +\infty$.*

Indeed, let us assume that $\mathbf{v}A = \lambda\mathbf{v}$ with $\mathbf{v} \geq \mathbf{0}, \sum v_i = 1$ and $\lambda > 0$. Let us consider the function

$$P(\mathbf{x}) = \prod_{i=1}^{n} x_i^{v_i}$$

and the sets $M_\alpha = \{\mathbf{x} \in \mathbf{R}^n_+ : P(\mathbf{x}) > \alpha\}$. From exercise 3.4.3 we have $P(\mathbf{x}) \leq \sum_{i=1}^{n} v_i x_i$. Thus on M_α

$$\frac{\dot{P}}{P} = \mathbf{v} \cdot (\mathbf{r} + A\mathbf{x}) = \mathbf{v} \cdot \mathbf{r} + \lambda \mathbf{v} \cdot \mathbf{x} \geq \mathbf{v} \cdot \mathbf{r} + \lambda\alpha \quad . \tag{21.3}$$

For α sufficiently large, we have $\dot{P}(\mathbf{x}) > 0$. The sets M_α are therefore forward invariant. All orbits in M_α escape to infinity, which proves the lemma.

Let us prove now the theorem.

(M1) $\Rightarrow$ (M2). Since A is mutualistic, we may write it in the form

$$A = B - cI \tag{21.4}$$

where $c > 0$ and B is a nonnegative matrix. The Perron-Frobenius theorem (cf. **11.4**) shows that there exists a dominant eigenvalue $\rho > 0$ of B with nonnegative left and right eigenvectors $\mathbf{v} \geq \mathbf{0}$ and $\mathbf{u} \geq \mathbf{0}$. Clearly $\mathbf{v}$ and $\mathbf{u}$ are also eigenvectors of A, corresponding to the eigenvalue $\lambda = \rho - c$. The previous lemma, together with (M1), shows that $\lambda \leq 0$.

If we suppose that $\lambda = 0$, we obtain $A\mathbf{u} = \mathbf{0}$. The line $\hat{\mathbf{x}} + t\mathbf{u}, t \in \mathbf{R}$, corresponds to fixed points of (21.1). This is a contradiction to (M1). Hence $\lambda < 0$. But no eigenvalue of A has real part larger than λ, and hence A is stable.

(M2) $\Rightarrow$ (M3). Since A is stable, det A has sign $(-1)^n$. The same is valid for every principal submatrix of A. Indeed, for every $J \subseteq \{1,\ldots,n\}$, the principal submatrix $B_J = (b_{ij})_{i,j\in J}$ of B has a dominant eigenvalue $\rho(J)$ which is not larger than the dominant eigenvalue ρ of B. Hence the submatrices $A_J = B_J - cI$ are also stable, and their determinant has sign $(-1)^{\text{card}\, J}$.

(M3) $\Rightarrow$ (M4). This can be proved by induction on n. We eliminate x_1 in (21.2) by multiplying the first equation

$$\sum_{k=1}^{n} a_{1k}x_k + c_1 = 0 \tag{21.5}$$

by $\dfrac{a_{i1}}{a_{11}}$ (note that $a_{11} < 0$), and subtracting it from the i-th equation. This yields a linear system in $x_2,\ldots,x_n$:

$$\sum_{k=2}^{n} \bar{a}_{ik}x_k + \bar{c}_i = 0 \tag{21.6}$$

with

$$\bar{a}_{ik} = a_{ik} - a_{1k}\frac{a_{i1}}{a_{11}} \quad \text{and} \quad \bar{c}_i = c_i - c_1\frac{a_{i1}}{a_{11}} \geq c_i > 0 \quad .$$

If we apply the corresponding operations to the leading principal minor, we obtain

$$\det(a_{ij})_{1\leq i,j\leq k} = \begin{vmatrix} a_{11} & a_{12} & \cdots & a_{1k} \\ 0 & \bar{a}_{22} & \cdots & \bar{a}_{2k} \\ \cdot & \cdot & & \cdot \\ \cdot & \cdot & & \cdot \\ 0 & \bar{a}_{k2} & \cdots & \bar{a}_{kk} \end{vmatrix} = a_{11}\det(\bar{a}_{ij})_{2\leq i,j\leq k} \quad .$$

Hence the $(n-1) \times (n-1)$ matrix $\bar{A}$ satisfies (M3). By induction hypothesis, (21.6) has a positive solution $\bar{x}_2, \dots, \bar{x}_n$. This yields a positive solution $x_1, \bar{x}_2, \dots, \bar{x}_n$ of (21.2), since $x_1 > 0$ is an obvious consequence of (21.5) and (21.6).
(M4) $\Rightarrow$ (M5) is trivial.
(M5) $\Rightarrow$ (M6). For mutualistic systems, (M5) means that A has a ***negative dominant diagonal***, i.e.

$$\exists d_i > 0: \ a_{ii} d_i + \sum_{j \neq i} |a_{ij}| d_j < 0 \quad . \tag{21.7}$$

We show that this condition ensures global stability of the interior rest point $\hat{\mathbf{x}}$ (which exists by assumption) for general, not necessarily mutualistic Lotka-Volterra systems. Indeed, let

$$V(\mathbf{x}) = \max_{i=1,\dots,n} \frac{|x_i - \hat{x}_i|}{d_i} \quad . \tag{21.8}$$

Then $V(\mathbf{x}) \geq 0$, with equality iff $\mathbf{x} = \hat{\mathbf{x}}$. The constant level sets of V are boxes of side lengths $2d_i$ centred in $\hat{\mathbf{x}}$. We claim that all these boxes are forward invariant.

Indeed, let i be any index for which $\left|\frac{x_i - \hat{x}_i}{d_i}\right|$ is maximal. Then

$$\begin{aligned}
|x_i - \hat{x}_i|^{\cdot} &= \dot{x}_i \operatorname{sgn}(x_i - \hat{x}_i) \\
&= x_i [a_{ii}(x_i - \hat{x}_i) + \sum_{j \neq i} a_{ij}(x_j - \hat{x}_j)] \operatorname{sgn}(x_i - \hat{x}_i) \\
&\leq x_i [a_{ii} |x_i - \hat{x}_i| + \sum_{j \neq i} |a_{ij}| |x_j - \hat{x}_j|] \\
&\leq x_i V(\mathbf{x}) [a_{ii} d_i + \sum_{j \neq i} |a_{ij}| d_j] < 0
\end{aligned}$$

for all $\mathbf{x} \neq \hat{\mathbf{x}}$ in int $\mathbf{R}^n_+$. Thus $V(\mathbf{x})$ is a strictly decreasing Ljapunov function. Hence all interior orbits converge to $\hat{\mathbf{x}}$. By the same token all boundary orbits are uniformly bounded.
(M6) $\Rightarrow$ (M1) is obvious.

Exercise 1: Show that under the assumptions of the theorem, the following properties are also equivalent to (M1) - (M6):
(M7) $\hat{\mathbf{x}}$ is globally stable;
(M8) (21.1) is permanent;
(M9) All interior orbits are uniformly bounded for $t \to +\infty$;
(M10) A^{-1} exists and has nonpositive elements;
(M11) For every $\mathbf{x} \geq \mathbf{0}$ (but $\mathbf{x} \neq \mathbf{0}$) there is an i such that $(A\mathbf{x})_i < 0$.

(Hint: (M6) $\Rightarrow$ (M7) $\Rightarrow$ (M8) $\Rightarrow$ (M9) $\Rightarrow$ (M2); (M10) $\Leftrightarrow$ (M4); (M11) $\Leftrightarrow$ (M5)).

Exercise 2: For $n = 3$, these conditions are equivalent to $a_{ii} < 0$ and $\det A < 0$.

Exercise 3: Study mutualistic Lotka-Volterra systems without interior fixed point (Hint: Construct a saturated fixed point $\hat{\mathbf{x}}$ on the boundary).

If the matrix A satisfies the conditions of the theorem, then $-A$ is said to be a (nonsingular) *M-matrix*. We have met this class in **20.5**. In **21.3**, we shall see that even if the mutualistic equation (21.1) has no interior rest point, the ecologically trivial assumption of uniform boundedness implies the existence of a globally stable equilibrium.

21.2 Boundedness and B-matrices

We now turn to the characterization of *uniform boundedness.*

Theorem 1. *The following conditions are equivalent for a matrix A.*

(B 1) *For every $\mathbf{r} \in \mathbf{R}^n$, the solutions of the Lotka-Volterra equation (21.1) are uniformly bounded for $t \to +\infty$.*

(B 2) *The origin $\mathbf{0}$ is globally asymptotically stable for the solutions of*

$$\dot{x}_i = x_i(A\mathbf{x})_i \tag{21.9}$$

in $\mathbf{R}^n_+$.

(B 3) *Whenever*

$$x_i(A\mathbf{x})_i = \lambda x_i \quad i = 1,\dots,n \tag{21.10}$$

holds for some $\mathbf{x} \geq \mathbf{0}$ ($\mathbf{x} \neq \mathbf{0}$), then $\lambda < 0$.

The matrix A is said to be a *B-matrix* if it satisfies one of the equivalent conditions (B 1) to (B 3).

Before proving the theorem, we recall the projective change in coordinates from **16.6** which brings infinity into view. With

$$z_k = \frac{x_k}{1+\sum x_i} \quad (k=1,\dots,n) \quad \text{and} \quad z_{n+1} = \frac{1}{1+\sum x_i}\ , \tag{21.11}$$

equation (21.1) transforms into the replicator equation

$$\begin{aligned} \dot{z}_k &= z_k\Big(\sum_{j=1}^{n} a_{kj}z_j + r_k z_{n+1} - \bar{a}(\mathbf{z})\Big) \quad (k = 1,\dots,n) \\ \dot{z}_{n+1} &= z_{n+1}\big(-\bar{a}(\mathbf{z})\big) \end{aligned} \tag{21.12}$$

on S_{n+1}, with

$$\bar{a}(\mathbf{z}) = \sum_{i,j=1}^{n} a_{ij} z_i z_j + z_{n+1} \sum_{k=1}^{n} r_k z_k \quad .$$

Obviously, (21.1) has uniformly bounded solutions for $t \to +\infty$, iff the closed invariant face $F_\infty = \{\mathbf{z} \in S_{n+1} : z_{n+1} = 0\}$ corresponding to the points at infinity is a *repellor* for (21.12) in the sense that there exists a $c > 0$ such that

$$\liminf_{t \to +\infty} z_{n+1}(t) > c$$

for all $\mathbf{z} \in S_{n+1}$ with $z_{n+1} > 0$. This will be useful for the proof of the theorem.

(B 3) $\Rightarrow$ (B 1). It is enough to show that $P(\mathbf{z}) = z_{n+1}$ is an *average Ljapunov function* in the vicinity of F_∞. By the same argument as in Theorem 19.6.1 this holds iff

$$\frac{\dot{P}}{P}(\bar{\mathbf{z}}) = \left.\frac{\dot{z}_{n+1}}{z_{n+1}}\right|_{\bar{\mathbf{z}}} = -\bar{a}(\bar{\mathbf{z}}) > 0 \tag{21.13}$$

for all fixed points $\bar{\mathbf{z}}$ of (21.12) with $\bar{z}_{n+1} = 0$. These fixed points $\bar{\mathbf{z}}$ at infinity are characterized by

$$\bar{z}_i (A\bar{\mathbf{z}})_i = \lambda \bar{z}_i \tag{21.14}$$

where λ is an arbitrary constant. Using $\bar{\mathbf{z}} \in S_{n+1}$, we see that λ is then just $\bar{a}(\bar{\mathbf{z}})$. Hence $P(\mathbf{z}) = z_{n+1}$ is an average Ljapunov function for (21.12) if $\lambda = \bar{a}(\bar{\mathbf{z}}) < 0$ for any such point $\bar{\mathbf{z}} \in F_\infty$.

(B 1) $\Rightarrow$ (B 2). By assumption there exists a constant $k > 0$ such that

$$B_k = \{\mathbf{x} \in \mathbf{R}_+^n : x_i \leq k \text{ for all } i\}$$

contains all ω-limits of solutions of (21.1) with $\mathbf{r} = \mathbf{0}$, i.e. of (21.9). But (21.9) is homogeneous and hence invariant under scalings $\mathbf{x} \to \alpha\mathbf{x}$ for $\alpha > 0$. So each set $B_{\alpha k}, \alpha > 0$, has the same property. For $\alpha \to 0$ this proves that all solutions converge to $\mathbf{0}$. The stability of $\mathbf{0}$ follows from the compactness of the sets $B_{\alpha k}$.

(B 2) $\Rightarrow$ (B 3). Let $\bar{\mathbf{x}} \geq \mathbf{0}$ satisfy (21.10). The line $\{t\bar{\mathbf{x}} : t > 0\}$ is obviously invariant for (21.9), since $(x_i/x_j)^\cdot = 0$ holds there. On this line, the flow reduces to $\dot{x}_i = \lambda x_i$. If $\lambda > 0$ the orbit grows to infinity. For $\lambda = 0$ the line consists of rest points. Since both possibilities contradict (B 2), $\lambda < 0$ follows.

We note that the vectors $\bar{\mathbf{x}}$ satisfying (21.10) correspond to asymptotic eigendirections of (21.1), and the points $\bar{\mathbf{z}}$ in (21.14) to rest points of (21.12). The corresponding "transversal" eigenvalue is $-\lambda = -\bar{a}(\bar{\mathbf{z}})$. The condition $\lambda < 0$ means that $\bar{\mathbf{z}}$ is not saturated.

Exercise 1: Show that the condition obtained from (B 1) by dropping "uniformly" still implies (B 1). (Hint: consider the case $r_i = r > 0$).

Exercise 2: Show that for symmetric matrices A, (B 3) is equivalent to

$$\mathbf{x} \cdot A\mathbf{x} < 0 \text{ for all } \mathbf{x} \geq \mathbf{0} \text{ with } \mathbf{x} \neq \mathbf{0} \quad . \tag{21.15}$$

In this case, $\sum x_i$ is decreasing for large $\mathbf{x}$ for (21.1). The matrix $-A$ is said to be *strictly copositive* (Hint: Show that the maxima of $\mathbf{x} \cdot A\mathbf{x}$ on S_n are given by (21.10)).

Theorem 2. *The matrix A is a B-matrix iff*
(B 4) *For all $\mathbf{x} \geq \mathbf{0}$ $(\mathbf{x} \neq \mathbf{0})$ there is an i such that $x_i > 0$ and $(A\mathbf{x})_i < 0$.*

Biologically, this means that for every state $\mathbf{x} \neq \mathbf{0}$, at least one species i has its growth rate reduced by the interaction of the species.

Proof: (B 4) $\Rightarrow$ (B 3) is obvious. For the converse we show that (B 1) to (B 3) imply that the transpose A^t of A satisfies (B 4). (The class of B-matrices is therefore closed under transposition).

So let us assume that (B 1) or (B 2) holds for A, but that A^t does not satisfy (B 4). Then we can find a $\mathbf{p} \geq \mathbf{0}$, $\mathbf{p} \neq \mathbf{0}$, such that $p_i(A^t\mathbf{p})_i \geq 0$ for all i. Since $(A^t\mathbf{p})_i = (\mathbf{p}A)_i \geq 0$ for all $i \in I = \operatorname{supp}(\mathbf{p})$, we obtain $\mathbf{p} \cdot A\mathbf{x} \geq 0$ for all $\mathbf{x}$ with $\operatorname{supp}(\mathbf{x}) \subseteq I$. Now $P(\mathbf{x}) = \prod x_i^{p_i}$ satisfies

$$\frac{\dot{P}}{P}(\mathbf{x}) = \mathbf{p} \cdot (\mathbf{r} + A\mathbf{x}) = \mathbf{p} \cdot \mathbf{r} + \mathbf{p} \cdot A\mathbf{x} \quad .$$

This expression is nonnegative for $\mathbf{r} = \mathbf{0}$ and even positive for $\mathbf{r} > \mathbf{0}$. Thus P increases along the orbit of $\mathbf{x}$ if $\operatorname{supp}(\mathbf{x}) = \operatorname{supp}(\mathbf{p})$. All these solutions converge to infinity, a contradiction to (B 1) and (B 2).

Exercise 3: If A is a B-matrix and D a positive diagonal matrix, then DA, AD and every principal submatrix of A are also B-matrices.

Exercise 4: For $n = 2$, A is a B-matrix iff $a_{ii} < 0$ and for mutualistic interactions additionally $\det A > 0$.

Exercise 5: The following conditions are equivalent to (B 1) - (B 4):
(B5) If $A\mathbf{x} = \lambda D\mathbf{x}$ for a positive diagonal matrix D and $\mathbf{x} \geq \mathbf{0}$ $(\mathbf{x} \neq \mathbf{0})$, then $\lambda < 0$; and every principal submatrix of A has the same property.

(B6) For all nonnegative diagonal matrices $D \geq 0, A - D$ has no zero eigenvalue in a direction $\mathbf{x} \geq \mathbf{0}$, and every principal submatrix of A has the same property.
(B7) There exists an $\mathbf{x} > \mathbf{0}$ such that $A\mathbf{x} < \mathbf{0}$, and every principal submatrix has the same property.
(B8) For some $\mathbf{r} \in \mathbf{R}^n$, the Lotka-Volterra equation (21.1) has all orbits bounded, as $t \to +\infty$, and this property is robust against small perturbations of the a_{ij}.

Exercise 6: An ecosystem described by (21.1) is said to be *hierarchically ordered* if the graph $G^+(A)$, obtained by drawing an arrow from j to i whenever $a_{ij} > 0$, has no directed cycle and additionally $a_{ii} < 0$ holds for all i. (In this case we can order the species hierarchically such that positive actions occur only from a lower to a higher level). Show that the ecosystem is hierarchically ordered iff A is a qualitative B-matrix, i.e. if every matrix $\tilde{A}$ with $\operatorname{sgn} \tilde{a}_{ij} = \operatorname{sgn} a_{ij}$ is a B-matrix.

Exercise 7: Show that the orbits of a general predator-prey system

$$\dot{x}_i = x_i(r_i - \sum a_{ij}x_j - \sum b_{ik}y_k)$$

$$\dot{y}_k = y_k(-s_k + \sum c_{ki}x_i - \sum d_{kl}y_l)$$

($i = 1,\dots,n$; $k = 1,\dots,m$; all $a_{ij}, b_{ij}, c_{ij}, d_{ij}, s_j$ and r_i nonnegative and $a_{ii} > 0$, $d_{ii} > 0$) are uniformly bounded for $t \to +\infty$.

When A is a B-matrix, the Lotka-Volterra equation (21.1) has uniformly bounded orbits for $t \to +\infty$ and so the index theorem applies (cf. exercise 19.4.3). One could also use the index theorem for the replicator equation (21.12) and argue that (B 3) prevents the existence of saturated fixed points with $z_{n+1} = 0$. Hence all saturated fixed points of (21.12) correspond to "finite" saturated fixed points of (21.1). Their indices have to add up to $(-1)^n$. In fact, even some sort of converse holds.

Exercise 8: Show that the following conditions are equivalent to (B 1) - (B 8).
(B 9) For all $\mathbf{r} \in \mathbf{R}^n$, (21.1) has a saturated rest point, and the same holds for all subsystems (Hint: (B 9) $\Rightarrow$ (B 7)).
(B 10) For every $\mathbf{r} \leq \mathbf{0}$, (21.1) has only one saturated rest point, namely $\mathbf{0}$. (Hint: (B 2) $\Rightarrow$ (B 10) $\Rightarrow$ (B 4)).

Theorem 3. *If all proper principal submatrices of A are B-matrices and* $\det(-A) > 0$ *then A itself is a B-matrix.*

Proof: Let us suppose that A is not a B-matrix. By (B 3) there exists a $\bar{\mathbf{z}} \in \operatorname{int} S_n$ such that

$$(A\bar{\mathbf{z}})_i = \lambda > 0 \quad . \tag{21.16}$$

Indeed, $\bar{\mathbf{z}} \in \operatorname{bd} S_n$ would contradict the B-property of the submatrices of A, and $\lambda = 0$ would imply $A\bar{\mathbf{z}} = \mathbf{0}$ and hence $\det A = 0$. The point $\bar{\mathbf{z}}$ corresponds to a regular saturated rest point of (21.12) with $z_{n+1} = 0$. Its index is the negative (for the transversal eigenvalue $-\lambda$) of the determinant of the Jacobian of (21.12) at $\bar{\mathbf{z}}$ restricted to the face $z_{n+1} = 0$, independently of $\mathbf{r}$. By exercise 19.5.2 this is

$$i\,(\bar{\mathbf{x}}) = -\operatorname{sgn}\det A \quad . \tag{21.17}$$

We shall show that for $\mathbf{r} = \mathbf{0}$, $\bar{\mathbf{z}}$ is the only saturated rest point of (21.12). Then by Theorem 19.4.1 its index equals $(-1)^n$. By (21.17), therefore, $\operatorname{sgn}\det A = (-1)^{n-1}$ or $\operatorname{sgn}\det(-A) = -1$, which is a contradiction.

For $\mathbf{r} = \mathbf{0}$, (21.9) has no interior fixed point since $\det A \neq 0$. Furthermore, (B 2) and the assumption imply that all boundary orbits converge to $\mathbf{0}$. So $\mathbf{0}$ is the only (finite) fixed point of (21.9), and since all of its transversal eigenvalues are 0, it is a (completely degenerate) saturated fixed point. Its index is zero, however, and hence it does not count in Theorem 19.4.1, since there are no near-by points $\mathbf{x} > \mathbf{0}$ with $A\mathbf{x} < \mathbf{0}$ (cf. exercise 19.4.4). This follows either directly from (B 6) or from the following argument: As A is not a B-matrix, neither is its transpose A^t. Hence (B 4) gives a $\mathbf{p} > \mathbf{0}$ with $A^t\mathbf{p} \geq \mathbf{0}$. But then $0 \leq \mathbf{p} \cdot A\mathbf{x} < 0$, a contradiction.

Exercise 9: Give an alternative proof of (21.17). Consider (21.1) resp. (21.12) with $r_i = -r < 0$. This system has three saturated rest points: $\mathbf{0}$, an interior fixed point $r\mathbf{z}$, and a point at infinity in direction $\mathbf{z}$. Compute the indices of $\mathbf{0}$ and $r\mathbf{z}$ and apply Theorem 19.4.1.

Exercise 10: Show that Theorem 3 can be strengthened as follows:

If all $(n-1) \times (n-1)$ principal submatrices of A are B-matrices, then the following statements are equivalent:

(a) A itself is not a B-matrix

(b) $\det(-A) \leq 0$ and $\operatorname{adj}(-A) > 0$.

Here $\operatorname{adj}(A)$ is the *adjoint* matrix of A. Its (i,j)-entry is the determinant of the $(n-1) \times (n-1)$ submatrix of A obtained by deleting the j-th row and the i-th column, multiplied by $(-1)^{i+j}$. Since for a nonsingular matrix

$$A^{-1} = \frac{\operatorname{adj} A}{\det A} \quad , \tag{21.18}$$

condition (b) means in this case that all entries of A^{-1} are positive. (Hint: (i) Applying (B 3) to the matrix $(b_i^{-1}a_{ij})$, we may see that for each $\mathbf{b} > \mathbf{0}$, there is an $\mathbf{x} > \mathbf{0}$ with $A\mathbf{x} = \lambda\mathbf{b}$ and $\lambda \geq 0$. (ii) If $\det A \neq 0$ then (i) implies that A^{-1} is nonnegative. It is even positive since a zero entry would contradict the assumption on submatrices. The converse is easy. (iii) If $\det A = 0$, the right and left eigenvectors $\mathbf{u}$ and $\mathbf{v}$ with $A\mathbf{u} = \mathbf{v}A = \mathbf{0}$ are positive. There are, then, no other eigenvectors for 0, and hence A has rank $n-1$. Then $(\operatorname{adj} A)_{ij} = \rho u_i v_j$ with $\operatorname{sgn} \rho = (-1)^{n-1}$. For the converse, observe that $\operatorname{adj} A \neq 0$ implies that the rank of A is $n-1$.)

Exercise 11: A matrix A with the sign pattern

$$\begin{bmatrix} - & + & - \\ - & - & + \\ + & - & - \end{bmatrix}$$

corresponding to a cyclic predator-prey system is a B-matrix iff $\det A < 0$.

21.3 VL-stability and global stability

In this section we turn to the strongest stability concept: global asymptotic stability.

The matrix A will be called *Volterra-Ljapunov stable* (VL-stable) if there exists a positive diagonal matrix $D > 0$ such that the symmetric matrix $DA + A^t D$ is negative definite, i.e. if there exist positive numbers d_i such that

$$\sum_{i,j} d_i a_{ij} x_i x_j < 0 \text{ for all } \mathbf{x} \neq \mathbf{0} \quad . \tag{21.19}$$

This means that for suitable $d_i > 0$, the function $V(\mathbf{x}) = \sum d_i x_i^2$ is a strict Ljapunov function for the linear ODE $\dot{\mathbf{x}} = A\mathbf{x}$, and hence that the ellipsoids $V(\mathbf{x}) \leq c$ are strictly forward invariant. This is related to Ljapunov's characterisation of stable matrices, see exercise 8.1.1. VL-stability is much stronger, as it requires the positive definite matrix Q to be diagonal, or the ellipsoid to be symmetric with respect to the coordinate planes.

Theorem. *If A is VL-stable, then for every $\mathbf{r} \in \mathbf{R}^n$ the Lotka-Volterra equation* (21.1) *has one globally stable fixed point.*

Proof: (21.19) implies (B 4) and hence by (B 9) the existence of a saturated rest point $\bar{\mathbf{x}}$. Let $V(\mathbf{x})$ be the standard Ljapunov function proposed by Volterra,

$$V(\mathbf{x}) = \sum_{i=1}^{n} d_i(\bar{x}_i \log x_i - x_i) \quad . \tag{21.20}$$

It has a unique global maximum at the point $\bar{\mathbf{x}}$ and is defined for all $\mathbf{x}$ with $\operatorname{supp}(\mathbf{x}) \supseteq \operatorname{supp}(\bar{\mathbf{x}})$. Now

$$\dot{V}(\mathbf{x}) = \sum_i d_i(\bar{x}_i - x_i)\frac{\dot{x}_i}{x_i} = \sum_i d_i(\bar{x}_i - x_i)(r_i + \sum a_{ij}x_j)$$

$$= -\sum_{i,j} d_i a_{ij}(x_i - \bar{x}_i)(x_j - \bar{x}_j) + \sum_i d_i(\bar{x}_i - x_i)(r_i + \sum_j a_{ij}\bar{x}_j) \ . \tag{21.21}$$

By (21.19) the first sum is positive for $\mathbf{x} \neq \bar{\mathbf{x}}$. In the second sum all terms with $\bar{x}_i > 0$ vanish since $\bar{\mathbf{x}}$ is a rest point, and the remaining terms are nonnegative since $\bar{\mathbf{x}}$ is saturated. Thus $\bar{\mathbf{x}}$ is globally stable.

Exercise 1: Show that A is VL-stable iff for every positive definite Q the matrix QA has a positive diagonal entry.

Exercise 2: If A is VL-stable and D,D' are positive diagonal matrices, then A^t, A^{-1}, DAD' and all its principal submatrices are VL-stable.

Exercise 3: The 2×2-matrix A is VL-stable iff $a_{ii} < 0$ $(i = 1,2)$ and $\det A > 0$.

Exercise 4: The Lotka-Volterra equation with immigration terms $\epsilon_i > 0$,

$$\dot{x}_i = x_i(r_i + (A\mathbf{x})_i) + \epsilon_i \quad , \tag{21.22}$$

(which corresponds to (19.17)) has a unique, globally stable interior fixed point, whenever A is a VL-stable matrix.

Exercise 5: Consider a "two patch" ecosystem with migration,

$$\begin{aligned} \dot{x}_i &= x_i(r_i + (A\mathbf{x})_i) + d_i(y_i - x_i) \\ \dot{y}_i &= y_i(r_i + (A\mathbf{y})_i) + d_i(x_i - y_i) \quad . \end{aligned} \tag{21.23}$$

Show that if A is VL-stable, then for arbitrary $d_i > 0$ and r_i, (21.23) has a unique globally stable fixed point.

Exercise 6: The function $V(\mathbf{x})$ from (21.20) is a constant of motion for the Lotka-Volterra equation (21.1) iff $d_i a_{ij} = -d_j a_{ji}$ holds for all i,j. (The matrix DA is then antisymmetric).

We have seen in **21.1** that matrices with negative dominant diagonal also lead to global stability (using a rectangular Ljapunov function). The next exercises compare this with VL-stability.

Exercise 7: The following conditions are equivalent:

(DD1) The matrix A is diagonally dominant, i.e. satisfies (21.7).

(DD2) The rectangles $\{\mathbf{x} \in \mathbf{R}^n\colon |x_i| \leq d_i\, const\}$ are strictly forward invariant for the linear ODE $\dot{\mathbf{x}} = A\mathbf{x}$.

(DD3) The matrix $\tilde{A}$ defined by $\tilde{a}_{ii} = -a_{ii}$ and $\tilde{a}_{ij} = -|a_{ij}|$ for $i \neq j$, is an M-matrix.

(DD4) A^t is diagonally dominant, i.e. there exist $c_i > 0$ with

$$a_{ii}c_i + \sum_{j \neq i} |a_{ji}| c_j < 0 \quad .$$

(DD5) The function $\sum c_i |x_i|$ is a strict Ljapunov function for $\dot{\mathbf{x}} = A\mathbf{x}$.

(DD6) $\sum c_i |\log(x_i/\hat{x}_i)|$ is a strict Ljapunov function for (21.1), if it has an interior fixed point $\hat{\mathbf{x}}$.

(DD7) There exists an $\mathbf{x} > \mathbf{0}$ such that $SAS\mathbf{x} < \mathbf{0}$ holds for every signature matrix S (a diagonal matrix with entries ± 1).

Exercise 8: Show that diagonal dominance implies VL-stability. (Hint: use c_i/d_i, with c_i from (DD4) and d_i from (DD1), as multipliers in (21.19)).

21.4 P-matrices

The matrix A is said to be a *P-matrix* if all its principal minors are positive.

Theorem 1. *The following properties are equivalent:*

(P1) *A is a P-matrix.*

(P2) *For every diagonal matrix $D \geq 0$, $A + D$ is a P-matrix.*

(P3) *For all $\mathbf{x} \neq \mathbf{0}$, there exists an i such that $x_i(A\mathbf{x})_i > 0$.*

(P4) *For all $\mathbf{x} \neq \mathbf{0}$, there exists a diagonal matrix $D > 0$ such that $\mathbf{x} \cdot DA\mathbf{x} = D\mathbf{x} \cdot A\mathbf{x} > 0$.*

(P5) *Every real eigenvalue of a principal submatrix of A is positive.*

Proof: (P1) $\Rightarrow$ (P2) is a consequence of the well known determinant formula

$$\det(A + D) = \sum_I \Big(\prod_{i \notin I} d_i\Big) \det(A_I)$$

where the sum runs over all 2^n subsets $I \subseteq \{1,...,n\}$ and A_I is the

corresponding principal submatrix.
(P2) $\Rightarrow$ (P3). Let us suppose that there exists an $\mathbf{x} \neq \mathbf{0}$ such that $x_i(A\mathbf{x})_i \leq 0$ for all i. Let $I = \text{supp}(\mathbf{x})$ and let $\mathbf{x}_I$ be the restricted vector. The components of $A_I\mathbf{x}_I$ and the corresponding components of $\mathbf{x}_I$ have opposite signs and hence $A_I\mathbf{x}_I$ may be written as $-D\mathbf{x}_I$ for a suitable nonnegative diagonal matrix D (over the index set I). Thus $(A_I + D)\mathbf{x}_I = \mathbf{0}$, and hence $A_I + D$ is a singular matrix, which is a contradiction.
(P3) $\Rightarrow$ (P4). If we choose a diagonal matrix D with a 1 in the i-th position given by (P3) and small positive entries elsewhere, we obtain $\mathbf{x} \cdot DA\mathbf{x} > 0$.
(P4) $\Rightarrow$ (P5). Let λ be a real eigenvalue of some submatrix A_I, i.e. $A_I\mathbf{x}_I = \lambda\mathbf{x}_I$. Let $\mathbf{x} \in \mathbf{R}^n$ have the same coordinates x_i as $\mathbf{x}_I$ for $i \in I$, and $x_i = 0$ for $i \notin I$. If we choose $D > 0$ according to (P4), we obtain

$$0 < A\mathbf{x} \cdot D\mathbf{x} = A_I\mathbf{x}_I \cdot D_I\mathbf{x}_I = \lambda\mathbf{x}_I \cdot D_I\mathbf{x}_I \quad .$$

Since this last inner product is positive, we obtain $\lambda > 0$.
(P5) $\Rightarrow$ (P1) follows from the fact that the determinant of a matrix is the product of its eigenvalues.

Exercise 1: Show that (P3) is equivalent to
(P6) For every signature matrix S there is an $\mathbf{x} > \mathbf{0}$ with $SAS\mathbf{x} > \mathbf{0}$.

Exercise 2: Show that the following implications hold for an $n \times n$ matrix A: (i) if A is VL-stable, then $-A$ is a P-matrix; (ii) if $-A$ is a P-matrix then A is a B-matrix.

Exercise 3: Compare the properties (P2), (P3), (P5), (P6) with (B6), (B4), (B5), (B7) and (DD7), respectively.

The relevance of P-matrices for ecological systems relies upon the fact that they guarantee the *uniqueness* of a saturated fixed point. In the generic situation, where all fixed points of (21.1) are regular, this is an immediate consequence of the index theorem 19.4.1: If $-A$ is a P-matrix, then all saturated fixed points have the same index $(-1)^n$, and therefore there exists only one.

Theorem 2. *The Lotka-Volterra equation* (21.1) *has a unique saturated rest point for every* $\mathbf{r} \in \mathbf{R}^n$ *iff* $-A$ *is a P-matrix.*

Proof: Let $-A$ be a P-matrix and $\mathbf{r} \in \mathbf{R}^n$. The existence of saturated rest points follows from (B9). Let us assume that there were two of them, $\mathbf{x}'$ and $\mathbf{x}''$. If $x_i' < x_i''$ then $x_i'' > 0$ and (since $\mathbf{x}''$ is a rest point) $(A\mathbf{x}'')_i + r_i = 0$. Since $\mathbf{x}'$ is saturated we have $(A\mathbf{x}')_i + r_i \leq 0$. Thus we obtain $(A(\mathbf{x}'-\mathbf{x}''))_i \le 0$ and $(\mathbf{x}'-\mathbf{x}'')_i < 0$. This shows that $-A$ reverses the sign of all components of $\mathbf{x}'-\mathbf{x}'' \neq \mathbf{0}$, a contradiction to (P3).

Now let us assume that $-A$ is not a P-matrix. By (P3) we obtain an $\mathbf{x} \neq \mathbf{0}$ such that $x_i y_i \le 0$ for all i (where $\mathbf{y} = -A\mathbf{x}$). Separating $\mathbf{x}$ and $\mathbf{y}$ into positive and negative parts, $\mathbf{x} = \mathbf{x}^+ - \mathbf{x}^-$, $\mathbf{y} = \mathbf{y}^+ - \mathbf{y}^-$ ($\mathbf{x}^+$, $\mathbf{y}^-$,$\mathbf{y}^+$,$\mathbf{y}^- \ge \mathbf{0}$) and defining $\mathbf{r} = -\mathbf{y}^+ - A\mathbf{x}^+ = -\mathbf{y}^- - A\mathbf{x}^-$ we obtain $\mathbf{r} + A\mathbf{x}^+ = -\mathbf{y}^+ \le \mathbf{0}$, $\mathbf{r} + A\mathbf{x}^- = -\mathbf{y}^- \le \mathbf{0}$ and $x_i^+ y_i^+ = x_i^- y_i^- = 0$, which shows that both $\mathbf{x}^+$ and $\mathbf{x}^-$ are saturated rest points of (21.1).

Exercise 4: Illustrate Theorem 21.4.2 for $n = 2$ (bistability versus global stability).

Exercise 5: The matrix

$$A = \begin{pmatrix} -1 & -\alpha & -\beta \\ -\beta & -1 & -\alpha \\ -\alpha & -\beta & -1 \end{pmatrix}$$

is VL-stable iff $-1 < \alpha + \beta < 2$, and $-A$ is a P-matrix iff $-1 < \alpha + \beta$ and $\alpha\beta < 1$ (cf. **9.5**).

Exercise 6: Let $-A$ be a P-matrix and $\bar{\mathbf{x}} = \bar{\mathbf{x}}(\mathbf{r})$ the unique saturated rest point of (21.1). If r_i increases and all other r_j remain constant, then $\bar{x}_i(\mathbf{r})$ increases. This seems biologically obvious, but does not hold if $-A$ is not a P-matrix.

Exercise 7: The equation $\dot{x}_i = x_i f_i(\mathbf{x})$ admits only one saturated rest point if its orbits are uniformly bounded and $-D_{\mathbf{x}}\mathbf{f}$ is a P-matrix for every $\mathbf{x} \in \mathbf{R}^n_+$.

If (21.1) has an interior fixed point $\hat{\mathbf{x}}$ and $-A$ is a P-matrix then Theorem 21.4.2 implies that there is no saturated rest point on the boundary. As the following exercises show, this implies some weak form of persistence, but it cannot prevent that most orbits spiral away from $\hat{\mathbf{x}}$ and towards the boundary.

Exercise 8: If (21.1) has an interior fixed point and $-A$ is a P-matrix, then no interior orbit $\mathbf{x}(t)$ can converge to a point on the boundary. Show that it is also impossible that $\omega(\mathbf{x})$ lies in the interior of one boundary face. Does it follow that (21.1) is persistent? (Hint: Show that the time-average of $\mathbf{x}(t)$ would converge to a saturated fixed point on the boundary).

Exercise 9: Find a P-matrix $-A$ such that (21.1) is not permanent although it has an interior rest point. (Hint: see exercise 5).

Exercise 10: Construct a system (21.1), with interior fixed point $\hat{\mathbf{x}}$ and $-A$ a P-matrix, which is not persistent. (Hint: add a fourth competitor to (9.18) with $\alpha\beta < 1$ and $\alpha + \beta > 2$ which obeys

$$\dot{y} = y(r - x_1 - x_2 - x_3 - y) \quad .$$

If $\frac{3}{1+\alpha+\beta} < r < 1$, then the system has an interior fixed point $\hat{\mathbf{x}}$ which attracts all orbits on the line $x_1 = x_2 = x_3 > 0$ and $y > 0$. Show that for all other interior orbits, however, $y(t) \to 0$ and $\omega(\mathbf{x})$ is the heteroclinic cycle of (9.18) again.)

Exercise 11: If $-A$ is a P-matrix, then the Lotka-Volterra system "with immigration" (21.22) has exactly one fixed point for every choice of $\epsilon_i > 0$. Is the converse also valid?

21.5 Communities with a special structure

Let A be the interaction matrix of an ecosystem modeled by (21.1). We obtain the *undirected graph* $G(A)$ by joining i and j with an edge whenever $a_{ij} \neq 0$ or $a_{ji} \neq 0$, and the *directed graph* $\vec{G}(A)$ by drawing an arrow $j \to i$ whenever $a_{ij} \neq 0$. A *cycle* of A is a *nonvanishing* product of the form $a_{i_1 i_2} a_{i_2 i_3} \cdots a_{i_k i_1}$, for a sequence of *pairwise distinct* indices $i_1, i_2, \ldots, i_k$. The *length* of this cycle is k.

Theorem 1. *Suppose that A has no cycles of length ≥ 3. Then A is VL-stable iff $-A$ is a P-matrix.*

Proof: We shall use induction on the number of species n. The case $n = 2$ was settled in exercise 21.3.3. Let us assume that $-A$ is a P-matrix. We have to find $d_i > 0$ such that (21.19) holds.
(a) Let us assume first that A is *qualitatively symmetric*, i.e. that $a_{ij} \neq 0$ implies $a_{ji} \neq 0$. In this case we may choose $d_i > 0$ such that

$$d_i |a_{ij}| = d_j |a_{ji}| \quad . \tag{21.24}$$

This is possible in a consistent way since the graph $G(A)$ has no cycles. Let D be the diagonal matrix with entries d_i. If $a_{ij} a_{ji} \geq 0$ holds for all i,j then $-DA$ is a symmetric P-matrix. Then the quadratic form $\sum d_i a_{ij} x_i x_j$ is negative definite. Otherwise, there are indices $k \neq l$ with $a_{kl} a_{lk} < 0$. Then the term $x_k x_l$ in the quadratic form vanishes and the community splits into two blocks if we drop the connection between k

and l. Applying the induction hypothesis to both subcommunities, both blocks of the quadratic form are positive definite and so is the whole form itself. Hence A is VL-stable.

(b) Let us assume now that there exist k,l with $a_{kl} \neq 0$ but $a_{lk} = 0$. Note that the undirected graph $G(A)$ may have cycles in this case. The matrix A turns out to be reducible now, if we split the community into two blocks as above. Using induction, we are left to show the following lemma.

Lemma. *Let A be a reducible matrix of the form*

$$A = \begin{bmatrix} A_1 & A_2 \\ 0 & A_3 \end{bmatrix}$$

Then A is VL-stable iff A_1 and A_3 are VL-stable.

Proof: Replacing each of the principal VL-stable submatrices A_i ($i = 1{,}3$) by $D_iA_i + A_i^tD_i$ with suitable $D_i > 0$, we may assume that A_1 and A_3 are symmetric and negative definite. Thus we have only to find an $\epsilon > 0$ such that the quadratic form

$$(\mathbf{x},\mathbf{y}) \begin{bmatrix} \epsilon A_1 & \epsilon A_2 \\ 0 & A_3 \end{bmatrix} \begin{pmatrix} \mathbf{x} \\ \mathbf{y} \end{pmatrix}$$

is negative definite. Rescaling $\mathbf{x}$ to $\mathbf{x}\epsilon^{-1/2}$, we are left with

$$\mathbf{x} \cdot A_1\mathbf{x} + \epsilon^{1/2}\mathbf{x} \cdot A_2\mathbf{y} + \mathbf{y} \cdot A_3\mathbf{y} \quad . \tag{21.25}$$

Since both A_1 and A_3 are negative definite, so is (21.25) for small $\epsilon > 0$.

If all pairwise interactions are of predator-prey type, the theorem simplifies even more:

Theorem 2. *Suppose that $\vec{G}(A)$ has no cycles of length ≥ 3, $a_{ii} < 0$ and $a_{ij}a_{ji} \leq 0$ for all $i \neq j$. Then A is VL-stable.*

Proof: By the construction in the proof of the previous theorem, for each irreducible block of A, $DA + A^tD$ yields already a negative diagonal matrix. Together with the lemma this implies VL-stability.

Exercise 1: Check under the above assumptions that $-A$ is a P-matrix.

A particular case of Theorem 2 are *food chains*, which are always VL-stable, as was seen in **9.3**. The systems satisfying the assumptions of Theorem 2 are of interest from another viewpoint. They represent those communities which guarantee stability when only the signs (+,0,−) of the interaction coefficients a_{ij} are known.

Exercise 2: Show that A is *qualitatively semi-stable* (i.e. every matrix with the same sign pattern has only eigenvalues with real part ≤ 0) iff the following three conditions are satisfied

(i) $a_{ii} \leq 0$;

(ii) $a_{ij}a_{ji} \leq 0$ for $i \neq j$;

(iii) no cycles of length ≥ 3.

Exercise 3: Show that a matrix A is qualitatively VL-stable iff

(i) $a_{ii} < 0$

(ii) $a_{ij}a_{ji} \leq 0$ for $i \neq j$

(iii) no cycles of length ≥ 3.

Exercise 4: Show that the matrix $-A$ is a qualitative P-matrix iff all cycles are negative: $a_{ii} < 0, a_{ij}a_{ji} \leq 0$ $(i \neq j)$, $a_{ij}a_{jk}a_{ki} \leq 0$ $(i,j,k$ pairwise distinct), etc ...

Another type of communities defined by sign conditions is the following:

$$\begin{aligned}&\text{(i)}\ a_{ii} \leq 0\ ;\\ &\text{(ii)}\ \text{all cycles of } A \text{ of length} \geq 2 \text{ are positive}\ .\end{aligned} \tag{21.26}$$

Exercise 5: Consider a matrix A describing the competition of two symbiotic systems (i.e. there is a partition of the species into two disjoint groups I and J, one of which could be empty, such that $a_{ij} \geq 0$ for $i \neq j$, $i,j \in I$ or $i,j \in J$ and $a_{ij} \leq 0, a_{ji} \leq 0$ for $i = j$ or $i \in I$ and $j \in J$). Show that such a matrix A satisfies (21.26), but that the converse is not true. Show that (21.1) generates a monotone flow in the variables $x_i (i \in I), -x_j (j \in J)$.

The following theorem generalizes the case of symbiosis, treated in **21.1.**

Theorem 3. *If A satisfies condition* (21.26) *then A is VL-stable iff $-A$ is a P-matrix.*

Proof: Let $\tilde{A}$ be the matrix defined by $\tilde{a}_{ii} = a_{ii} \leq 0$ and $\tilde{a}_{ij} = |a_{ij}|$ for $i \neq j$. Then every principal minor of $\tilde{A}$ equals the corresponding minor of A, since in every cycle the number of negative entries of A (which change their sign in $\tilde{A}$) is even by (21.26). Thus when $-A$ is a P-matrix then so is $-\tilde{A}$. By (M3) and (21.7), $\tilde{A}$ and hence also A have a negative dominant diagonal. By exercise 21.3.8 A is VL-stable.

Exercise 6: A matrix A is qualitatively diagonally dominant iff $a_{ii} < 0$ and there are no cycles of length ≥ 2.

(21.26) is the most general condition on signs under which P-matrices can be characterized by diagonal dominance. Both Theorem 21.5.1 and Theorem 21.5.3 give conditions on the sign patterns of A which guarantee that being a P-matrix already implies VL-stability. It would be interesting to find a common generalization of these two results.

We conclude this section with a class of systems which is not defined by qualitative conditions on the signs only. (21.1) is said to be a *symmetric system* if sgn $a_{ij} =$ sgn a_{ji} for all i,j and if every cycle is equal to its reversed cycle, i.e.

$$a_{i_1 i_2} a_{i_2 i_3} \cdots a_{i_k i_1} = a_{i_1 i_k} a_{i_k i_{k-1}} \cdots a_{i_2 i_1} \quad . \tag{21.27}$$

Exercise 7: Show that it suffices that condition (21.27) holds for cycles of length 3.

Exercise 8: Show that (21.1) is a symmetric system iff there exist $d_i > 0$ such that $(d_i a_{ij})$ is a symmetric matrix.

Exercise 9: Show that a symmetric system admits a global Ljapunov function of the form

$$\sum_{i,j} d_i a_{ij} (x_i - \bar{x}_i)(x_j - \bar{x}_j) \quad . \tag{21.28}$$

(The case $n = 2$ was treated in exercise 8.3.4. In exercise 24.4.7 we will see that these systems are gradients.)

Exercise 10: A symmetric system is VL-stable iff $-A$ is a P-matrix.

Exercise 11: Analyse the symmetric competition system where $a_{ij} = -f(|i-j|)$ with f a positive function, e.g. $f(x) = a^{x^2}$. Under which conditions on f is the system stable?

Exercise 12: Analyse competition systems with $-a_{ij} = \epsilon_i \delta_{ij} + c_i d_j$ (for $\epsilon_i, c_i, d_i > 0$).

21.6 D-stability and total stability

Let $\bar{\mathbf{x}}$ be an interior fixed point of the Lotka-Volterra equation (21.1). The Jacobian at $\bar{\mathbf{x}}$ is given by $(\bar{x}_i a_{ij})$, and hence depends on the rates r_i. In order to ensure that $\bar{\mathbf{x}}$ is always asymptotically stable, the interaction matrix A should be *D-stable*, which means that DA is stable for every diagonal matrix $D > 0$.

It turns out that this concept is rather strong, and - unexpectedly - even has some *global* meaning for Lotka-Volterra equations.

Exercise 1: If A is D-stable, then every principal submatrix is D-semistable (all eigenvalues have real part ≤ 0). (Hint: Choose some d_i very small.)

Exercise 2: If A is a D-stable matrix, then $-A$ has nonnegative principal minors.

Exercise 3: If A is VL-stable, then it is D-stable.

Exercise 4: Show that a 2×2 matrix is D-stable iff $a_{11} \leq 0$ and $a_{22} < 0$ (or vice versa) and $\det A > 0$.

The last example shows that D-stable matrices need not be B-matrices (in some marginal cases). Thus we consider a slightly stronger stability concept: A is said to be *totally stable* if every principal submatrix of A is D-stable.

This concept has again some global meaning for the Lotka-Volterra equations.

Theorem. *A is totally stable iff for every* $\mathbf{r}$, (21.1) *has exactly one saturated fixed point and this point is asymptotically stable within its face.*

Exercise 5: Prove this. (Hint: use Theorem 21.4.2).

Exercise 6: Show that qualitative D-stability is the same as qualitative stability.

Exercise 7: Show that qualitative total stability is equivalent to qualitative VL-stability (compare exercise 21.5.3.).

A 30 year old open problem in the theory of ODE's is the *"Jacobian Conjecture"*: Let $\dot{\mathbf{x}} = \mathbf{f}(\mathbf{x})$ be an ODE in $\mathbf{R}^n$. Suppose that

(i) $\mathbf{0}$ is a fixed point, i.e. $\mathbf{f}(\mathbf{0}) = \mathbf{0}$,

(ii) The Jacobian is a stable matrix at every point $\mathbf{x} \in \mathbf{R}^n$.

Then $\mathbf{0}$ is a *globally stable* fixed point.

If this conjecture is true, it will imply the following:

If the Lotka-Volterra equation (21.1) *has an interior fixed point* $\bar{\mathbf{x}}$ *and the interaction matrix is D-stable, then* $\bar{\mathbf{x}}$ *is globally stable.*

Indeed, let us make the change of coordinates $x_i = \bar{x}_i e^{u_i}$ or $u_i = \log(x_i/\bar{x}_i)$. Then

$$\dot{u}_i = r_i + \sum_{j=1}^{n} a_{ij}\bar{x}_j e^{u_j}$$

and $(a_{ij}\bar{x}_j e^{u_j})$ is the Jacobian matrix, which is stable by assumption for all $\mathbf{u} \in \mathbf{R}^n$.

Exercise 8: Show that a 3×3 matrix A is stable iff

$$D < 0,\ T < 0 \text{ and } D > MT \quad , \tag{21.29}$$

where $D = \det A$, $T = \operatorname{trace} A$, and M is the sum of the three principal 2×2-minors of A. This is the *Routh-Hurwitz* criterion. Show that for $D = MT$, A has a pair of imaginary eigenvalues.

Exercise 9: Show that a 3×3 matrix A is D-stable iff $a_{ii} \leq 0$, $\det A < 0$, the 2×2 minors $A_i = a_{jj}a_{kk} - a_{kj}a_{jk}$ are nonnegative and

$$\sqrt{-a_{11}A_1} + \sqrt{-a_{22}A_2} + \sqrt{-a_{33}A_3} > \sqrt{-\det A} \quad .$$

Exercise 10: A two-prey, one-predator system (or any other three species system with both 3-cycles nonnegative) whose interaction matrix is a P-matrix is totally stable.

Exercise 11: Construct matrices A which are totally stable but not VL-stable: Consider two-prey, one-predator systems with

$$-A_\epsilon = \begin{bmatrix} a & b & 1 \\ c & d & 1 \\ -1 & -1 & \epsilon \end{bmatrix} \qquad (a,b,c,d,\epsilon > 0) \quad .$$

Show that A_ϵ is VL-stable for every $\epsilon > 0$ iff $(b-c)^2 < 4ad$; $-A_\epsilon$ is a P-matrix for every $\epsilon > 0$ iff A_ϵ is totally stable for every $\epsilon > 0$ iff $ad > bc$ and $a + d > b + c$.

Exercise 12: Show that for $n = 3$, D-stability together with the existence of an interior fixed point implies permanence. (Hint: wait for **22.1** and **22.2**).

21.7 Notes

Many of the matrix stability concepts occurring in this chapter were originally motivated by mathematical economics, see Nikaido (1968), Arrow and Hahn (1971). Their importance for ecological systems was stressed by Svirezhev and Logofet (1983) and Šiljak (1978). An excellent book on matrix theory is Berman and Plemmons (1979). It contains a unique survey on M-matrices as well as a chapter on the "linear complementarity problem". Its importance for the study of Lotka-Volterra equations (it is equivalent to the problem of finding saturated fixed points) was realized by Takeuchi and Adachi (1980). Theorems 21.4.2 and the equivalence of (B4) and (B9) are two basic results of this theory, due to Gale, see Berman and Plemmons (1979) and Cottle (1980). Mutualistic systems were studied by Goh (1979) and Smith (1986a). The characterization of boundedness in terms of B-matrices goes back to Moltchanov (1961), see also Hofbauer (1987a). The explicit characterization of B-matrices (exercise 21.2.10) generalizes a criterion for copositive matrices of Hadeler (1983). The concept of VL-stability goes back to Volterra, see Scudo and Ziegler (1978), who called these matrices "dissipative". For recent results see Redheffer (1985). The global stability results for diagonally dominant and VL-stable matrices extend to more general models allowing time-dependent interaction terms, diffusion or time delays, see Gopalsamy (1984, 1985), Hastings (1978), Takeuchi (1986) and Wörz-Busekros (1978). Theorem 21.4.1 on P-matrices is due to Fiedler and Ptak (1962) and Gale and Nikaido (1965). **21.5** follows Takeuchi et al. (1978). Theorem 21.5.1 was also shown by Berman and Hershkowitz (1983). Competition between two symbiotic subcommunities (exercise 21.5.5) were studied by Smith (1986b) using monotone flows. The Ljapunov function for symmetric systems was found by MacArthur (1970). Exercise 21.5.11 arises in a theory of species packing and niche overlap, see MacArthur (1970) or May (1973). Exercise 21.5.12 is from Shigesada et al. (1984). D-stable matrices were studied by Johnson (1974), Cross (1978) and Clark and Hallam (1982). For the "Jacobian Conjecture" see Meisters (1982). Sign stable matrices were emphasized by May (1973), for a survey on qualitative (or sign) matrices see Quirk (1981).

22. Some Low Dimensional Ecological Systems

22.1 Heteroclinic cycles

We shall now investigate permanence and persistence for *three-species systems*, starting with three-dimensional Lotka-Volterra equations

$$\dot{x}_i = x_i(r_i + \sum_{j=1}^{3} a_{ij}x_j) \qquad i = 1,2,3 \quad . \tag{22.1}$$

In this context, the most remarkable phenomenon is the occurrence of *heteroclinic cycles*, i.e. cyclic arrangements of saddle equilibria and saddle connections (orbits having one saddle as α-limit and the next one as ω-limit). We have met with them already in **9.5**. The saddles were the one-species fixed points $\mathbf{F}_i$, which can occur (in a biologically reasonable way) only if $r_i > 0$ and $a_{ii} < 0$. Under this assumption, we may write (22.1) in the form

$$\dot{x}_i = r_i x_i(1 - \sum_{j=1}^{3} c_{ij}x_j) \qquad i = 1,2,3 \tag{22.2}$$

(with $c_{ii} > 0$). The transversal eigenvalue at $\mathbf{F}_i$ in direction $\mathbf{e}_j$ is given by

$$\left.\frac{\dot{x}_j}{x_j}\right|_{F_i} = r_j(1 - \frac{c_{ji}}{c_{ii}}) \quad . \tag{22.3}$$

In order to obtain a heteroclinic cycle $\mathbf{F}_1 \to \mathbf{F}_2 \to \mathbf{F}_3 \to \mathbf{F}_1$, each $\mathbf{F}_i$ must have one positive and one negative transversal eigenvalue, and they must be arranged in a cyclic pattern. This means that

$$\begin{aligned} c_{31} &> c_{11} > c_{21} \\ c_{12} &> c_{22} > c_{32} \\ c_{23} &> c_{33} > c_{13} \quad . \end{aligned} \tag{22.4}$$

To stress the analogy with the special case treated in **9.5**, we introduce the abbreviations

$$\alpha_i = \frac{c_{i-1,i}}{c_{ii}} \qquad \beta_i = \frac{c_{i+1,i}}{c_{ii}} \tag{22.5}$$

(where the indices are counted modulo 3). Then (22.4) reads as

$$\alpha_i > 1 > \beta_i \quad . \tag{22.6}$$

We note that the β_i may be negative.

From the discussion in **8.3** we know that on the face $x_3 = 0$, the isoclines do not intersect and hence that an orbit starting in $(x_1,x_2,0)$ (with $x_1,x_2 > 0$) converges to $\mathbf{F}_2$. In particular the unstable manifold of $\mathbf{F}_1$ converges to $\mathbf{F}_2$. Similarly, there is an orbit from $\mathbf{F}_2$ to $\mathbf{F}_3$ and one from $\mathbf{F}_3$ to $\mathbf{F}_1$. These together with the $\mathbf{F}_i$ form the heteroclinic cycle γ.

Theorem. (a) *If* $\det C > 0$ *and*

$$\prod_{i=1}^{3}(\alpha_i - 1) < \prod_{i=1}^{3}(1-\beta_i) \tag{22.7}$$

then (22.2) *is permanent.*
(b) *If*

$$\prod_{i=1}^{3}(\alpha_i - 1) > \prod_{i=1}^{3}(1-\beta_i) \tag{22.8}$$

then the heteroclinic cycle is an attractor.

Proof: (a) We shall prove first that $\det C > 0$ implies that all orbits are uniformly bounded. Indeed, the two-species subsystems have uniformly bounded orbits and the corresponding interaction matrices are B-matrices. Together with $\det(-A) > 0$, this implies by Theorem 21.2.3 that A is a B-matrix.

In order to prove permanence, it is enough to show that there exists a positive solution (p_1,p_2,p_3) of the following system of inequalities, which is just (19.45) at the boundary fixed points $\mathbf{0},\mathbf{F}_1,\mathbf{F}_2,\mathbf{F}_3$:

$$\begin{aligned} r_1p_1 + r_2p_2 + r_3p_3 &> 0 \\ r_2(1-\beta_1)p_2 - r_3(\alpha_1 - 1)p_3 &> 0 \\ -r_1(\alpha_2 - 1)p_1 + r_3(1-\beta_2)p_3 &> 0 \\ r_1(1-\beta_3)p_1 - r_2(\alpha_3 - 1)p_2 &> 0. \end{aligned} \tag{22.9}$$

The first inequality is satisfied for every positive $\mathbf{p}$. The other inequalities can be written as

$$\frac{p_{i+1}}{p_{i-1}} > \frac{r_{i-1}}{r_{i+1}}\,\frac{\alpha_i - 1}{1-\beta_i} \tag{22.10}$$

for $i = 1,2,3 \pmod 3$. By multiplying those inequalities, it is easy to see that (22.10) has a positive solution $\mathbf{p}$ iff (22.7) is satisfied. This proves (a).

(b) If (22.8) is valid, then one can find $p_i > 0$ such that the inequalities in (22.10) are reversed. The corresponding function $P(x) = \prod x_i^{p_i}$ is then exponentially decreasing near $\mathbf{F}_1$, $\mathbf{F}_2$ and $\mathbf{F}_3$. The heteroclinic cycle γ is an attractor in bd $\mathbf{R}_+^3$. Indeed,the function

$$\psi(\mathbf{x}) = \sum_{i=1}^{3} p_i r_i (1 - \sum_{j=1}^{3} c_{ij} x_j)$$

satisfies $\frac{\dot{P}}{P} = \psi$ in int $\mathbf{R}_+^3$ and

$$\int_0^T \psi(\mathbf{x}(t))\,dt < 0 \tag{22.11}$$

for every $\mathbf{x} \in \gamma$. Thus P is an average Ljapunov function (but a decreasing one) near γ. The same argument as in the proof of Theorem 13.2.1 shows that all orbits in $\mathbf{R}_+^3$ near γ converge to γ, i.e. that the heteroclinic cycle is an attractor.

Exercise 1: Show that (22.2) is always persistent under assumptions (22.4).

Exercise 2: Show that the following conditions are equivalent:

(a) (22.2) has uniformly bounded orbits;
(b) (22.2) has a unique interior fixed point;
(c) $\det A < 0$;
(d) $\det C > 0$.

(Hint: (d) $\Rightarrow$ (a) was shown in the previous proof; (a) $\Rightarrow$ (b) follows from the index theorem 19.4.1 and (b) $\Rightarrow$ (c) from Cramer's rule and (22.6)).

Exercise 3: Assume $\det C > 0$ and show that (22.7) is equivalent to the unique interior fixed point lying above the plane through $\mathbf{F}_1$,$\mathbf{F}_2$ and $\mathbf{F}_3$. Check that the geometric condition from Theorem 19.6.2 is satisfied in this case.

Exercise 4: For $r_i = 1$, (22.2) is equivalent (in which sense?) to the general rock-scissors-paper game (16.13).

Exercise 5: Determine the asymptotic behaviour of the time average of (22.2) under the assumption (22.8). (Hint: recall **9.5**).

Exercise 6: Suppose γ is a heteroclinic cycle for a differential equation in the plane, consisting of n saddle points $\mathbf{P}_i$ and the connecting orbits. Let $\lambda_i > 0$ and $-\mu_i < 0$ be the eigenvalues at $\mathbf{P}_i$ and $\rho = \prod_{i=1}^{n} \frac{\lambda_i}{\mu_i}$. Show that γ is attracting if $\rho < 1$ and repelling if $\rho > 1$. (Hint: choose coordinates x_i such that $x_i = 0$ along the orbit connecting $\mathbf{P}_{i-1}$ with $\mathbf{P}_i$ and $x_i > 0$ inside γ such that $\dot{x}_i \approx \lambda_i x_i$ and $\dot{x}_{i+1} \approx -\mu_i x_{i+1}$ holds near $\mathbf{P}_i$. Then use a suitable average Ljapunov function of the form $\prod_{i=1}^{n} x_i^{p_i}$).

Exercise 7: Apply the above result to the Battle of the Sexes (17.23) and the "rock-scissors-paper" game (exercise 16.5.2).

22.2 Permanence for three dimensional Lotka-Volterra systems

In **19.5** the following conditions were shown to be necessary both for permanence and robust persistence of the n-dimensional Lotka-Volterra equation (22.1):

(i) there exists an interior rest point $\hat{\mathbf{x}}$;

(ii) $\det(-A) > 0$;

(iii) if the truncated system without species k admits a unique interior rest point, then the corresponding principal minor of $-A$ is positive.

Indeed, (i) is a consequence of Theorem 19.5.1, (ii) is just (19.21) and (iii) is Theorem 19.5.4.

In dimension three, (iii) excludes those two-species subsystems where the interior fixed point is a saddle. By the index theorem such a system would have two saturated and hence stable boundary fixed points. Thus (iii) states that *the bistable case* (see Fig. 8.3) *cannot occur as a subsystem of a persistent three species system.*

We will now show that for a uniformly bounded three dimensional system, (i) to (iii) are sufficient conditions for persistence and - up to the occurrence of heteroclinic cycles - even for permanence. We shall strengthen (iii) to

(iii)' the two species subsystems are uniformly bounded and not bistable. More explicitly: the determinant of a two species subsystem (say $x_3 = 0$) must be positive ($a_{11}a_{22} > a_{12}a_{21}$) if it is symbiotic ($a_{12}$ and $a_{21} \geq 0$) or if it (is competitive and) has a fixed point $\bar{\mathbf{x}}$ with $\bar{x}_1 \geq 0$ and $\bar{x}_2 \geq 0$.

Note that (iii)' excludes also the degenerate cases of a line of fixed points or of the two isoclines intersecting at an unstable fixed point on a coordinate axis. Both cases cannot occur in robustly persistent systems.

(iv) If the system admits a heteroclinic cycle, then (22.7) holds.

With those conditions we obtain

Theorem. *Consider the three dimensional Lotka-Volterra equation with intraspecific competition* ($a_{ii} < 0$)

$$\dot{x}_i = x_i(r_i + \sum_{j=1}^{3} a_{ij}x_j) \qquad i = 1,2,3 \quad . \tag{22.12}$$

(a) *This system is persistent and uniformly bounded, and both properties are robust, iff* (i), (ii) *and* (iii)' *hold.*
(b) *This system is robustly permanent and uniformly bounded iff* (i), (ii), (iii)' *and* (iv) *hold.*

Proof: We are left to show that (i), (ii) and (iii)' guarantee permanence if there is no heteroclinic cycle.

First we note that Theorem 21.2.3 together with (ii) and (iii)' implies that A is a B-matrix. Hence the orbits of (22.12) are uniformly bounded for $t \to +\infty$.

Next we show that none of the boundary rest points of (22.12) is saturated. For the two species equilibria this follows from (iii)' and (19.36). So we are left with the origin $\mathbf{0}$ and the one species equilibria $\mathbf{F}_1$,$\mathbf{F}_2$ and $\mathbf{F}_3$ (if they exist). If any of these were regular and saturated, it would be a sink and hence have index -1. By the index theorem 19.4.1, however, only one saturated rest point can exist: this must be the interior fixed point. If one of the fixed points on the boundary had a zero eigenvalue, a suitable small perturbation of the r_i would make all its transversal eigenvalues negative. Since (i), (ii) and (iii)' are open conditions, the above argument applies to the perturbed system.

Now we apply exercise 19.6.2. We have to find p_1,p_2,$p_3 > 0$ such that

$$\sum_{i:x_i=0} p_i(r_i + \sum_{j=1}^{3} a_{ij}x_j) > 0 \tag{22.13}$$

at every boundary fixed point $\mathbf{x}$. Again we need not worry about the two species rest points, since their unique transversal eigenvalue is positive by the above considerations, so that (22.13) is trivially satisfied. At $\mathbf{0}$,$\mathbf{F}_1$,$\mathbf{F}_2$ and $\mathbf{F}_3$, (22.13) yields:

$$r_1p_1 + r_2p_2 + r_3p_3 > 0 \tag{22.14}$$

$$(r_2 - r_1\frac{a_{21}}{a_{11}})p_2 + (r_3 - r_1\frac{a_{31}}{a_{11}})p_3 > 0 \tag{22.15}$$

$$(r_1 - r_2\frac{a_{12}}{a_{22}})p_1 + (r_3 - r_2\frac{a_{32}}{a_{22}})p_3 > 0 \tag{22.16}$$

$$(r_1 - r_3\frac{a_{13}}{a_{33}})p_1 + (r_2 - r_3\frac{a_{23}}{a_{33}})p_2 > 0 \quad . \tag{22.17}$$

Since none of these fixed points is saturated, at least one of the coefficients of the p_i is positive in each inequality. Note that $\mathbf{F}_i$ exists iff $r_i > 0$, since $a_{ii} < 0$ by assumption. We distinguish three cases concerning the number of one species equilibria in the system. Since $\mathbf{0}$ is not saturated, at least one of the r_i is positive.

Case A: $r_1 > 0$ and $r_2, r_3 \le 0$. Then only $\mathbf{F}_1$ exist. We choose first p_2 and $p_3 > 0$ such that (22.15) holds. Then for large p_1, (22.14) holds too.

Case B: $r_1, r_2 > 0$ and $r_3 \le 0$. Then $\mathbf{F}_1$ and $\mathbf{F}_2$ exist. We note that either $r_2 - r_1\frac{a_{21}}{a_{11}} > 0$ or $r_1 - r_2\frac{a_{12}}{a_{22}} > 0$. Indeed we would otherwise have $a_{12}, a_{21} < 0$ and $\frac{a_{21}}{a_{11}} \ge \frac{r_2}{r_1} \ge \frac{a_{22}}{a_{12}}$ and so by (8.12) a bistable two dimensional competition system, in contradiction to (iii)'. If $r_1 - r_2\frac{a_{12}}{a_{22}} > 0$, say, then we choose $p_2, p_3 > 0$ to fulfill (22.15), and then some p_1 which is suitably large.

Case C: $r_1, r_2, r_3 > 0$. Then all $\mathbf{F}_i$ exist, but (22.14) is trivially satisfied. If there exists one species i which invades both other species, i.e. if the transversal eigenvalues at both $\mathbf{F}_j (j \neq i)$ in direction $\mathbf{F}_i$ is positive, then choosing p_i very large will give a solution of (22.15) to (22.17). If at one of the $\mathbf{F}_j$, both transversal eigenvalues are nonnegative, then it is again straightforward to find a solution $p_1, p_2, p_3 > 0$. The only remaining case is that of a heteroclinic cycle, which was treated in the previous section.

Exercise 1: Suppose that (22.12) has uniformly bounded orbits. Then it is robustly permanent iff the two sets C and D from Theorem 19.6.2 are disjoint.

Exercise 2: Try to give a purely algebraic proof (without referring to the index theorem) that (i), (ii) and (iii) exclude the possibility of saturated rest points on the boundary.

Exercise 3: Show that (22.12) with

$$A = \begin{bmatrix} -1 & 2 & 2 \\ 1 & -1 & 0 \\ 0 & -1 & -1 \end{bmatrix}$$

and $\hat{\mathbf{x}} = (1,1,1)$ satisfies (i), (ii), (iii) but not (iii)' since there is a stable fixed point at infinity in the subsystem $x_3 = 0$. Thus (i) , (ii) and (iii) do not imply permanence or persistence.

Exercise 4: Show that the competition model with

$$-A = \begin{bmatrix} 8 & 5 & 7 \\ 10 & 2 & 3 \\ 2 & 10 & 9 \end{bmatrix}$$

and $\hat{\mathbf{x}} = (1,1,1)$ is not permanent, although $\hat{\mathbf{x}}$ is asymptotically stable. Thus permanence and local stability share the conditions (ii) on the determinant and $T < 0$ on the trace, but the conditions (iii)' and (21.29) concerning the 2×2-minors are different.

Exercise 5: If the growth rates r_i are nonzero, (22.12) can be written in the form

$$\dot{x}_i = r_i x_i (1 - \sum_{j=1}^{3} c_{ij} x_j) \quad . \tag{22.18}$$

Show that the criterion on permanence and persistence given by Theorem 22.2 depends only on the competition matrix c_{ij}, but not on the growth rates r_i (as long as they keep the same signs), whereas local stability of the interior fixed point depends both on the r_i and the c_{ij}. Construct an example where a change in the r_i results in the loss of the stability of the interior rest point $\hat{\mathbf{x}}$.

Exercise 6: Simplify the permanence criterion from Theorem 22.2 as much as possible for (a) a food chain; (b) one predator and two preys; (c) two predators and one prey.

Exercise 7: Classify the possible phase portraits of permanent three species Lotka-Volterra systems according to their boundary behaviour (ignoring the behaviour in int $\mathbf{R}^3_+$).

Exercise 8: Show that the two-prey one-predator system with $r_1 = r_2 = -r_3 = 1$ and

$$A = \begin{bmatrix} -1 & -1 & -10 \\ -1.5 & -1 & -1 \\ 5 & 0.5 & 0 \end{bmatrix}$$

is permanent. Prove that the interior fixed point is not stable. Run the system on a computer. You will find "chaotic" motion.

Exercise 9: Show that every two-species competition system which is not bistable may turn into a permanent system by the introduction of a suitable (a) common predator, or (b) third competitor.

22.3 General three species systems

For *general three species ecological equations*

$$\dot{x}_i = x_i f_i(x_1,x_2,x_3) \qquad i = 1,2,3 \tag{22.19}$$

we cannot expect as precise criteria as in the Lotka-Volterra case. A biologically intuitive, coarse statement which can be expected to hold under rather general circumstances, however, is that the system is permanent iff there are no saturated equilibria on the boundary. It is clear that this condition is necessary for robust permanence. For the converse, we need a few biologically reasonable restrictions to prevent limit cycles or attracting heteroclinic cycles on the boundary.

(H1) The solutions of (22.19) in $\mathbf{R}^3_+$ are uniformly bounded for $t \to +\infty$.

(H2) The ω-limit of every orbit on $\mathrm{bd}\,\mathbf{R}^3_+$ consists of fixed points.

(H3) The origin is hyperbolic, i.e. $f_i(\mathbf{0}) \neq 0$, and there is at most one further equilibrium on each axis.

Thus there are two possibilities for each species. Take for example species 1:

(a) $f_1(x_1,0,0) < 0$ for all $x_1 > 0$. Then there is no equilibrium on the x_1-axis. Species 1 is *not self-supporting*.

(b) There is a $k_1 > 0$ such that $(x_1 - k_1)f(x_1,0,0) < 0$ for all $x_1 \geq 0$ with $x_1 \neq k_1$. Then species 1 is *self-supporting*, and k_1 is its *carrying capacity*. By $\mathbf{F}_1$ we denote the corresponding equilibrium $(k_1,0,0)$ and by λ_{ij} the transversal eigenvalues $f_j(\mathbf{F}_i)$ at $\mathbf{F}_i$ in direction $\mathbf{e}_j (j \neq i)$.

It follows that the three-species system is of one of the three following types.

Case A: Only one species is self-supporting: for example, $f_1(\mathbf{0}) > 0$, $f_2(\mathbf{0}) < 0$, $f_3(\mathbf{0}) < 0$.

Case B: Two species are self-supporting, e.g. $f_1(\mathbf{0}) > 0, f_2(\mathbf{0}) > 0$ and $f_3(\mathbf{0}) < 0$.

Case C: All three species are self-supporting: $f_i(\mathbf{0}) > 0$ for $i = 1,2,3$.

If none of the $\mathbf{F}_i$ is saturated, then one of the following possibilities holds:

(C1) There exists a species i which invades both other species: $\lambda_{ji} > 0$ and $\lambda_{ki} > 0$

(C2) There exists a species i which can be invaded by both other species: $\lambda_{ij} > 0$ and $\lambda_{ik} > 0$.

(C3) One has $\lambda_{12},\lambda_{23},\lambda_{31} > 0$ and $\lambda_{21},\lambda_{32},\lambda_{13} \leq 0$ (or the other way round).

This last case gives rise to a heteroclinic cycle on the boundary (if there are no two-species equilibria which break up the connection between saddle points). In this case we need an additional hypothesis:

(H4) If (C3) applies, then

$$\lambda_{12}\lambda_{23}\lambda_{31} + \lambda_{21}\lambda_{32}\lambda_{13} > 0 \quad . \tag{22.20}$$

Theorem. *Let* (22.19) *describe a three-species community satisfying* (H1)-(H4). *If none of the boundary fixed points is saturated, then the system is permanent.*

Proof: We repeat essentially the proof from the last section, except that our average Ljapunov function will be of the form

$$P(\mathbf{x}) = (x_1 + x_2)^{p_0} x_1^{p_1} x_2^{p_2} x_3^{p_3} \quad . \tag{22.21}$$

In int $\mathbf{R}^3_+$, one has

$$\frac{\dot{P}}{P} = \psi(\mathbf{x}) = p_0 \frac{x_1 f_1(\mathbf{x}) + x_2 f_2(\mathbf{x})}{x_1 + x_2} + \sum_{i=1}^{3} p_i f_i(\mathbf{x}) \quad . \tag{22.22}$$

We observe that the first term in $\psi(\mathbf{x})$ admits no continuous extension to the x_3-axis. We have therefore to take the lower semicontinuous extension $\psi(0,0,x_3) = p_0 \min(f_1(0,0,x_3),\ f_2(0,0,x_3))$.

By exercise 13.2, condition (13.3) on the extension of ψ still guarantees permanence. By (H2), this condition reduces to $\psi(\mathbf{x}) > 0$ for all rest points $\mathbf{x} \in \operatorname{bd} \mathbf{R}^3_+$, for a suitable choice of $p_0 \geq 0, p_1,p_2,p_3 > 0$.

The condition is trivially satisfied for all two-species rest points, since these points are not saturated. We are left with up to four conditions corresponding to the fixed points $\mathbf{0},\mathbf{F}_1,\mathbf{F}_2$ and $\mathbf{F}_3$:

$$p_0 \min(f_1(\mathbf{0}), f_2(\mathbf{0})) + \sum_{i=1}^{3} p_i f_i(\mathbf{0}) > 0 \tag{22.23}$$

$$\lambda_{12} p_2 + \lambda_{13} p_3 > 0 \tag{22.24}$$

$$\lambda_{21} p_1 + \lambda_{23} p_3 > 0 \tag{22.25}$$

$$\lambda_{31} p_1 + \lambda_{32} p_2 > 0 \quad . \tag{22.26}$$

In case **A**, where only $\mathbf{0}$ and $\mathbf{F}_1$ exist, we proceed as in **22.2**, setting $p_0 = 0$. In case B, since neither $\mathbf{F}_1$ nor $\mathbf{F}_2$ is saturated, one of the $\lambda_{12}, \lambda_{13}$ and one of the $\lambda_{21}, \lambda_{23}$ is positive. Hence there is a choice of $p_1, p_2, p_3 > 0$ such that (22.24) and (22.25) are satisfied. Since $f_1(\mathbf{0})$ and $f_2(\mathbf{0})$ are both positive, (22.23) holds whenever p_0 is sufficiently large. Case C runs as in the Lotka-Volterra case (cf. **22.1**).

Exercise 1: Apply the above results to
- a) a general one-predator two-prey system;
- b) a general two-predator one-prey system;
- c) a food chain;
- d) two mutualists invaded by an "aprovechado" (a species which takes advantage of an association of mutualists without providing any benefits in return).

Show that (C3) does not apply in any of these cases. As long as there are no periodic orbits in the coordinate planes, permanence is therefore essentially equivalent to the absence of saturated boundary rest points.

Exercise 2: Derive permanence criteria for a two-predator, one-prey model of the form

$$\dot{x} = rx\left(1 - x - \frac{a_1 y_1}{1+b_1 x} - \frac{a_2 y_2}{1+b_2 x}\right)$$

$$\dot{y}_1 = s_1 y_1\left(-1 + \frac{c_1 x}{1+b_1 x} - y_1\right)$$

$$\dot{y}_2 = s_2 y_2\left(-1 + \frac{c_2 x}{1+b_2 x} - y_2\right)$$

assuming that there are no periodic orbits on the boundary planes.

Exercise 3: Construct examples of a permanent one-predator two-prey system with nonlinear growth rates where the competition of the two preys is bistable. (Hint: observe the slight difference in argumentation in sections **22.2** and **22.3**).

Exercise 4: Let $\mathbf{x}(t)$ be a periodic orbit with period T in the (x_1,x_2)-

plane for (22.19). Show that this periodic orbit is attracting (resp. repelling) in the x_3-direction if the integral $\int_0^T f_3(\mathbf{x}(t))\,dt$ is negative (resp. positive). (Hint: Use x_3 as average Ljapunov function).

22.4 A two-prey two-predator system

The occurrence of heteroclinic cycles in three dimensional systems may be dismissed as an artefact. In higher dimensions, however, such cycles should be taken more seriously. They occur, for instance, in the following simple Lotka-Volterra model, which also illustrates how bistable systems of two competitors can be made permanent by the introduction of *two* predating species (from **22.2** we know that *one* such species could never do this job).

We shall denote the densities of the preys by x_1 and x_2, and those of the predators by y_1 and y_2. For the sake of simplicity we shall assume that the predators do not interfere with each other, that they specialize on different prey and that they show no crowding effects. This leads to

$$\begin{aligned}
\dot{x}_1 &= x_1(r_1 - a_{11}x_1 - a_{12}x_2 - b_1y_1)\\
\dot{x}_2 &= x_2(r_2 - a_{21}x_1 - a_{22}x_2 - b_2y_2)\\
\dot{y}_1 &= y_1(-c_1 + d_1x_1)\\
\dot{y}_2 &= y_2(-c_2 + d_2x_2) \quad .
\end{aligned} \tag{22.27}$$

We denote the fixed points of (22.27) with $\mathbf{F}_1$,$\mathbf{F}_2$,$\mathbf{F}_{12}$,$\mathbf{F}_1^1$, $\mathbf{F}_2^2$ etc. ..., the subscripts referring to the prey species and the superscripts to the predator species present.

Theorem. *Suppose that the interior fixed point $\mathbf{F}_{12}^{12} = (\bar{x}_1,\bar{x}_2,\bar{y}_1,\bar{y}_2)$ of (22.27) exists and that the (x_1,x_2) -subsystem is bistable. Then (22.27) is permanent if*

$$\frac{a_{21}}{r_2}\bar{x}_1 + \frac{a_{12}}{r_1}\bar{x}_2 < 1 \quad . \tag{22.28}$$

Proof: Let us start by assuming only that $\mathbf{F}_{12}^{12}$ exists. (Then $a_{ij}\bar{x}_j < r_i$ for $i,j = 1,2$, so that both terms on the left hand side of (22.28) are less than 1). Suppose that initially only prey 1 is present, i.e. that the state is at $\mathbf{F}_1$, and that a small amount of predator 1 is introduced. Its rate of growth is

$$\left.\frac{\dot{y}_1}{y_1}\right|_{\mathbf{F}_1} = -c_1 + d_1\frac{r_1}{a_{11}} = \frac{d_1}{a_{11}}(r_1 - a_{11}\bar{x}_1) = \frac{d_1}{a_{11}}(a_{12}\bar{x}_2 + b_1\bar{y}_1) > 0 \quad .$$

So predator 1 will invade at $\mathbf{F}_1$ and the system will converge to $\mathbf{F}_1^1$. At that point, prey 2 can invade, because $x_1(\mathbf{F}_1^1) = c_1/d_1 = \bar{x}_1$ and hence

$$\left.\frac{\dot{x}_2}{x_2}\right|_{\mathbf{F}_1^1} = r_2 - a_{21}\bar{x}_1 = a_{22}\bar{x}_2 + b_2\bar{y}_2 > 0 \quad .$$

Biologically, this means that prey 1 is weakened by its predator 1 to such an extent that it loses the competition against prey 2. Because of bistability, prey 2 will take over. Obviously, $\mathbf{F}_2$ is a stable fixed point in the three species system corresponding to the face $y_2 = 0$. In particular, if $\mathbf{F}_{12}^1$ does not exist, Theorem 9.2.1 implies that the unstable manifold of $\mathbf{F}_1^1$ is an orbit converging to $\mathbf{F}_2$. At $\mathbf{F}_2$, by the same argument, predator 2 will invade, so that the system switches to $\mathbf{F}_2^2$. Again, prey 1 will invade and, if $\mathbf{F}_{21}^2$ does not exist, eventually will take over. Thus if neither $\mathbf{F}_{21}^2$ nor $\mathbf{F}_{12}^1$ exist, then a heteroclinic cycle $\mathbf{F}_1 \to \mathbf{F}_1^1 \to \mathbf{F}_2 \to \mathbf{F}_2^2 \to \mathbf{F}_1$ arises. We shall presently see that (22.28) is the condition needed to make this cycle unstable, and the system permanent.

Indeed, the hyperplane H spanned by $\mathbf{F}_1,\mathbf{F}_1^1,\mathbf{F}_2,\mathbf{F}_2^2$ is given by

$$\frac{a_{11}}{r_1}x_1 + \frac{b_1}{r_1}y_1 + \frac{a_{22}}{r_2}x_2 + \frac{b_2}{r_2}y_2 = 1 \quad . \tag{22.29}$$

The remaining boundary fixed points $\mathbf{0},\mathbf{F}_{12}$ and (if they exist) $\mathbf{F}_{12}^1$ and $\mathbf{F}_{21}^2$ lie below the plane H. We shall now use Theorem 19.6.2. Let (x_1,x_2,y_1,y_2) be in the cone D given by

$$\begin{aligned} a_{11}x_1 + a_{12}x_2 + b_1y_1 &\geq r_1 \\ a_{21}x_1 + a_{22}x_2 + b_2y_2 &\geq r_2 \\ x_1 &\leq \bar{x}_1 \\ x_2 &\leq \bar{x}_2 \quad . \end{aligned} \tag{22.30}$$

These inequalities imply

$$\frac{a_{11}x_1 + b_1y_1}{r_1} + \frac{a_{22}x_2 + b_2y_2}{r_2} \geq 1 - \frac{a_{12}\bar{x}_2}{r_1} + 1 - \frac{a_{21}\bar{x}_1}{r_2} \quad .$$

The latter expression is larger than 1 iff (22.28) holds. (22.28) implies that the set D is disjoint from the convex hull C of the boundary rest points. By Theorem 19.6.2 thus, (22.27) is permanent.

Exercise 1: Show that if (22.27) admits an interior fixed point and the (x_1,x_2)-subsystem is not bistable, then (22.27) is permanent.

Exercise 2: If $a_{11}a_{22} > a_{12}a_{21}$ then $\mathbf{F}_{12}^{12}$ is globally stable (Hint: compare with Theorem 21.5.1.)

Exercise 3: Show that if $\mathbf{F}_{12}^{12}$ exists, but neither $\mathbf{F}_{12}^{1}$ nor $\mathbf{F}_{21}^{2}$, and if $\frac{a_{21}}{r_2}\bar{x}_1 + \frac{a_{12}}{r_1}\bar{x}_2 > 1$, then the heteroclinic cycle is an attractor. The system is persistent but not permanent.

Exercise 4: Find examples where

a) $\mathbf{F}_{12}^{12}$ is asymptotically stable, but the heteroclinic cycle is attracting.

b) $\mathbf{F}_{12}^{12}$ is unstable but (22.27) is permanent.

Exercise 5: Study the two-prey two-predator system

$$\dot{x}_1 = x_1(1 - \frac{2}{3}x_1 - \frac{1}{3}x_2 - by_1 - y_2)$$

$$\dot{x}_2 = x_2(1 - \frac{1}{3}x_1 - \frac{2}{3}x_2 - y_1 - by_2)$$

$$\dot{y}_1 = y_1(-1 + 2x_1 + x_2)$$

$$\dot{y}_2 = y_2(-1 + x_1 + 2x_2) \quad .$$

Show that for $b > 2$ there is a heteroclinic cycle $\mathbf{F}_1^1 \to \mathbf{F}_2^1 \to \mathbf{F}_2^2 \to \mathbf{F}_1^2 \to \mathbf{F}_1^1$. It is attracting for $b > b_0 \approx 6.4$, whereas for $b < b_0$ it is repelling and the system is permanent. (Use an average Ljapunov function as in **22.3**). However, the inequalities (19.44) have a positive solution only for $b < 5$. For $b \geq 5$, the interior fixed point lies in the convex hull C of the boundary fixed points. Thus the sufficient conditions for permanence given by Theorems 19.6.1 and 19.6.2 are in general not necessary for $n \geq 4$.

22.5 Notes

Theorems 22.1 and 22.2 are due to Hofbauer. Theorem 22.2 generalizes the characterization of permanence of two-prey one-predator systems given by Hutson and Vickers (1983). Other criteria for persistence of three dimensional Lotka-Volterra systems were derived by Hallam et al. (1979). The example for chaotic motion (exercise 22.2.8) is from

Gilpin (1979). Other examples were found by Arneodo et al. (1980) and Takeuchi and Adachi (1983, 1984). **22.3** follows Hutson and Law (1985). Similar results were obtained by Freedman and Waltman (1984, 1985) using a completely different approach. Two-prey two-predator systems have been studied by Takeuchi and Adachi (1984); the condition for permanence is due to Kirlinger (1986).

PART VI: BACK TO THE GENE POOL: GRADIENTS AND CYCLES

23. The Continuous Selection Model

23.1. The selection equation

As in chapter 3, we shall consider one gene locus with n alleles $A_1,\dots,A_n$, but this time assume that generations blend into each other. If the population is very large and births and deaths occur more or less continuously, we may treat the number $N_i(t)$ of alleles A_i at time t as a differentiable function of t. By $2N$ we shall denote the total number of genes and by $x_i = N_i/2N$ the relative frequency of A_i. The total number of individuals in the population is then given by N.

Now let us assume that the population is at any time in Hardy-Weinberg equilibrium. Then the frequency of the gene pair (A_i,A_j) is given by $x_i x_j = \dfrac{N_i N_j}{4N^2}$. Let d_{ij} be the death rate and b_{ij} the birth rate of (A_i,A_j)-individuals. The difference $m_{ij} = b_{ij} - d_{ij}$ is said to be the *Malthusian fitness parameter*. For the present, we shall assume that it is a constant.

The allele A_i occurs in the gene pairs (A_i,A_j) and (A_j,A_i) (with $j = 1,\dots,n$). Its increase in the gene pool is given by

$$\dot{N}_i = \sum_j m_{ij}\frac{N_i N_j}{4N^2}N + \sum_j m_{ji}\frac{N_j N_i}{4N^2}N$$

i.e. since $m_{ij} = m_{ji}$ by

$$\dot{N}_i = \frac{N_i}{2N}\sum_j m_{ij}N_j \quad .$$

Hence the growth rate of the total population is given by

$$\dot{N} = \frac{1}{2}\sum_i \dot{N}_i = \frac{1}{4N}\sum_{i,j} m_{ij}N_i N_j = N\sum_{i,j} m_{ij}x_i x_j \quad . \tag{23.1}$$

For the relative frequencies we obtain

$$\dot{x}_i = \left(\frac{N_i}{2N}\right)^{\cdot} = \frac{\dot{N}_i N - \dot{N}N_i}{2N^2} = x_i\left(\sum_j m_{ij}x_j - \sum_{r,s} m_{rs}x_r x_s\right) \quad .$$

In matrix notation this reads

$$\dot{x}_i = x_i[(M\mathbf{x})_i - \mathbf{x} \cdot M\mathbf{x}] \quad . \tag{23.2}$$

(23.2) is the ***continuous time selection equation of population genetics***, defined on the probability simplex S_n. Obviously this is again a replicator equation, but now with the additional condition $m_{ij} = m_{ji}$.

We shall stress right away that while (23.2) is a classic, its derivation is rather shaky. First of all, (23.1) would imply exponential growth for the whole population. But this is a minor point, which is repaired in **23.6**. More seriously, the assumption of Hardy-Weinberg equilibrium can be justifiably challenged. Moreover, the notion of a birth rate for A_iA_j-genotypes depends crucially on this assumption. We shall discuss this in detail in Chapter 26. Another justification of (23.2) is given in the exercise below.

Exercise 1: Show that (23.2) arises as a limiting case of the discrete selection model (3.1). (Set $w_{ij} = 1 + hm_{ij}$, interpret h as generation length, and let $h \to 0$).

Exercise 2: Discuss the case of 2 alleles and compare the results with those of **3.5.**

Exercise 3: Derive a similar selection model for asexually reproducing populations. Where have these equations appeared earlier in this book? Identify them as a special case of (23.2) by choosing special fitness matrices M.

23.2. The Fundamental Theorem

The mean fitness of the population at time t is

$$\bar{m}(t) = \mathbf{x}(t) \cdot M\mathbf{x}(t) = \sum_{i,j} m_{ij} x_i(t) x_j(t) \quad . \tag{23.3}$$

We have seen in **3.3** that the mean fitness is never decreasing for the difference equation (3.1). A similar result holds for the differential equation (23.2).

Theorem. *The average fitness $\bar{m}$ is a strict Ljapunov function for* (23.2).

Proof: Clearly

$$\bar{m}(t)^{\cdot} = (\mathbf{x} \cdot M\mathbf{x})^{\cdot} = \dot{\mathbf{x}} \cdot M\mathbf{x} + \mathbf{x} \cdot M\dot{\mathbf{x}} \quad .$$

Since M is symmetric, both terms are equal. Hence

$$\frac{1}{2}\dot{\bar{m}}(t) = \sum \dot{x}_i (M\mathbf{x})_i = \sum x_i [(M\mathbf{x})_i - \mathbf{x}\cdot M\mathbf{x}](M\mathbf{x})_i$$

$$= \sum x_i (M\mathbf{x})_i^2 - (\sum x_i (M\mathbf{x})_i)^2 \tag{23.4}$$

$$= \sum x_i [(M\mathbf{x})_i - \mathbf{x}\cdot M\mathbf{x}]^2 \geq 0 \quad . \tag{23.5}$$

Equality holds iff all terms $(M\mathbf{x})_i$ for which $x_i > 0$ take the same value. This condition, as we have seen in **16.3**, characterizes the rest points of (23.2).

Let us note that the terms in (23.4) and (23.5) represent the variance of the fitness $(M\mathbf{x})_i$ of allele A_i. Thus we have obtained the continuous version of the Fundamental Theorem of natural selection, with an additional information: *the increase in mean fitness of the population is proportional to the variance of the fitness.* As long as the alleles A_i which are present in the population do not all have the same fitness values, natural selection will change the state of the gene pool. The theorem of Ljapunov implies that all orbits converge to the set of fixed points.

Exercise: Prove an analogue of the Fundamental Theorem for the asexual model derived in Exercise 23.1.3.

23.3. Evolutionary stability and the selection equation

Among the replicator equations, the selection equation (23.2) is characterized by the symmetry condition $m_{ij} = m_{ji}$. In the context of game dynamics, symmetry of the payoff matrix means that both players share their payoff equally. Such games are called *partnership games.* There is no clash of interest, no discrepancy between individual benefit and the welfare of the whole population, and no decline in average payoff.

How can selection be interpreted as an evolutionary game? The alleles $A_1, \cdots, A_n$ have to be viewed as strategies. They interact by forming an individual of genotype A_iA_j. The fitness of A_iA_j is then the common payoff for A_i and A_j. It measures the increase of both alleles due to the reproduction of the A_iA_j-individual. Viewed in this way, it is the genes who play the game: the living beings are their playing boards.

As shown in the Fundamental Theorem, the asymptotically stable rest points are just the strict maxima of the mean fitness $\bar{m} = \mathbf{x} \cdot M\mathbf{x}$. The connection with game dynamics is given by the

Theorem. *The asymptotically stable rest points of* (23.2) *are precisely the evolutionarily stable states.*

Proof: Let $\mathbf{p}$ be asymptotically stable. Then by Theorem 23.2

$$\mathbf{x} \cdot M\mathbf{x} < \mathbf{p} \cdot M\mathbf{p} \tag{23.6}$$

for all $\mathbf{x} \neq \mathbf{p}$ in some suitable neighbourhood of $\mathbf{p}$. Replacing $\mathbf{x}$ by $2\mathbf{x} - \mathbf{p}$ (which is also near $\mathbf{p}$) we get

$$\mathbf{x} \cdot M\mathbf{x} < \mathbf{x} \cdot M\mathbf{p} \tag{23.7}$$

or, since M is symmetric,

$$\mathbf{p} \cdot M\mathbf{x} > \mathbf{x} \cdot M\mathbf{x} \tag{23.8}$$

which is just condition (15.15) for an ESS. The converse follows from Theorem 16.4.

As a corollary, one obtains that *an asymptotically stable rest point of* (23.2) *in* int S_n *is globally stable.*

Exercise 1: Show that a rest point $\mathbf{p} \in \operatorname{int} S_n$ is asymptotically stable iff

$$\xi \cdot M\xi < 0 \tag{23.9}$$

for all $\xi \neq \mathbf{0}$ with $\sum \xi_i = 0$. Show that (23.9) is equivalent to the concavity of $\bar{m}$ on S_n.

Exercise 2: Show that (23.9) in turn implies the existence of a unique globally stable rest point (which may lie on the boundary). (This is a special case of exercise 16.4.3, but there exists a simpler direct proof here).

Exercise 3: Suppose $\mathbf{p}$ is an interior fixed point of (23.2) or (3.1) with $m_{ij} \geq 0$. Show that each of the following conditions is equivalent to the asymptotic stability of $\mathbf{p}$.

(a) M has one positive and $n-1$ negative eigenvalues.

(b) The leading principal minors of M alternate in sign.

Exercise 4: Show that for $n = 3$ and $m_{ij} \geq 0$, the existence of an asymptotically stable interior rest point implies that the heterozygote fitnesses m_{12}, m_{23}, m_{31} satisfy the triangle inequalities $m_{12} < m_{13} + m_{23}$ etc... Show that the converse holds if $m_{11} = m_{22} = m_{33} = 0$.

23.4. Convergence to equilibrium

The Fundamental Theorem implies that the orbits of (23.2) converge to the set of rest points. But does each orbit converge to one rest point? Under the generic assumption that the rest points are isolated, this is a simple consequence of the fact that each ω-limit is connected. In degenerate cases, it may happen that linear manifolds of rest points occur. It is conceivable that an orbit approaches such a manifold without zooming in on one of its points. That this cannot happen for (23.2) is shown in the following

Theorem. *Each orbit of the selection equation* (23.2) *converges to some equilibrium.*

The main idea in the proof is to exploit the coincidence of asymptotically stable rest points and ESS. This suggests the use of our standard Ljapunov function P. Let $\mathbf{p}$ be an arbitrary point in $\omega(\mathbf{x})$ (with $\mathbf{p} \neq \mathbf{x}$). Suppose first that $\mathbf{p}$ lies in $\operatorname{int} S_n$. Then with $P = \prod x_i^{p_i}$ we have for all t

$$\frac{\dot{P}}{P} = \mathbf{p} \cdot M\mathbf{x} - \mathbf{x} \cdot M\mathbf{x} = \mathbf{x} \cdot M\mathbf{p} - \mathbf{x} \cdot M\mathbf{x} > 0 \tag{23.10}$$

from the Fundamental Theorem. Since $\mathbf{p} \in \omega(\mathbf{x})$ implies that $\mathbf{x}(t)$ comes arbitrarily close to $\mathbf{p}$, which is the unique maximum of P, the orbit $\mathbf{x}(t)$ must converge to $\mathbf{p}$.

The general case is slightly more technical. For convenience we replace P by its log, or more precisely by

$$L(\mathbf{x}) = - \sum_{i:p_i>0} p_i \log \frac{x_i}{p_i} \quad . \tag{23.11}$$

Obviously L is a continuous extended real valued function on S_n which is differentiable on the set

$$\{\mathbf{x}: L(\mathbf{x}) < \infty\} = \{\mathbf{x}: \operatorname{supp}(\mathbf{x}) \supseteq \operatorname{supp}(\mathbf{p})\} \quad .$$

As we have seen in the proof of Theorem 16.4, it attains its minimal value 0 only at $\mathbf{p}$. Now let $J = \{i: (M\mathbf{p})_i \neq \mathbf{p} \cdot M\mathbf{p}\}$. Since $\mathbf{p}$ is a rest point, $p_i = 0$ for $i \in J$. If $J = \emptyset$, the same proof as for $\mathbf{p} \in \operatorname{int} S_n$ is valid. Thus we assume $J \neq \emptyset$. We define

$$S(\mathbf{x}) = \sum_{i \in J} x_i \quad \text{and} \quad Z(\mathbf{x}) = \min_{i \in J} [(M\mathbf{x})_i - \mathbf{x} \cdot M\mathbf{x}]^2 \quad . \tag{23.12}$$

Then (23.5) implies

$$\frac{1}{2}(\bar{m})^{\cdot} = \sum x_i [(M\mathbf{x})_i - \mathbf{x} \cdot M\mathbf{x}]^2 \geq S(\mathbf{x}) Z(\mathbf{x}) \quad . \tag{23.13}$$

By the definition of J, $Z(\mathbf{p}) =: z > 0$ and so $\{\mathbf{x} : Z(\mathbf{x}) > \frac{z}{2}\}$ is an open neighbourhood of $\mathbf{p}$. Now

$$\{\mathbf{p}\} = \bigcap_{\epsilon > 0} \{\mathbf{x} : L(\mathbf{x}) \leq \epsilon\}$$

is a decreasing intersection of compact sets. Thus there exists an $\bar{\epsilon} > 0$ such that

$$L(\mathbf{x}) \leq \bar{\epsilon} \;\Rightarrow Z(\mathbf{x}) > \frac{z}{2} \quad . \tag{23.14}$$

From $\mathbf{p} \in \omega(\mathbf{x})$ follows supp $(\mathbf{x}(t)) \supseteq$ supp $(\mathbf{p})$, and so we can differentiate L to obtain

$$L^{\cdot}(\mathbf{x}) = -\sum p_i \frac{\dot{x}_i}{x_i} = -\sum p_i [(M\mathbf{x})_i - \mathbf{x} \cdot M\mathbf{x}] = \mathbf{x} \cdot M\mathbf{x} - \mathbf{p} \cdot M\mathbf{x}$$

$$= \mathbf{x} \cdot M\mathbf{x} - \mathbf{p} \cdot M\mathbf{p} + \mathbf{p} \cdot M\mathbf{p} - \mathbf{x} \cdot M\mathbf{p}$$

$$= \mathbf{x} \cdot M\mathbf{x} - \mathbf{p} \cdot M\mathbf{p} + \sum_i [\mathbf{p} \cdot M\mathbf{p} - (M\mathbf{p})_i] x_i \quad .$$

Because $\mathbf{x} \cdot M\mathbf{x} < \mathbf{p} \cdot M\mathbf{p}$ on the orbit, we have

$$L^{\cdot}(\mathbf{x}) < K S(\mathbf{x}) \tag{23.15}$$

where

$$K = \max_{i \in J} |\mathbf{p} \cdot M\mathbf{p} - (M\mathbf{p})_i| \quad .$$

We choose a sequence $t_n \to +\infty$ such that $\mathbf{x}(t_n) \to \mathbf{p}$. Then $L(\mathbf{x}(t_n)) \to L(\mathbf{p}) = 0$. For $\epsilon < \bar{\epsilon}$, we choose $N = N(\epsilon)$ such that

$$L(\mathbf{x}(t_N)) < \frac{\epsilon}{2} \quad \text{and} \quad \frac{K}{z}[\bar{m}(\mathbf{p}) - \bar{m}(\mathbf{x}(t_N))] < \frac{\epsilon}{2} \quad . \tag{23.16}$$

As long as $t > t_N$ and $L(\mathbf{x}(s)) < \epsilon$ for $s \in [t_N, t]$, the estimates (23.13) to

(23.15) at $\mathbf{x}(s)$ imply

$$L^{\cdot}(\mathbf{x}(s)) \leq \frac{K}{2}\frac{[\bar{m}(\mathbf{x}(s))]^{\cdot}}{Z(\mathbf{x}(s))} < \frac{K}{z}[\bar{m}(\mathbf{x}(s))]^{\cdot} \ .$$

Integrating and applying (23.16) we have

$$0 \leq L(\mathbf{x}(t)) \leq L(\mathbf{x}(t_N)) + \frac{K}{z}[\bar{m}(\mathbf{x}(t)) - \bar{m}(\mathbf{x}(t_N))] < \epsilon \ . \quad (23.17)$$

Because $\epsilon < \bar{\epsilon}$, $\mathbf{x}(t)$ remains in the neighbourhood described by (23.14). Hence (23.17) holds for any $t > t_N$.

Since ϵ may be chosen arbitrarily small, $L(\mathbf{x}(t))$ approaches 0 as $t \to +\infty$, i.e. $\mathbf{x}(t)$ converges to $\mathbf{p}$.

Exercise: Give an alternative proof by showing that instead of (23.11) the function

$$L(\mathbf{x}) + 2\sum_{i\in J_+} x_i \quad \text{where} \quad J_+ = \{i\colon \mathbf{p}\cdot M\mathbf{p} > (M\mathbf{p})_i\}$$

increases along the orbit of $\mathbf{x}$ in a neighbourhood of $\mathbf{p}$.

23.5 The location of stable equilibria

The dynamics of the selection equation is essentially determined by the stable equilibria. How many can there be? We know that if one such equilibrium lies in the interior of S_n, then it is the only one. But several such equilibria may coexist if they all lie on the boundary, each one having some open subset of S_n as its basin of attraction. The simplest example is obtained if M is the identity matrix.

In order to get an idea on the variety of different "patterns of stable equilibria", we study a special class of fitness matrices which can be described by graphs: Let G be a (nonoriented) graph on the set of vertices $\{1,\dots,n\}$. Then we associate to G a symmetric matrix M, the *incidence matrix* of the graph:

$$m_{ij} = \begin{cases} 1 & \text{if } i \text{ and } j \text{ are joined } (i \neq j) \\ 0 & \text{if } i \text{ and } j \text{ are not joined} \\ \frac{1}{2} & \text{if } i = j \end{cases}$$

A *clique* of G is a maximal subset $I \subseteq \{1,\ldots,n\}$ such that any two elements i and j of I are joined by an edge.

We show now that the cliques of G correspond to the asymptotically stable equilibria of the selection equation (23.2) with the above fitness matrix M. Without restricting generality, we may assume $I = \{1,\ldots,k\}$. Let $\mathbf{p} = (\frac{1}{k},\ldots,\frac{1}{k},0,\ldots,0)$ be the barycenter of the face $S_n(I)$. The average fitness restricted to this face is

$$\mathbf{x}\cdot M\mathbf{x} = \sum_{i\neq j} x_i x_j + \frac{1}{2}\Sigma x_i^2 = (\sum_{i\in I} x_i)^2 - \frac{1}{2}\sum_{i\in I} x_i^2 = 1 - \frac{1}{2}\sum_{i\in I} x_i^2 \quad .$$

It takes its maximal value $1 - \frac{1}{2k}$ at $\mathbf{p}$. This shows that $\mathbf{p}$ is asymptotically stable within the restriction of (23.2) to $S_n(I)$. If $j \notin I$, there is some $i \in I$ with $m_{ij} = 0$. Hence

$$(M\mathbf{p})_j \leq \frac{1}{k}(k-1) = 1 - \frac{1}{k} < 1 - \frac{1}{2k} = \mathbf{p}\cdot M\mathbf{p} \quad .$$

Hence $\mathbf{p}$ is saturated and therefore asymptotically stable in S_n (cf. **23.3**).

Exercise 1: Show that all stable equilibria arise in this way.

Exercise 2: Draw all possible graphs for $n = 3$ and determine the corresponding pattern of stable equilibria.

Exercise 3: For any partition of $n = k_1 + k_2 + \cdots + k_s$ $(s, k_i = 1,2,\ldots)$, construct an $n \times n$ matrix M where the number of asymptotically stable equilibria of (23.2) is the product $k_1 \cdots k_s$.

Exercise 4: Construct an $n \times n$ matrix M where the number of coexisting stable equilibria is $g(n) = r3^k$ with $n = 3k + r$ and $r = 2,3$ or 4. (It has been shown that $g(n)$ is the maximal number of cliques in a graph with n vertices).

Theorem. *Suppose that for an incidence matrix M there exists an asymptotically stable equilibrium $\mathbf{p}$ of (3.1) or (23.2) involving $n-i$ alleles. Then 2^i is an upper bound for the total number of asymptotically stable equilibria.*

Proof: Let I (with cardinality i) be the complement of supp $(\mathbf{p})$. The map $C \to C \cap I$ is a one-to-one map of cliques to subsets of I: indeed, if $C_1 \cap I = C_2 \cap I$ for two cliques C_1 and C_2, then $C_1 \cup C_2$ is also complete and hence $C_1 = C_2$. Therefore the number of cliques is bounded by the number 2^i of subsets of I. The theorem then is implied by the correspondence of asymptotically stable equilibria and cliques.

It is natural to ask whether general fitness matrices allow for essentially other patterns of stable equilibria which cannot be described by cliques. An example for $n = 4$ is the following

$$M = \begin{pmatrix} 1 & 14 & 18 & 6 \\ 14 & 10 & 9 & 10.2 \\ 18 & 9 & 6 & 11 \\ 6 & 10.2 & 11 & 10 \end{pmatrix} .$$

With this matrix, (23.2) has 3 stable equilibria $(\frac{1}{3}, \frac{1}{3}, \frac{1}{3}, 0)$, $(0, 0, \frac{1}{6}, \frac{5}{6})$ and $(0, \frac{1}{2}, 0, \frac{1}{2})$. But the graph induced by these equilibria (by joining any two vertices in the same support) has only two cliques: 123 and 234.

Exercise 5: Check this.

Similarly the number of coexisting stable equilibria in general fitness matrices may exceed the maximal number of cliques given in exercise 4.

Exercise 6: Show that the circulant fitness matrix generated by the row $(0, 8, 13, 2, 2, 13, 8)$ has 14 stable equilibria: the cyclic permutations of $(8, 3, 8, 0, 0, 0, 0)$ and $(13, 0, 24, 0, 13, 0, 0,)$. But $g(7) = 12$.

23.6 Density dependent fitness

The classical selection equation (23.2) describes the changes in relative frequencies in a population. But our derivation started with absolute numbers. What happened, then, to the size of the whole population? The corresponding differential equation (23.1) for the total number N of individuals, $\dot{N} = N\bar{m}$, says that once an equilibrium $\mathbf{p}$ of the relative frequencies is established, the population will grow (or decay) exponentially, with the mean fitness $\bar{m} = \mathbf{p} \cdot M\mathbf{p}$ as growth rate. This is a rather unrealistic feature of the model, and the reason that (23.1) is usually ignored in population genetics.

The simplest way to repair this defect of the genetic model is to make the fitness values m_{ij} dependent on the population number N. In order to obtain saturation, we assume that due to overpopulation the fitness values $m_{ij}(N)$ become negative for large N, or more precisely that there exist constants K_{ij} (the *carrying capacities* of the genotypes A_iA_j), such that

$$m_{ij}(N) > 0 \text{ for } N < K_{ij} \text{ and } m_{ij}(N) < 0 \text{ for } N > K_{ij} \quad (23.18)$$

(The simplest choice would be

$$m_{ij}(N) = r_{ij}\left(1 - \frac{N}{K_{ij}}\right) \tag{23.19}$$

with some positive constants r_{ij}). Equations (23.1-2) read then

$$\begin{aligned} \dot{N} &= N\bar{m}(N,\mathbf{x}) \\ \dot{x}_i &= x_i[\sum_j m_{ij}(N)x_j - \bar{m}(N,\mathbf{x})] \end{aligned} \tag{23.20}$$

with

$$\bar{m}(N,\mathbf{x}) = \sum_{r,s} m_{rs}(N)x_r x_s \quad .$$

The state space is $\mathbf{R}_+ \times S_n$, a prism with base S_n. Since (23.18) implies that

$$\bar{m}(N,\mathbf{x}) < 0 \text{ for } N > \max_{i,j} K_{ij} = K_0 \quad ,$$

all orbits are bounded and end up in the region $N \leq K_0$. Let us consider the regions I (resp. II) where $\bar{m}(N,\mathbf{x})$ is positive (resp. negative) and the separating manifold $H = \{(N,\mathbf{x}) : \bar{m}(N,\mathbf{x}) = 0\}$ (see Fig. 23.1). Clearly N increases in I and decreases in II. This suggests that all orbits will end up on H. In order to exclude the possibility of oscillations between I and II we consider the mean fitness

$$\begin{aligned} (\bar{m})^{\cdot} &= \sum_{i=1}^{n} \frac{\partial \bar{m}}{\partial x_i}\dot{x}_i + \frac{\partial \bar{m}}{\partial N}\dot{N} \\ &= 2\sum_i x_i[(M\mathbf{x})_i - \mathbf{x}\cdot M\mathbf{x}]^2 + \sum_{i,j} x_i x_j (\frac{d}{dN}m_{ij}(N))N\bar{m}(N,\mathbf{x}) \quad . \end{aligned}$$

On the set H, the second term vanishes and we are left with the result of the Fundamental Theorem: $(\bar{m})^{\cdot} \geq 0$. Hence H is semipermeable. The region I is forward invariant. At all points of H, the orbits enter horizontally from II into I, with the exception of the points where $(\bar{m})^{\cdot} = 0$, which are the rest points of (23.20).

Thus there are two possibilities for a nonstationary solution: (a) either it remains all the time in the upper region II, where $\dot{N} < 0$ along the whole orbit, and converges to an invariant subset of H, which is just the set of rest points; (b) or - and this is the generic case - it reaches region I and remains there, converging again to the set of fixed points.

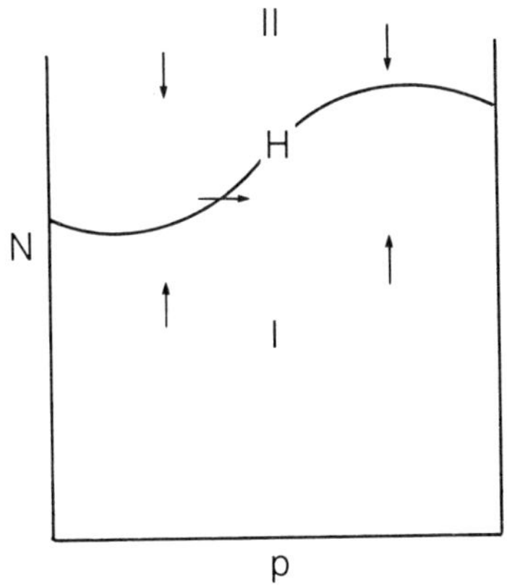

Figure 23.1

If all m_{ij} are decreasing functions of N, then $\bar{m}(N)$ is decreasing too and H is the graph of a function from S_n to $\mathbf{R}_+$ assigning to each frequency distribution $\mathbf{x}$ a carrying capacity $N(\mathbf{x})$. The rest points of (23.20) are exactly the critical points of this function and the stable rest points are its local maxima.

Exercise 1: Prove this last statement.

Exercise 2: Give a complete discussion of the case $n = 2$, when m_{ij} is given by (23.19). Show that a stable polymorphism exists iff the carrying capacity K_{12} of the heterozygote is larger than K_{11} and K_{22}.

Exercise 3: Show that when $m_{ij}(N) = r_{ij} - c_{ij}N$, with $c_{ij} = 0$ for $i \neq j$ and $c_{ii} > 0$ (heterozygotes have infinite carrying capacity), then there exists a unique polymorphism. Prove its global stability. (Hint: use the theorem of Perron-Frobenius).

23.7 Notes

The continuous time selection equation (23.2) goes back to the pioneers of population genetics, R.A. Fisher, J.B.S. Haldane and S. Wright. They used it at least implicitly as an approximation for the discrete time model (3.1) when selective differences are small, to make the mathematics simpler. Our derivation follows Crow and Kimura

(1970) and Hadeler (1974). Some population geneticists refuse to use it since its derivation and in particular the Hardy-Weinberg assumption is dubious and not correct in general (see **26.5**). The stability criteria in exercises 23.3.3-4 are due to Kingman (1961) and Mandel (1959). The convergence result of **23.4** is from Akin and Hofbauer (1982). For the discrete case, we refer to Lyubich et al. (1980) and Losert and Akin (1983). The connection between cliques of graphs and stable equilibria was discovered several times. A similar result appears e.g. as an exercise in Comtet (1974, p. 300). The formula for the maximal number of cliques in a graph (exercise 23.5.4) was conjectured by Erdös and proved by Moon and Moser (1965). Theorem 23.5 was announced by Karlin (1980) for arbitrary matrices (where it is false). The counterexamples in exercises 23.5.5-6 are due to Vickers and Cannings (1987). Selection with density dependent fitness was studied by MacArthur (1962), Roughgarden (1971), Anderson (1971) and Ginzburg (1977, 1983).

24. Gradient Systems

24.1. Kimura's Maximum Principle and gradient vector fields

The Fundamental Theorem (see **23.2**) states that the average fitness increases as long as the population is not in equilibrium. A refinement of this is Kimura's *Maximum Principle: The change in gene frequencies occurs in such a way that the increase in average fitness is maximal.*

Let us explain this in a general context. Assume that V is a real-valued continuously differentiable function on an open subset U of $\mathbf{R}^n$. In order to determine the direction of the steepest ascent of V at a point $\mathbf{x} \in U$, consider a vector $\mathbf{y}$ of length 1. The increase of V in the direction $\mathbf{y}$ is given by the *directional derivative*:

$$\lim_{h\to 0} \frac{V(\mathbf{x}+h\mathbf{y}) - V(\mathbf{x})}{h} = \sum_{i=1}^{n} \frac{\partial V}{\partial x_i} y_i = \text{grad } V(\mathbf{x}) \cdot \mathbf{y} \quad . \tag{24.1}$$

By the Cauchy-Schwarz inequality (see exercise 3.4.1) this quantity is bounded by

$$\left(\sum_{i=1}^{n} \left(\frac{\partial V}{\partial x_i} \right)^2 \right)^{\frac{1}{2}} = \|\text{grad } V(\mathbf{x})\|$$

and this bound is attained precisely if $\mathbf{y}$ points into the same direction as grad $V(\mathbf{x})$. Thus the vector field

$$\mathbf{x} \to \text{grad } V(\mathbf{x}) = \left(\frac{\partial V}{\partial x_1}(\mathbf{x}), \ldots, \frac{\partial V}{\partial x_n}(\mathbf{x}) \right) \tag{24.2}$$

points into the direction of greatest increase of V. The orbits of the ODE

$$\dot{\mathbf{x}} = \text{grad } V(\mathbf{x}) \tag{24.3}$$

follow the lines of steepest ascent of the "landscape" given by V. (24.3) is called a *gradient vector field*, and V is its *potential*.

Let us now collect some properties of gradient vector fields. The potential V is obviously a strict Ljapunov function:

$$\dot{V} = \sum_{i=1}^{n} \frac{\partial V}{\partial x_i} \dot{x}_i = \sum_{i=1}^{n} \left(\frac{\partial V}{\partial x_i} \right)^2 \geq 0 \tag{24.4}$$

with equality only at the rest points. This observation settles the qualitative behaviour of gradient vector fields: all orbits converge to the set of rest points.

Moreover, the orbits cross the constant level sets orthogonally. (This is an easy consequence of (24.1), and intuitively obvious to every hiker). The rest points are the critical points of V. The stable rest points are the local maxima of V.

Orbits converging to a particular rest point do so with a distinguished direction. They do not spiral in towards the rest point. This geometric fact corresponds to the algebraic result that the eigenvalues of the linearization are all real: indeed, the linearization yields the matrix $\left(\frac{\partial^2 V}{\partial x_i \partial x_j} \right)$ which is symmetric.

Exercise: For which $n \times n$ matrices A is the linear equation $\dot{\mathbf{x}} = A\mathbf{x}$ a gradient vector field? Determine its potential.

Now let us return to the selection equation (23.2). If the average fitness $\mathbf{x} \cdot M\mathbf{x}$ were its potential, as the Maximum Principle claims, this selection equation should plainly be a linear differential equation. But it isn't! What is wrong? It turns out, not that Kimura made a mistake, but that our mathematical approach was too narrow-minded.

The notion of a gradient vector field depends, in fact, on the metric of the underlying space: on *distances* (the direction of steepest ascent depends on the shape of the unit sphere) and on *angles* (the vector field is orthogonal to the constant level sets). Rather than give up Kimura's principle, we shall give up Euclid's metric.

24.2 The Euclidean metric and others

Let $\mathbf{x}$ be a point in int S_n. Any path $\mathbf{x}(t)$ in S_n passing through $\mathbf{x}$ at time 0 has a velocity $\xi = \dot{\mathbf{x}}(0)$ which, since $\sum x_i(t) = 1$ for all t, satisfies $\sum \xi_i = 0$. Conversely, to any vector ξ in

$$\mathbf{R}_0^n = \{\xi = (\xi_1,...,\xi_n) \in \mathbf{R}^n : \sum_{i=1}^{n} \xi_i = 0\} \tag{24.5}$$

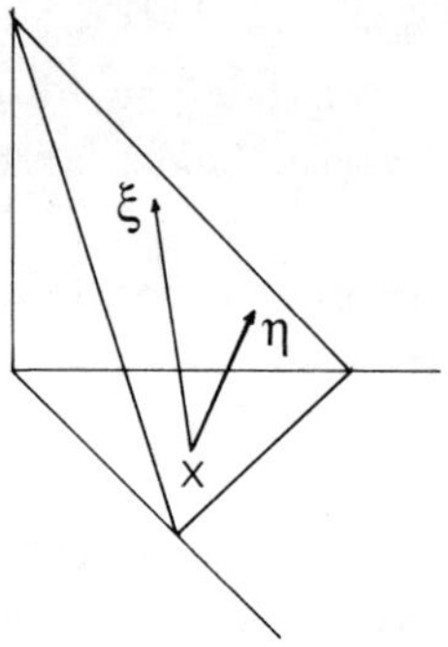

Figure 24.1

we can assign a path $t \to \mathbf{x}(t)$ in S_n with $\dot{\mathbf{x}}(0) = \xi$ (by simply setting $\mathbf{x}(t) = \mathbf{x} + t\,\xi$). Thus the *tangent space* of S_n at $\mathbf{x}$, which is denoted by $T_{\mathbf{x}}S_n$, is just $\mathbf{R}_0^n$. We picture the vector $\xi \in T_{\mathbf{x}}S_n$ as being "attached" to the point $\mathbf{x}$ (see Fig. 24.1).

The usual *Euclidean inner product* of two vectors ξ and η in $T_{\mathbf{x}}S_n$ is given by

$$\xi \cdot \eta = \sum_{i=1}^{n} \xi_i \eta_i \quad . \tag{24.6}$$

But there exist many other inner products in $T_{\mathbf{x}}S_n$. A particularly interesting one, the *Shahshahani inner product*, is defined by

$$<\xi,\eta>_{\mathbf{x}} = \sum_{i=1}^{n} \frac{1}{x_i} \xi_i \eta_i \quad . \tag{24.7}$$

(It is obviously linear in ξ and η and satisfies $<\xi,\xi>_{\mathbf{x}} \geq 0$, with equality iff $\xi = \mathbf{0}$). Just as the Euclidean product defines the length of a vector ξ (by $\|\xi\| = \sqrt{\xi \cdot \xi}$) and the angle Θ between two nonzero vectors ξ and η

(by $\cos\Theta = \xi\cdot\eta/\|\xi\|\|\eta\|$) so does the Shahshahani product. It differs from the Euclidean one by attaching more weight to changes near to the boundary, where x_i is small and the term $\xi_i\eta_i$ therefore more important.

Now let $L:\xi \to L(\xi)$ be a linear map from $T_{\mathbf{x}}S_n$ into $\mathbf{R}$. There exists a unique vector $l \in T_{\mathbf{x}}S_n$ such that $L(\xi) = l\cdot\xi$ for all $\xi \in T_{\mathbf{x}}S_n$. By the same token there exists a unique vector $l_{\mathbf{x}} \in T_{\mathbf{x}}S_n$ such that $L(\xi) = \langle l_{\mathbf{x}},\xi\rangle_{\mathbf{x}}$ for all $\xi \in T_{\mathbf{x}}S_n$. Let us apply this to the linear map $D_{\mathbf{x}}V$ (where V is, as before, a map from S_n to $\mathbf{R}$). Since $D_{\mathbf{x}}V(\xi) = \sum \frac{\partial V(\mathbf{x})}{\partial x_i}\xi_i$, we see that the vector l corresponding to $D_{\mathbf{x}}V(\xi)$ is just $\operatorname{grad} V(\mathbf{x})$: indeed,

$$D_{\mathbf{x}}V(\xi) = \operatorname{grad} V(\mathbf{x})\cdot\xi \quad . \tag{24.8}$$

The vector $\operatorname{grad} V(\mathbf{x})$ is said to be the *Euclidean gradient*, since it relates to the Euclidean inner product via (24.8). The vector $l_{\mathbf{x}}$ with the corresponding property for the Shahshahani product is said to be the *Shahshahani gradient* and denoted by $\operatorname{Grad} V(\mathbf{x})$:

$$D_{\mathbf{x}}V(\xi) = \langle \operatorname{Grad} V(\mathbf{x}),\xi\rangle_{\mathbf{x}} \quad . \tag{24.9}$$

Before showing in the next section that the Shahshahani gradient does the job for Kimura's Maximum Principle, let us note that there are many other inner products and hence gradients on $T_{\mathbf{x}}S_n$. More generally, for any manifold M, a Riemannian metric is an assignment (in a smooth way) of an inner product to each tangent space $T_{\mathbf{x}}M$. The corresponding gradient for a function $V:M\to\mathbf{R}$ is defined in the obvious way. It yields a vector field on M which depends both on the potential V and on the Riemannian metric.

Exercise 1: If U is an open subset of $\mathbf{R}^n$, then $T_{\mathbf{x}}U$ is just $\mathbf{R}^n$, for every $\mathbf{x}\in U$. A Riemannian metric is defined by a symmetric positive definite matrix $G(\mathbf{x})$ depending smoothly on $\mathbf{x}$:

$$(\xi,\eta)_{\mathbf{x}} = \xi\cdot G(\mathbf{x})\eta = \sum_{ij} g_{ij}(\mathbf{x})\xi_i\eta_j \quad .$$

The corresponding gradient of V at $\mathbf{x}$ is given by $G(\mathbf{x})^{-1}\operatorname{grad} V(\mathbf{x})$.

Exercise 2: If G and A are two symmetric $n\times n$-matrices, then GA is symmetric iff $GA = AG$. Prove that GA is symmetric for all positive definite symmetric G iff A is diagonal. If G is positive definite and A symmetric, then GA is similar to a symmetric matrix and thus has all eigenvalues real. (Hint: Every positive definite matrix has a positive definite square root).

Exercise 3: Show that the Jacobian matrix of a gradient vector field at any fixed point has only real eigenvalues.

Exercise 4: For which $n \times n$ matrices A is the linear differential equation $\dot{\mathbf{x}} = A\mathbf{x}$ a gradient vector field with respect to some Riemannian metric on $\mathbf{R}^n$?

24.3 The Shahshahani gradient

Consider now the selection equation (23.2)

$$\dot{x}_i = x_i[(M\mathbf{x})_i - \mathbf{x} \cdot M\mathbf{x}] \ . \tag{24.10}$$

Since $m_{ij} = m_{ji}$, the linear differential equation $\dot{\mathbf{x}} = M\mathbf{x}$ is an Euclidean gradient on $\mathbf{R}^n$ with $\frac{1}{2}\mathbf{x} \cdot M\mathbf{x}$ as its potential. The following theorem shows that (24.10) is therefore the Shahshahani gradient of $\frac{1}{2}\mathbf{x} \cdot M\mathbf{x}$ on S_n.

Theorem. *Let $\dot{x}_i = f_i(\mathbf{x}) = \frac{\partial V}{\partial x_i}$ be an Euclidean gradient vector field on $\mathbf{R}^n$. Then the corresponding replicator equation*

$$\dot{x}_i = \hat{f}_i(\mathbf{x}) = x_i(f_i(\mathbf{x}) - \bar{f}(\mathbf{x})) \tag{24.11}$$

is a Shahshahani gradient with the same potential function V.

Proof: For $\mathbf{x} \in \operatorname{int} S_n$ and $\xi \in \mathbf{R}^n_0$ we compute

$$\begin{aligned} <\hat{\mathbf{f}}(\mathbf{x}),\xi>_{\mathbf{x}} &= \sum_{i=1}^{n} \frac{1}{x_i}\hat{f}_i(\mathbf{x})\xi_i = \sum_{i=1}^{n} \frac{1}{x_i} x_i(f_i(\mathbf{x}) - \bar{f}(\mathbf{x}))\xi_i \\ &= \sum_{i=1}^{n} f_i(\mathbf{x})\xi_i - \bar{f}(\mathbf{x}) \sum_{i=1}^{n} \xi_i \\ &= \sum_{i=1}^{n} \frac{\partial V}{\partial x_i}\xi_i = D_{\mathbf{x}} V(\xi) \ . \end{aligned}$$

By (24.9), $\hat{\mathbf{f}}(\mathbf{x})$ is the Shahshahani gradient of V.

In particular, the proof shows that

$$\dot{V}(\mathbf{x}) = D_{\mathbf{x}} V(\dot{\mathbf{x}}) = <\hat{\mathbf{f}}(\mathbf{x}),\hat{\mathbf{f}}(\mathbf{x})>_{\mathbf{x}} = \sum_{i=1}^{n} x_i[f_i(\mathbf{x}) - \bar{f}(\mathbf{x})]^2 \ . \tag{24.12}$$

Thus - as we have shown for the selection equation already in (23.5) - the change of V is equal to the variance of the values $f_i(\mathbf{x})$.

Another important special case is when the functions $f_i(\mathbf{x})$ are constants. This leads to equation (12.5) modelling the growth of selfreplicating macromolecules or selection in a haploid population. The potential then is given by $\bar{f}(\mathbf{x}) = \sum x_i f_i$. Since the linear function $\bar{f}$ attains its maximal values at the corners of S_n, this shows again that only those species i with maximal f_i will survive.

Exercise 1: If the potential $V(\mathbf{x}): \mathbf{R}^n_+ \to \mathbf{R}$ is *homogeneous of degree s*, i.e. if $V(\alpha x_1,...,\alpha x_n) = \alpha^s V(x_1,...,x_n)$ for all $\alpha > 0$, then $\bar{f} = sV$. (Hint: Use Euler's theorem). Hence the "average fitness" $\bar{f}$ is increasing at maximal rate. Show also that every function $V: S_n \to \mathbf{R}$ is the restriction of a homogeneous function (of any prescribed degree) on $\mathbf{R}^n_+$.

Exercise 2: Show that equations of the form

$$\dot{x}_i = x_i(f_i(x_i) - \bar{f}) \tag{24.13}$$

are Shahshahani gradients. If the f_i are monotonically decreasing functions, then the potential is strictly concave on S_n and hence there is a unique, globally attracting equilibrium point $\mathbf{p}$ for (24.13). Show that $P(\mathbf{x}) = \prod_{i=1}^{n} x_i^{p_i}$ is also a global Ljapunov function in this case.

Exercise 3: Show that the change of coordinates $x_i = y_i^2/4$ transforms S_n with the Shahshahani metric into the part of the $(n-1)$ dimensional sphere of radius 2 lying in the positive orthant, equipped with the usual Euclidean metric.

Exercise 4: Show that the geodesic distance, i.e. the length of the shortest path between two states $\mathbf{p},\mathbf{q} \in \operatorname{int} S_n$ with respect to the Shahshahani metric is given by

$$d(\mathbf{p},\mathbf{q}) = 2\arccos\left(\sum_{i=1}^{n} \sqrt{p_i q_i}\right) .$$

(Hint: The shortest path between two points on a sphere is on a great circle).

24.4 A criterion for Shahshahani gradients

Euclidean gradients $\mathbf{f}(\mathbf{x}) = \operatorname{grad} V(\mathbf{x})$ can be easily recognized by the symmetry of the Jacobian matrix $D_{\mathbf{x}}\mathbf{f}$, i.e. by the *integrability conditions*

$$\frac{\partial f_i}{\partial x_j} = \frac{\partial f_j}{\partial x_i}$$

(at least if the domain of $\mathbf{f}$ has no holes). We are looking for a similar criterion for Shahshahani gradients.

If the vector field $\hat{\mathbf{f}}$ given by (24.11) is defined on a neighbourhood of int S_n (in $\mathbf{R}^n$), we may compute

$$\frac{\partial \hat{f}_i}{\partial x_j} = \delta_{ij}(f_i - \bar{f}) + x_i\left(\frac{\partial f_i}{\partial x_j} - \frac{\partial \bar{f}}{\partial x_j}\right) .$$

Since only the action of the Jacobian $D_{\mathbf{x}}\hat{\mathbf{f}}$ on vectors from the tangent space $T_{\mathbf{x}}S_n = \mathbf{R}^n_0$ is of relevance, we study the bilinear form $H_{\mathbf{x}}\hat{\mathbf{f}}$ given by

$$H_{\mathbf{x}}\hat{\mathbf{f}}(\xi,\eta) = <\xi, D_{\mathbf{x}}\hat{\mathbf{f}}(\eta)>_{\mathbf{x}} \tag{24.14}$$

for $\xi,\eta \in \mathbf{R}^n_0$. Clearly

$$\begin{aligned} H_{\mathbf{x}}\hat{\mathbf{f}}(\xi,\eta) &= \sum_{i,j=1}^{n} \frac{1}{x_i}\frac{\partial \hat{f}_i}{\partial x_j}(\mathbf{x})\xi_i\eta_j \\ &= \sum_{i=1}^{n} \frac{1}{x_i}(f_i - \bar{f})\xi_i\eta_i + \sum_{i,j=1}^{n} \frac{\partial f_i}{\partial x_j}(\mathbf{x})\xi_i\eta_i \quad . \end{aligned} \tag{24.15}$$

At the interior rest points of (24.11), the first term vanishes and the bilinear form $H_{\mathbf{x}}\hat{\mathbf{f}}$ reduces to the Jacobian $D_{\mathbf{x}}\mathbf{f}$.

Theorem. *For a vector field* $\hat{\mathbf{f}}$ *defined by* (24.11) *in a neighbourhood* U *of* intS_n, *the following conditions are equivalent:*

(a) $\hat{\mathbf{f}}$ *is a Shahshahani gradient on* int S_n;

(b) *There exist functions* $V,G: U \to \mathbf{R}$ *such that*

$$f_i(\mathbf{x}) = \frac{\partial V}{\partial x_i} + G(\mathbf{x}) \tag{24.16}$$

holds on int S_n;

(c) *The Jacobian bilinear form* $H_{\mathbf{x}}\hat{\mathbf{f}}$ *is symmetric at every* $\mathbf{x} \in$ int S_n;

(d) *The relation*

$$\frac{\partial f_i}{\partial x_j} + \frac{\partial f_j}{\partial x_k} + \frac{\partial f_k}{\partial x_i} = \frac{\partial f_i}{\partial x_k} + \frac{\partial f_k}{\partial x_j} + \frac{\partial f_j}{\partial x_i} \tag{24.17}$$

holds on int S_n *for all* i,j,k.
A function V *satisfying* (24.16) *is a Shahshahani potential for* $\hat{\mathbf{f}}$.

Proof: (a) $\Rightarrow$ (b) If $\hat{\mathbf{f}} = \text{Grad } V$, then (24.9) implies

$$\sum_{i=1}^{n} \frac{1}{x_i} \hat{f}_i(\mathbf{x}) \xi_i = \sum_{i=1}^{n} f_i(\mathbf{x}) \xi_i = \sum_{i=1}^{n} \frac{\partial V}{\partial x_i} \xi_i$$

for all $\boldsymbol{\xi} \in \mathbf{R}_0^n$ and $\mathbf{x} \in \text{int } S_n$. With $\xi_i = 1, \xi_n = -1$ and $\xi_j = 0$ for all $j \neq i,n$ this yields

$$f_i(\mathbf{x}) - f_n(\mathbf{x}) = \frac{\partial V}{\partial x_i}(\mathbf{x}) - \frac{\partial V}{\partial x_n}(\mathbf{x}) \quad .$$

It follows that $\frac{\partial V}{\partial x_i}(\mathbf{x}) - f_i(\mathbf{x})$ is independent of i, which proves (b).
(b) $\Rightarrow$ (c) Since the f_i are of the form

$$f_i(\mathbf{x}) = \frac{\partial V}{\partial x_i}(\mathbf{x}) + G(\mathbf{x}) + (\sum_{j=1}^{n} x_j - 1)\varphi_i(\mathbf{x})$$

for $\mathbf{x} \in U$, the φ_i being suitable functions on U, the partial derivatives are given by

$$\frac{\partial f_i}{\partial x_j} = \frac{\partial^2 V}{\partial x_i \partial x_j} + \frac{\partial G}{\partial x_j} + \varphi_i(\mathbf{x})$$

for $\mathbf{x} \in \text{int } S_n$. Inserting this into (24.15), we see that the terms with G and φ_i disappear because $\sum \xi_i = \sum \eta_j = 0$. What remains is a symmetric bilinear form.
(c) $\Rightarrow$ (d) The symmetry of $H_{\mathbf{x}}\hat{\mathbf{f}}(\boldsymbol{\xi},\boldsymbol{\eta})$ implies

$$\sum_{i,j=1}^{n} \left(\frac{\partial f_i}{\partial x_j} - \frac{\partial f_j}{\partial x_i} \right) \xi_i \eta_j = 0$$

for all $\boldsymbol{\xi},\boldsymbol{\eta} \in \mathbf{R}_0^n$. With $\boldsymbol{\xi} = \mathbf{e}_i - \mathbf{e}_k$ and $\boldsymbol{\eta} = \mathbf{e}_j - \mathbf{e}_k$ we obtain (d).
(d) $\Rightarrow$ (b) Define for $\mathbf{x} \in \text{int } S_n$

$$g_i(x_1,\ldots,x_{n-1}) = f_i(x_1,\ldots,x_{n-1},1 - x_1 - \cdots - x_{n-1}) \quad . \tag{24.18}$$

Then g_i coincides with f_i on S_n and $\frac{\partial g_i}{\partial x_j} = \frac{\partial f_i}{\partial x_j} - \frac{\partial f_i}{\partial x_n}$ by the chain rule. So (d) implies (with $k = n$)

$$\frac{\partial g_i}{\partial x_j} - \frac{\partial g_n}{\partial x_j} = \frac{\partial g_j}{\partial x_i} - \frac{\partial g_n}{\partial x_i} \quad .$$

These are just the Euclidean integrability conditions for

$g_i - g_n \, (1 \le i \le n-1)$ on R^{n-1}. Thus we find a potential $V = V(x_1,...,x_{n-1})$ with

$$g_i - g_n = \frac{\partial V}{\partial x_i} \quad (i = 1,...,n-1) \quad .$$

By (24.18) this implies (b) with $G = g_n$.
(b) ⇒ (a) is Theorem 24.3.

The *triangular integrability condition* (24.17) is very useful to find out whether a given system is a Shahshahani gradient. Some applications of it will be given soon.

Corollary. *The equation* (16.4) *is a Shahshahani gradient iff for all* i,j,k

$$a_{ij} + a_{jk} + a_{ki} = a_{ik} + a_{kj} + a_{ji} \tag{24.19}$$

Exercise 1: Show some more characterizations, equivalent to (24.19):

(a) There exists a symmetric matrix $b_{ij} = b_{ji}$ and numbers c_j, such that $a_{ij} = b_{ij} + c_j$.

(b) There exists constants c_i such that $a_{ij} - a_{ji} = c_i - c_j$ for all i,j.

(c) There exist vectors $\mathbf{u},\mathbf{v} \in \mathbf{R}^n$ such that $a_{ij} - a_{ji} = u_i + v_j$ for all i,j.

Exercise 2: Show that (24.19) implies that analogous conditions hold for p-cycles, $p \ge 3$.

Exercise 3: It was shown in exercise 16.7.1 that a replicator equation (16.1) without interior fixed point has a global Ljapunov function of the form $V(\mathbf{x}) = \sum_{i=1}^{n} c_i \log x_i$, with $\mathbf{c} \in \mathbf{R}_0^n$. Find a Riemannian metric on int S_n which makes (16.1) the gradient of $V(\mathbf{x})$.

Exercise 4: We know that if $\mathbf{p}$ is an interior ESS, then $V(\mathbf{x}) = \sum p_i \log x_i$ is a Ljapunov function for the replicator equation. Show that one cannot, in general, find a Riemannian metric on int S_n which makes (16.4) the gradient of $V(\mathbf{x})$.

Exercise 5: Show that an ecological differential equation

$$\dot{x}_i = x_i f_i(\mathbf{x}) \qquad \mathbf{x} \in \mathbf{R}_+^n$$

is a gradient with respect to the Riemannian metric

$$<\xi,\eta>_{\mathbf{x}} = \sum_{i=1}^{n} \frac{1}{x_i} \xi_i \eta_i \tag{24.20}$$

in int $\mathbf{R}^n_+$ iff $\dot{x}_i = f_i(\mathbf{x})$ is an Euclidean gradient.

Exercise 6: The Lotka-Volterra equation (9.1) is a gradient with respect to (24.20) iff the interaction matrix is symmetric. Compute its potential.

Exercise 7: Extend the above result to matrices satisfying

$$a_{ij}c_j = a_{ji}c_i$$

for some $c_i > 0$. (Modify (24.20) or make a change of variables). In particular every two-species Lotka-Volterra system which is competitive or symbiotic is a gradient. (Compare **8.3, 18.7, 21.5**).

24.5 Mixed strategists and gradient systems

In **15.5** and **16.4** we considered games with N pure strategies R_1 to R_N and populations consisting of n phenotypes E_1 to E_n, where E_i plays strategy R_j with probability p^i_j and hence is characterized by a vector $\mathbf{p}^i \in S_N$. If we denote the frequencies of E_i in the population by x_i, then

$$p_k = \sum_{i=1}^{n} x_i p^i_k \tag{24.21}$$

is the frequency of the strategy R_k in the population. Thus $\mathbf{x} \in S_n$ denotes the state of the population and $\mathbf{p} = \mathbf{p}(\mathbf{x}) \in S_N$ the distribution of the pure strategies. We shall, as usual, assume that the payoff F_k for strategy R_k depends on $\mathbf{x}$ only through the strategy mix $\mathbf{p}$ in the population. Then the payoff for the phenotype E_i is given by

$$f_i(\mathbf{x}) = \mathbf{p}^i \cdot \mathbf{F}(\mathbf{p}) = \sum_{j=1}^{N} p^i_j F_j(\mathbf{p}) \tag{24.22}$$

while the average payoff in the population is

$$\bar{f}(\mathbf{x}) = \sum x_i(\mathbf{p}^i \cdot \mathbf{F}(\mathbf{p})) = \mathbf{p} \cdot \mathbf{F}(\mathbf{p}) \quad . \tag{24.23}$$

The game dynamical Ansatz leading to (16.1) now yields

$$\dot{x}_i = x_i[(\mathbf{p}^i - \mathbf{p}) \cdot \mathbf{F}(\mathbf{p})] \tag{24.24}$$

which we shall call the *mixed strategist dynamics.*

We may also consider a *pure strategist dynamics* for a fictitious population whose phenotypes correspond to the pure strategies. It is given by

$$\dot{y}_i = y_i(F_i(\mathbf{y}) - \bar{F}(\mathbf{y})) \tag{24.25}$$

on S_N.

Theorem. *If the pure strategist game (24.25) is a Shahshahani gradient with potential $V(\mathbf{y})$, then the mixed strategist game (24.24) is a Shahshahani gradient with potential $V(\mathbf{p}(\mathbf{x}))$.*

Proof: Indeed, if

$$F_i(\mathbf{y}) = \frac{\partial V}{\partial y_i}(\mathbf{y}) + G(\mathbf{y})$$

(cf. (24.16)), then

$$f_i(\mathbf{x}) = \mathbf{p}^i \cdot \mathbf{F}(\mathbf{p}) = \sum_j p_j^i(F_j(\mathbf{p}) - G(\mathbf{p})) + G(\mathbf{p}) = \frac{\partial V(\mathbf{p})}{\partial x_i} + G(\mathbf{p}) \; .$$

Obviously, every 2×2-game is a Shahshahani gradient, since the triangular integrability condition (24.17) is trivially satisfied. Hence every mixed strategist dynamics based on two pure strategies is a Shahshahani gradient. Denoting by $p_1,...,p_n$ and p the first component of $\mathbf{p}^1,...,\mathbf{p}^n$ resp. $\mathbf{p}$, we obtain from (24.24)

$$\dot{x}_i = x_i(p_i - p)[F_1(p) - F_2(p)] \tag{24.26}$$

where

$$p = \sum_{i=1}^{n} p_i x_i \quad .$$

The Shahshahani potential is given by

$$V(p) = \int [F_1(p) - F_2(p)]\,dp \quad . \tag{24.27}$$

If the 2×2-game is linear (as in the Hawk-Dove-example **15.2**) and described by the matrix A, then we obtain in the generic case where $a = a_{11} - a_{21} + a_{22} - a_{12} \neq 0$:

$$F_1(p) - F_2(p) = a(p - \hat{p}) \tag{24.28}$$

(where $\hat{p} = a^{-1}(a_{22} - a_{12})$) and hence

$$V(p) = \frac{1}{2}a(p - \hat{p})^2 \quad . \tag{24.29}$$

For the sex ratio game (see **15.4**) we have

$$F_1(m) - F_2(m) = \frac{1}{m} - \frac{1}{1-m} \tag{24.30}$$

(we write now m, the frequency of males, instead of p). Hence

$$V(m) = \log m(1-m) \quad . \tag{24.31}$$

It follows that the product $m(1-m)$ of the frequencies of males and females increases to its maximal value, obtained for $m = \frac{1}{2}$: furthermore, the population evolves in such a way that the product increases at a maximal rate.

Exercise 1: When does the frequency p in (24.26) converge to an evolutionarily stable strategy? (Hint: Start out with the linear case (24.28)).

Exercise 2: For any three phenotypes p_1, p_2 and p_3, (24.26) admits a constant of motion:

$$Q(\mathbf{x}) = (p_2 - p_3)\log x_1 + (p_3 - p_1)\log x_2 + (p_1 - p_2)\log x_3 \tag{24.32}$$

Show that

$$\dot{p} = [F_1(p) - F_2(p)]\mathrm{Var}\,\tilde{P} \tag{24.33}$$

where $\tilde{P}$ is the random variable taking the value p_i with probability x_i. Determine the rest points and the phase portrait of (24.26) (cf. 16.15).

Exercise 3: Find invariants of motion for (24.24) if $n > N$. (Hint: There is one for every nontrivial linear relation $\sum c_i \mathbf{p}^i = \mathbf{0}$; note that $\sum c_i = 0$). Find an analogue of (24.33) for $\dot{\mathbf{p}}$ (cf. (16.10) and **28.4**).

Exercise 4: Set up a discrete equivalent for the evolution of the sex ratio, and verify that m converges to $\frac{1}{2}$.

Exercise 5: Consider the case of $N > 2$ mating types: every "sex" R_i can mate with any other $R_j (j \neq i)$. Show that the payoff for an R_i-individual is

$$F_i(\mathbf{m}) = \frac{1 - m_i}{\sum_k m_k(1 - m_k)}$$

and that

$$V(\mathbf{m}) = \frac{1}{2}\log \sum_{k=1}^{N} m_k(1-m_k) = \frac{1}{2}\log[1 - \frac{1}{N} - \sum_k (m_k - \frac{1}{N})^2]$$

is a Shahshahani potential for the corresponding equation (24.24). Hence the "sex ratio" converges to $(\frac{1}{N}, \dots, \frac{1}{N})$. So why are there only two mating types?

24.6 A selective Hopf bifurcation theorem

More advanced models in population genetics, like those including recombination or mutation, which will be considered in the next chapter, often take the form

$$\dot{x}_i = x_i[(M\mathbf{x})_i - \mathbf{x}\cdot M\mathbf{x}] + \hat{f}_i(\mathbf{x}) \tag{24.34}$$

where $\hat{\mathbf{f}}(\mathbf{x})$ is again a vector field on S_n which we way write in replicator form (24.11).

Theorem. *Whenever $\hat{\mathbf{f}}(\mathbf{x})$ is not a Shahshahani gradient there exist symmetric matrices M such that the combined vector field (24.34) has periodic orbits.*

Thus the flow of the combined field (24.34) is not gradient-like. Not all orbits converge to equilibrium states.

Proof: We prove first the following statement:

Given a point $\mathbf{x} \in \text{int } S_n$ and a linear map $S: \mathbf{R}_0^n \to \mathbf{R}_0^n$ which is symmetric with respect to the Shahshahani metric, there exists a symmetric matrix M such that $\mathbf{x}$ is a fixed point of (24.34) and S is the symmetric part of its Jacobian at $\mathbf{x}$.

Indeed, for given $\mathbf{x}$, we can split up any fitness matrix M as

$$m_{ij} = \bar{m} + (m_i - \bar{m}) + (m_j - \bar{m}) + \Theta_{ij} \tag{24.35}$$

where $m_i = (M\mathbf{x})_i$, $\bar{m} = \sum x_i m_i = \mathbf{x}\cdot M\mathbf{x}$, and the dominance terms Θ_{ij} satisfy

$$\sum_{j=1}^{n} \Theta_{ij} x_j = 0 \quad . \tag{24.36}$$

Conversely, arbitrary numbers m_i and $\Theta_{ij} = \Theta_{ji}$ obeying (24.36) determine a fitness matrix M. Now choose m_i such that

$$x_i m_i + \hat{f}_i(\mathbf{x}) = 0 \quad . \tag{24.37}$$

Since $\hat{\mathbf{f}}(\mathbf{x}) \in \mathbf{R}_0^n$ one then has $\sum x_i m_i = \bar{m} = 0$. Thus $\mathbf{x}$ is a fixed point of (24.34). Obviously, after specifying m_i and $\bar{m}$ by (24.37), every symmetric operator S in the $(n-1)$ dimensional space $\mathbf{R}_0^n$ can be realized by suitably chosen Θ_{ij} obeying (24.36).

We may now prove the theorem by viewing the Jacobian of (24.34) at $\mathbf{x}$ as a linear map on $T_{\mathbf{x}}S_n = \mathbf{R}_0^n$ which we split into its symmetric and antisymmetric part with respect to the Shahshahani inner product: $S + A$. By assumption (and using Theorem 24.4 (c)) there exists a point $\mathbf{x} \in \text{int } S_n$ where $A \neq 0$. We choose an orthogonal basis $\xi^1,\dots,\xi^{n-1}$ of $\mathbf{R}_0^n$ such that $A\xi^j = i\omega_j\xi^{n-j}$ and $A\xi^{n-j} = -i\omega_j\xi^j$ for $j = 1,\dots,k$ where $\pm i\omega_j (1 \le j \le k)$ are the pairs of nonzero imaginary eigenvalues of A $(k \ge 1)$ and $A\xi^j = \mathbf{0}$ for $k < j < n-k$. Define the symmetric operators S_λ by $S_\lambda\xi^j = \lambda\xi^j$ for $j = 1$ and $j = n-1$ and $S_\lambda\xi^j = -\xi^j$ for $1 < j < n-1$. Then $S_\lambda + A$ has eigenvalues $\lambda \pm i\omega_1$, $-1 \pm i\omega_j$ $(2 \le j \le k)$ and maybe -1. By the above statement we can realize S_λ by symmetric matrices M_λ. Then (24.34) with M_λ undergoes a Hopf bifurcation at the rest point $\mathbf{x}$, as λ varies from -1 to 1. Hence there exist nontrivial periodic solutions.

24.7 Notes

The Maximum Principle was stated by Kimura (1958), see also Crow and Kimura (1970). The Riemannian metric which makes the selection equation the gradient of mean fitness was introduced by Shahshahani (1979) and further investigated by Akin (1979). The characterization (b) from Theorem 24.4 is from Sigmund (1984), (c) from Akin (1979) and (d) from Hofbauer (1985b). The linear case of the triangular integrability conditions is from Sigmund (1984). **24.5** follows Sigmund (1987a). **24.6** as well as exercises 24.3.3-4 are due to Akin (1979).

25. Selection, Mutation and Recombination

25.1 The selection mutation model

In this section we extend the model introduced in **4.2**. Consider now one gene locus with n alleles $A_1,\dots,A_n$ in a population with discrete generations and let $x_1,\dots,x_n$ be their relative frequencies in the gene pool at the time of mating. Assuming random mating, the relative frequency of new gene pairs (A_i,A_j) will be x_ix_j. Due to natural selection only a proportion of $w_{ij}x_ix_j$ will survive to mature age ($w_{ij} = w_{ji} \ge 0$ being the fitness parameters). So the number of newly produced genes A_j is proportional to $\sum_k w_{jk}x_jx_k = x_j(W\mathbf{x})_j$. Now let ϵ_{ij} be the *mutation rate* from A_j to A_i, so that

$$\epsilon_{ij} \ge 0 \quad \text{and} \quad \sum_{i=1}^n \epsilon_{ij} = 1 \tag{25.1}$$

for all $j = 1,...,n$ (where the ϵ_{ii} are suitably defined). Then the frequency x_i' of genes A_i in the gene pool of the new generation is proportional to $\sum_j \epsilon_{ij} x_j (W\mathbf{x})_j$. More precisely it is given by

$$x_i' = \frac{1}{\bar{w}(\mathbf{x})} \sum_{j=1}^{n} \epsilon_{ij} x_j (W\mathbf{x})_j \tag{25.2}$$

with $\bar{w}(\mathbf{x}) = \mathbf{x} \cdot W\mathbf{x}$ (the mean fitness of the population) as the usual normalization factor. This is the *discrete time selection mutation equation.* We shall always assume $w_{ii} > 0$ for all i in order to guarantee that $\bar{w}(\mathbf{x}) > 0$ for all $\mathbf{x} \in S_n$.

In order to get a differential equation we replace as usual $x_i' - x_i$ by $\dot{x}_i = dx_i/dt$:

$$\dot{x}_i = \bar{w}(\mathbf{x})^{-1} \sum_{j,k} \epsilon_{ij} x_j w_{jk} x_k - x_i \quad . \tag{25.3}$$

A biologically more satisfactory way to derive a *continuous time model* might be the following. Consider a population with overlapping generations and let selection act in the way described in **23.1** with Malthusian fitness parameters m_{ij}. Mutation effects, being small in general, will change the gene frequencies in a linear way. Now a simultaneous action of selectional and mutational forces in a small time interval Δt will be of smaller order $(\Delta t)^2$, since both forces are independent. Thus we arrive at the following continuous time model with separate selection and mutation terms:

$$\dot{x}_i = x_i[(M\mathbf{x})_i - \mathbf{x} \cdot M\mathbf{x}] + \sum_{j=1}^{n} (\epsilon_{ij} x_j - \epsilon_{ji} x_i) \quad . \tag{25.4}$$

All three equations (25.2) to (25.4) describe dynamical systems on S_n.

Exercise 1: To model weak selection and weak mutation, replace ϵ_{ij} by $\delta\epsilon_{ij}$ (for $i \neq j$) in (25.2) and set $w_{ij} = 1 + \delta m_{ij}$, with δ a small parameter. Show that both the discrete time model (25.2) and its continuous time counterpart (25.3) reduce to the differential equation (25.4) as $\delta \to 0$.

Exercise 2: Analyse (25.3) and (25.4) for $n = 2$ alleles as in **4.2**.

25.2 Mutation and additive selection

In the absence of selection, i.e. if all w_{ij} (and m_{ij}) are equal, equations (25.3) and (25.4) reduce to the *continuous time mutation equation*

$$\dot{x}_i = \sum_{j=1}^{n} \epsilon_{ij} x_j - x_i \tag{25.5}$$

and (25.2) to the *discrete time mutation equation*

$$x_i' = \sum_{j=1}^{n} \epsilon_{ij} x_j \quad . \tag{25.6}$$

These equations are linear and they are easily analysed using Perron-Frobenius theory. As we have treated even more general equations in **11.4-5**, the proof of the following statement is left as an exercise to the reader (note that **1** is a left eigenvector of the mutation matrix by (25.1)).

Theorem 1. *If the mutation matrix* (ϵ_{ij}) *is primitive (resp. irreducible) then the discrete (resp. continuous) time model* (25.6) *(resp.* (25.5)) *has a unique equilibrium point* $\mathbf{p} \in \operatorname{int} S_n$. *This point is globally stable.*

This simple dynamics carries over to the case where "dominance free" selection is allowed. Suppose in (25.2) or (25.3) that selection is multiplicative (i.e. $w_{ij} = w_i w_j$). The corresponding assumption for (25.4) is that Malthusian fitness is additive (i.e. $m_{ij} = m_i + m_j$): one has only to take the limit as in exercise 25.1. Equations (25.2) to (25.4) simplify then to

$$\bar{w} x_i' = \sum_{j=1}^{n} \epsilon_{ij} w_j x_j \tag{25.7}$$

$$\dot{x}_i = \sum_{j=1}^{n} \epsilon_{ij} w_j x_j - x_i \bar{w} \tag{25.8}$$

$$\dot{x}_i = x_i(m_i - \bar{m}) + \sum_{j=1}^{n} \epsilon_{ij} x_j - x_i \tag{25.9}$$

with $\bar{w} = \sum_{i=1}^{n} x_i w_i$ and $\bar{m} = \sum_{j=1}^{n} x_i m_i$.

Theorem 2. *Under the assumptions of the previous theorem, the models* (25.7) *to* (25.9) *give rise to a unique equilibrium* $\mathbf{p} \in \operatorname{int} S_n$. *This point is globally stable.*

The proof is again left as an exercise, as one has only to repeat the argument in **11.4-5**.

Exercise 1: Show that equations (25.7) - (25.9) arise also as selection mutation models for haploid populations.

Exercise 2: If the mutation matrix (ϵ_{ij}) is irreducible and $\mathbf{p}$ is the interior fixed point, then

$$V(\mathbf{x}) = \sum_{i=1}^{n} \frac{1}{p_i}(x_i - p_i)^2 \quad \text{and} \quad P(\mathbf{x}) = \prod_{i=1}^{n} x_i^{p_i}$$

are global Ljapunov functions for (25.5). (Hint. $(\delta_{ij} - \epsilon_{ij})$ is a singular M-matrix. Proceed as in exercise 21.3.8).

Exercise 3: Let A be the matrix $(\epsilon_{ij} - \delta_{ij})$. If A is symmetric and $\mathbf{p}$ is the interior fixed point then each of the following functions is a Ljapunov function for (25.5):

$$(\mathbf{x}-\mathbf{p})^2, \quad \mathbf{x}\cdot A\mathbf{x}, \quad \frac{\mathbf{x}\cdot A\mathbf{x}}{\mathbf{x}\cdot\mathbf{x}}, \quad (\mathbf{x}-\mathbf{p})\cdot A(\mathbf{x}-\mathbf{p}) \quad .$$

Exercise 4: Define for $\mathbf{x},\mathbf{y} \in \operatorname{int} S_n$

$$g_1(\mathbf{x},\mathbf{y}) = \max_i \frac{x_i}{y_i} \qquad g_2(\mathbf{x},\mathbf{y}) = \min_i \frac{x_i}{y_i} \quad .$$

Show that for $\epsilon_{ij} > 0$ the mutation equations (25.6) and its generalizations (25.7) and (11.13) are contractions for the metric $d(\mathbf{x},\mathbf{y}) = \log\left(g_1(\mathbf{x},\mathbf{y})/g_2(\mathbf{x},\mathbf{y})\right)$.

Exercise 5: Prove the Perron-Frobenius theorem **11.4** for positive matrices M. Consider the map (11.13) on S_n. The positive eigenvector $\mathbf{u} > \mathbf{0}$ corresponds to the interior fixed point $\mathbf{p} > \mathbf{0}$ of (11.13) whose existence follows from fixed point arguments as in exercise 4 or **19.2**. The stability of $\mathbf{p}$ implies the dominance of the positive eigenvalue λ.

25.3 Special mutation rates

An important question is to which extent the Fundamental Theorem and the Maximum Principle generalize to the selection mutation model. Under the assumption of theorem 25.2.2 there will exist some Ljapunov function for (25.7) to (25.9). No biologically satisfactory one has been singled out yet, however.

Concerning the Maximum Principle, we cannot hope to extend it to general mutation rates ϵ_{ij}. While ϵ_{ij}, being a stochastic matrix, has leading eigenvalue 1 and thus all eigenvalues of the restriction to S_n are of absolute value less than 1 (which implies stability), there remains the possibility of complex eigenvalues excluding gradient behaviour.

Exercise 1: Find 3×3 mutation matrices, with ϵ_{ij} $(i \neq j)$ arbitrarily small, which have complex eigenvalues.

This suggests to single out those mutation matrices which are *Shahshahani gradients*. For this we write the mutation equation (25.5) as a replicator equation (24.11). This implies on int S_n:

$$f_i(\mathbf{x}) = \sum_{j=1}^{n} \epsilon_{ij} \frac{x_j}{x_i}$$

and hence $\dfrac{\partial f_i}{\partial x_j} = \dfrac{\epsilon_{ij}}{x_i}$, $i \neq j$. The triangular integrability condition (24.17) then reads (for pairwise different i,j,k)

$$\frac{\epsilon_{ij}}{x_i} + \frac{\epsilon_{jk}}{x_j} + \frac{\epsilon_{ki}}{x_k} = \frac{\epsilon_{ik}}{x_i} + \frac{\epsilon_{kj}}{x_k} + \frac{\epsilon_{ji}}{x_j} \tag{25.10}$$

for all $\mathbf{x} \in$ int S_n. Taking the limit $x_i \to 0$, we find that $\epsilon_{ij} = \epsilon_{ik}$ for all $j \neq k$. Thus *the mutation rates must depend on the target gene only*: they can be written as

$$\epsilon_{ij} = \epsilon_i \ (i \neq j) \quad . \tag{25.11}$$

(25.1) implies for these special mutation rates

$$\epsilon_{ii} = 1 + \epsilon_i - \epsilon \quad \text{with} \quad \epsilon = \epsilon_1 + \cdots + \epsilon_n \quad . \tag{25.12}$$

(25.3) reduces to

$$\begin{aligned} \bar{w}(\mathbf{x})\dot{x}_i &= x_i[(W\mathbf{x})_i - \bar{w}(\mathbf{x})] + \epsilon_i \bar{w}(\mathbf{x}) - \epsilon x_i (W\mathbf{x})_i \\ &= x_i[(1-\epsilon)(W\mathbf{x})_i - \bar{w}(\mathbf{x})] + \epsilon_i \bar{w}(\mathbf{x}) \quad . \end{aligned} \tag{25.13}$$

This gives a replicator equation (24.11) with

$$f_i(\mathbf{x}) = (1-\epsilon)\frac{(W\mathbf{x})_i}{\bar{w}(\mathbf{x})} + \frac{\epsilon_i}{x_i} \quad \text{and} \quad \bar{f}(\mathbf{x}) = 1 \quad . \tag{25.14}$$

Obviously the functions $f_i(\mathbf{x})$ fulfill the integrability condition and thus by Theorem 24.4, (25.13) is a Shahshahani gradient. The *potential* is easily computed to be

$$V(\mathbf{x}) = \frac{1-\epsilon}{2}\log \bar{w}(\mathbf{x}) + \sum_{i=1}^{n} \epsilon_i \log x_i \quad . \tag{25.15}$$

A more appealing Ljapunov function is obtained by exponentiating $2V$:

$$W(\mathbf{x}) = \bar{w}(\mathbf{x})^{1-\epsilon} \prod_{i=1}^{n} x_i^{2\epsilon_i} \quad . \tag{25.16}$$

For $\epsilon = 0$ (i.e. no mutation), $W(\mathbf{x})$ reduces to the mean fitness $\bar{w}(\mathbf{x})$. (25.13) implies

$$\dot{V}(\mathbf{x}) = \frac{1}{2}\frac{\dot{W}(\mathbf{x})}{W(\mathbf{x})} = \sum_{i=1}^{n} x_i (f_i - \bar{f})^2 \geq 0 \quad . \tag{25.17}$$

This gives a direct analogue of the Fundamental Theorem: *The rate of growth of the modified mean fitness function $W(\mathbf{x})$ is proportional to the variance of the terms $f_i(\mathbf{x})$.*

It is easy to derive a similar result for the uncoupled version of the continuous time equation (25.4). One finds

$$f_i(\mathbf{x}) = (M\mathbf{x})_i + \frac{\epsilon_i}{x_i} \quad \text{and} \quad \bar{f}(\mathbf{x}) = \mathbf{x}\cdot M\mathbf{x} + \epsilon \tag{25.18}$$

and the potential is given by

$$V(\mathbf{x}) = \frac{1}{2}\mathbf{x}\cdot M\mathbf{x} + \sum_{i=1}^{n} \epsilon_i \log x_i \quad . \tag{25.19}$$

Exercise 2: Work this out in detail (either directly or using the weak selection limit argument of exercise 25.1.1).

Summarizing, we have shown

Theorem 1. *The mutation equation (25.5) is a Shahshahani gradient iff the mutation rates take the special form (25.11). In this case, for arbitrary selection terms, both continuous time selection mutation equations (25.3) and (25.4) are Shahshahani gradients too.*

A nice application of this result is

Theorem 2. *Suppose that the model without mutation (i.e. $\epsilon_{ij} = 0$ for $i \neq j$) admits a stable interior equilibrium. Then for every choice of mutation rates satisfying (25.11) with $\epsilon \leq 1$, the equations (25.3) and (25.4) have a unique equilibrium point in S_n, which is globally attracting.*

Exercise 3: Prove this. (Hint: show that $\bar{w}(\mathbf{x})$ and $V(\mathbf{x})$ are strictly concave and will therefore attain a unique maximum.)

Exercise 4: Show that the conclusion need not hold if selection (in the absence of mutation) leads to a globally stable equilibrium on the boundary. (Hint: Take a two-allelic model where the fitter allele is recessive).

Exercise 5: For $n = 2$, the selection mutation equation is always a Shahshahani gradient.

25.4 Limit cycles for the selection mutation equation

What happens if mutation rates are not of the special form (25.11)? The mutation equation (25.5) is not a Shahshahani gradient, then, and Theorem 24.6 shows that for some selection matrices (m_{ij}) the continuous time selection mutation equation (25.4) has periodic orbits. Since (25.4) is a limiting case of (25.2) and (25.3), this result carries over to these more general equations. It is therefore *impossible* to find an analogue of the Fundamental Theorem in this general situation. The following example even displays *stable* limit cycles.

Assume that all homozygotes A_iA_i have the same fitness and all heterozygotes too. When working with (25.4) this means $m_{ij} = s\delta_{ij}$ where s measures the selective advantage of the homozygotes. Motivated by **9.5** and **20.2** we assume the mutation rates to be *cyclically symmetric*, that is,

$$\epsilon_{ij} = \epsilon_{j-i} \tag{25.20}$$

Then $\sum_{i=0}^{n-1} \epsilon_i = 1$ where the index i is again considered modulo n. Now (25.4) reads

$$\dot{x}_i = sx_i(x_i - Q(\mathbf{x})) + \sum_{j=1}^{n} \epsilon_{j-i}x_j - x_i \tag{25.21}$$

with $Q(\mathbf{x}) = \sum_{i=1}^{n} x_i^2$. Obviously the barycenter $\mathbf{m} = \frac{1}{n}\mathbf{1}$ is a rest point of (25.21). The Jacobian of (25.21) at a point $\mathbf{x}$ is given by

$$\partial\dot{x}_i/\partial x_j = s\,\delta_{ij}(x_i - Q(\mathbf{x})) + s\,x_i(\delta_{ij} - 2x_j) + \epsilon_{j-i} - \delta_{ij} \quad .$$

The divergence is then

$$\text{div} = \sum_{i=1}^{n} \partial\dot{x}_i/\partial x_i = s(2 - (n+2)Q(\mathbf{x})) + n(\epsilon_0 - 1) \quad .$$

After subtracting the eigenvalue orthogonal to S_n, which is given by

$-\bar{f}(\mathbf{x}) = -sQ(\mathbf{x})$, we obtain the divergence within S_n,

$$\mathrm{div}_0 = s(2-(n+1)Q(\mathbf{x})) + n(\epsilon_0 - 1) \quad . \tag{25.22}$$

Since $Q(\mathbf{x}) = \sum x_i^2 \geq \frac{1}{n}(\sum x_i)^2 = \frac{1}{n}$, we have for positive s

$$\mathrm{div}_0 \leq s(1-\frac{1}{n}) + n(\epsilon_0 - 1) \quad . \tag{25.23}$$

So the divergence is negative on $S_n \backslash \{\mathbf{m}\}$ whenever

$$s \leq \frac{n^2}{n-1}(1-\epsilon_0) \quad . \tag{25.24}$$

Now we specialize to $n = 3$ alleles. Then we can compute the eigenvalues $\gamma, \bar{\gamma}$ at $\mathbf{m}$ within S_3, using formula (9.20), to

$$\gamma = \frac{s}{3} - 1 + \epsilon_0 + \epsilon_1\lambda + \epsilon_2\bar{\lambda}$$

with $\lambda = \exp(2\pi i/3)$. They are complex if $\epsilon_1 \neq \epsilon_2$ and their real part is

$$\mathrm{Re}\,\gamma = \frac{s}{3} - \frac{3}{2}(\epsilon_1 + \epsilon_2) \quad .$$

For $s = \frac{9}{2}(\epsilon_1 + \epsilon_2)$ the eigenvalues are purely imaginary and a *Hopf bifurcation* occurs, with s as parameter. Since for all $s \leq \frac{9}{2}(\epsilon_1 + \epsilon_2)$, $\mathrm{div}_0 < 0$ holds on $S_3 \backslash \{\mathbf{m}\}$ by (25.23) and (25.24), the Bendixson-Dulac test shows that there are no periodic orbits in this case, and hence that $\mathbf{m}$ is globally asymptotically stable. As this holds even in the critical case $s = \frac{9}{2}(\epsilon_1 + \epsilon_2)$, *stable* limit cycles appear for s slightly larger than $\frac{9}{2}(\epsilon_1 + \epsilon_2)$ (see Fig. 25.1).

25.5 Selection at two loci

In this section we extend the selection recombination model described in **4.3** to an arbitrary number of alleles at the two loci. Let A_i $(1 \leq i \leq n)$ and B_j $(1 \leq j \leq m)$ be the possible alleles at the A- and B-locus. Then there are nm different types of gametes A_iB_j, the frequencies of which we denote by x_{ij}. Assuming random mating and discrete generations, the proportion of A_iB_j/A_kB_l individuals will change by natural selection from $x_{ij}x_{kl}$ in zygotes to $w_{ij,kl}x_{ij}x_{kl}$ at adult age. When haploid gametes are produced during meiosis, not only the parental combinations A_iB_j and A_kB_l but also the *recombinants* A_iB_l and

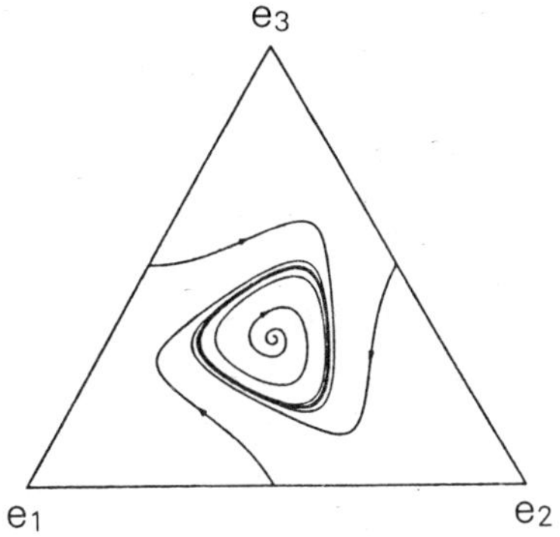

Figure 25.1

$A_k B_j$ will appear. Such crossovers happen with a certain probability r depending on the distance between the two loci. (Remember that this recombination fraction r takes its maximal value $\frac{1}{2}$ if the two loci are on different chromosomes.) This leads to the following *discrete two-loci selection model* for the gamete frequencies x'_{ij} in the next generation:

$$\begin{aligned} \bar{w}x'_{ij} &= (1-r)x_{ij}\sum_{k,l} w_{ij,kl}x_{kl} + r\sum_{k,l} w_{il,kj}x_{il}x_{kj} \\ &= x_{ij}\sum_{k,l} w_{ij,kl}x_{kl} - rD_{ij} \end{aligned} \tag{25.25}$$

with

$$D_{ij} = \sum_{k,l}(w_{ij,kl}x_{ij}x_{kl} - w_{il,kj}x_{il}x_{kj}) \quad . \tag{25.26}$$

The expressions D_{ij} are called *linkage disequilibrium functions.* The continuous time version of the two loci selection equation takes the form

$$\dot{x}_{ij} = x_{ij}(\sum_{k,l} m_{ij,kl}x_{kl} - \bar{m}) - rD_{ij} \tag{25.27}$$

with the disequilibrium functions now given by

$$D_{ij} = \sum_{k,l}(b_{ij,kl}x_{ij}x_{kl} - b_{il,kj}x_{il}x_{kj}) \quad . \tag{25.28}$$

These equations are obtained similarly to (23.2) or (25.4). Changes in the frequencies x_{ij} of $A_i B_j$ gametes in a small time interval $\triangle t$ are due to either births or deaths. With $b_{ij,kl}$, $d_{ij,kl}$ and $m_{ij,kl} = b_{ij,kl} - d_{ij,kl}$ denoting birthrates, deathrates and fitnesses of A_iB_j/A_kB_l individuals, this leads to the usual selection equation. A fraction r of the newborn has undergone a crossover, however, which gives the *recombination terms* $-rD_{ij}$ in (25.27). Note that the continuous time model has more parameters than the discrete time model.

Now usually $w_{ij,kl} = w_{il,kj}$ holds since both an A_iB_j/A_kB_l and A_iB_l/A_kB_j individual have A_iA_k on their A-locus and B_jB_l on the B-locus. Given the composition of the genotype, the arrangement of the alleles is irrelevant. With this assumption of "no position effects", $D_{ij} = 0$ holds if the frequencies x_{ij} can be written as a product of the gene frequencies $p_i = \sum_{j=1}^{m} x_{ij}$ of A_i and $q_j = \sum_{i=1}^{n} x_{ij}$ of B_j. The set where this linkage equilibrium holds is again called the *Wright manifold*

$$\mathbf{W} = \{\mathbf{x} \in S_{nm}: x_{ij} = p_i q_j \text{ for all } i,j\} \quad .$$

If there is no selection, i.e. if all birth- and deathrates are equal to 1 and d respectively, then (25.27) reduces to

$$\dot{x}_{ij} = -rD_{ij} \tag{25.29}$$

with

$$D_{ij} = x_{ij} - p_i q_j \quad . \tag{25.30}$$

Then

$$\dot{p}_i = \sum_j \dot{x}_{ij} = -r\sum_j D_{ij} = 0$$

and similarly $\dot{q}_j = 0$. Thus - as is biologically obvious - recombination does not alter the gene frequencies. Furthermore,

$$\dot{D}_{ij} = -rD_{ij} \quad .$$

Therefore the linkage disequilibrium terms all tend to zero and the population state converges to the Wright manifold, where genes at the two loci are independently distributed.

Exercise 1: Show that entropy $H(\mathbf{x}) = -\sum_{i,j} x_{ij} \log x_{ij}$ is also a Ljapunov function for the recombination equation (25.29).

Exercise 2: Analyse in a similar way the discrete time version of (25.29): Show that $x'_{ij} = x_{ij} - rD_{ij}$ and $D'_{ij} = (1-r)D_{ij}$.

Since the convergence to the Wright manifold is of a uniform exponential rate, namely $-r$, general theorems show that for weak selection, all orbits converge to a manifold near $\mathbf{W}$, where the state is in *quasi linkage equilibrium.*

On the other extreme, if $r = 0$ (which means no recombination or very tight linkage), (25.27) can be interpreted as a one-locus selection equation for nm "alleles" A_iB_j. Thus the selection part of (25.27) is again a gradient system with respect to the Shahshahani metric on S_{nm}.

Exercise 3: Assume that in (25.27) fitnesses are *additive* among the loci: $m_{ij,kl} = a_{ij} + b_{kl}$. Show that the gene frequencies satisfy

$$\dot{p}_i = p_i(\sum_j a_{ij}p_j - \bar{m}) + \sum_j (B\mathbf{q})_j x_{ij}$$

$$\dot{q}_j = q_j(\sum_i b_{ij}q_i - \bar{m}) + \sum_i (A\mathbf{p})_i x_{ij}$$

with $\bar{m} = \bar{a} + \bar{b}$. Observe that the mean fitness depends only on the gene frequencies. Show by an argument like that in **4.5** that the mean fitness increases. Show that on invariant sets of constant mean fitness, entropy increases. Conclude that each solution converges to some equilibrium point and that at equilibrium points both the selection and the recombination terms vanish.

Exercise 4: Assume that fitnesses are *multiplicative* in (25.25): $w_{ij,kl} = a_{ij}b_{kl}$. Show that the Wright manifold $\mathbf{W}$ is an invariant manifold, i.e. linkage equilibrium is preserved. On $\mathbf{W}$, the gene frequencies evolve independently on the two loci. On $\mathbf{W}$, mean fitness increases. Is $\mathbf{W}$ necessarily attracting?

In the case of only two alleles at each of the two loci, the difference equation (25.25) reduces to (4.10) and the differential equation (25.27) to

$$\dot{x}_i = x_i(\sum_{j=1}^{4} m_{ij}x_j - \bar{m}) + \epsilon_i bD \quad i = 1,\dots,4 \tag{25.31}$$

where we adopt the notation of **4.3**: the birthrate of the double heterozygote is denoted by b and $\epsilon_1 = -\epsilon_2 = -\epsilon_3 = \epsilon_4 = -1$, $D = x_1x_4 - x_2x_3$.

Exercise 5: Consider the recombination equation $\dot{x}_i = \epsilon_i D$. Show that the triangular integrability conditions (24.17) are satisfied on the Wright manifold $\mathbf{W}$ but not on the whole simplex S_4.

This implies that the recombination equation is not a Shahshahani gradient. Theorem 24.6 shows then the existence of a 4×4 selection matrix such that (25.31) has periodic orbits. Thus the two-loci two-alleles system (25.31) is not gradient-like in general, and there is no chance to extend the Fundamental Theorem to two-loci systems.

25.6 Notes

The selection mutation models were formulated by Crow and Kimura (1970). The pure mutation equation and mutation with dominance free selection were studied by Moran (1976) for discrete and Akin (1979) for continuous time. **25.3** and **25.4** are based on Hofbauer (1985b). Theorem 23.3.2 is essentially due to Hadeler (1981). Special mutation rates were studied also by Kingman (1980), Turelli (1984), and Bürger (1988). **25.5** only scratches the vast and difficult area of multilocus systems. We refer to Nagylaki (1977) and Karlin (1975). Kun and Lyubich (1979) proved the convergence to equilibrium if fitness is additive, using Ewens' (1969) observation that mean fitness is still increasing in this case. Akin (1979, 1983, 1987) demonstrated that even in the simplest case of two alleles at two loci, stable limit cycles can occur. A particularly interesting application of multilocus systems is the theory of modifier genes, see Karlin and McGregor (1974) for a survey, Bürger (1983a,b) and Wagner and Bürger (1985) for a global analysis of the evolution of dominance modifiers, and Hutson and Law (1981) for a model for the evolution of recombination.

26. Fertility Selection

26.1 The fertility equation

All selection models treated so far relied on two basic assumptions: random mating (or rather random union of genes) resulting in Hardy-Weinberg proportions for the zygote population, and selection by means of viability differences in the genotypes. With these assumptions we could express the selection equations in terms of gene frequencies rather than genotype frequencies.

However, natural selection generally works in a more complex way. Random mating is not the rule in nature (tall women, for example, usually prefer tall men), and some types of pairing may be more fertile than others. In this chapter we will consider selection models taking into account nonrandom mating, different fertilities of mating pairs and different viabilities of genotypes. We start with a simple two-allelic case focussing on fertility and derive the general equations for n alleles afterwards.

With two alleles A_1 and A_2 there are three genotypes A_1A_1, A_1A_2 and A_2A_2 and hence 9 different mating types. We assign to each of these mating types an average fertility, as indicated in Table 26.1. (F_{12}, for example, denotes the expected number of offspring of an A_1A_1 father and an A_1A_2 mother). As in **2.1**, we denote by x,y and z the frequencies of A_1A_1, A_1A_2 and A_2A_2 in the parental generation. With random mating, xy counts the frequency of matings of an A_1A_1 male and an A_1A_2 female. The number of their offspring is proportional to xyF_{12}; half of them will be of type A_1A_1, the other half of type A_1A_2. Carrying this through for all mating types as indicated in Table 26.1, we obtain the frequencies x',y',z' of the genotypes of the next generation.

father	mother	frequency of matings	offspring A_1A_1	A_1A_2	A_2A_2	fertility
	A_1A_1	x^2	1	0	0	F_{11}
A_1A_1	A_1A_2	xy	1/2	1/2	0	F_{12}
	A_2A_2	xz	0	1	0	F_{13}
	A_1A_1	xy	1/2	1/2	0	F_{21}
A_1A_2	A_1A_2	y^2	1/4	1/2	1/4	F_{22}
	A_2A_2	yz	0	1/2	1/2	F_{23}
	A_1A_1	xz	0	1	0	F_{31}
A_2A_2	A_1A_2	yz	0	1/2	1/2	F_{32}
	A_2A_2	z^2	0	0	1	F_{33}

Table 26.1 Genotype frequencies in a fertility selection model for two alleles.

$$x' = \bar{F}^{-1}[F_{11}x^2 + \frac{1}{2}(F_{12} + F_{21})xy + \frac{1}{4}F_{22}y^2]$$

$$y' = \bar{F}^{-1}[\frac{1}{2}F_{22}y^2 + \frac{1}{2}(F_{12} + F_{21})xy + \frac{1}{2}(F_{23} + F_{32})yz + (F_{31} + F_{13})xz]$$

$$z' = \bar{F}^{-1}[F_{33}z^2 + \frac{1}{2}(F_{23} + F_{32})yz + \frac{1}{4}F_{22}y^2] \quad . \tag{26.1}$$

The factor $\bar{F}$ has to be chosen in such a way that the relation $x' + y' + z' = 1$ holds. Thus $\bar{F}$ is given by the sum of all terms in the brackets on the right hand side of (26.1). $\bar{F}$ is the *mean fertility* of the population.

For simplicity we assumed random mating and equal viabilities in our derivation of the *fertility equation* (26.1). These assumptions are not crucial, however. Dropping them leads to the same type of equations, with different biological meanings for the constants F_{ij}.

Let us see this for the general case of n alleles $A_1,...,A_n$. Consider a life cycle from zygote to adult age (with sex-dependent selection), followed by (possibly nonrandom) mating with different fertilities and production of the next zygote generation. Denoting the frequencies of A_iA_i at the zygote stage with x_{ii} and that of A_iA_j $(i \neq j)$ with $2x_{ij}$, one has

$$\sum_{i,j=1}^{n} x_{ij} = 1 \quad .$$

The frequency of the allele A_i is given by

$$x_i = \sum_{j=1}^{n} x_{ij} \quad . \tag{26.2}$$

We let $h(ij,rs)$ be the probability for a mating of an A_iA_j male with an A_rA_s female, multiplied with the fecundity (i.e. number of progeny) of such a pairing. Furthermore we denote by $m(ij)$ (resp. $f(ij)$) the viability (i.e. the probability of surviving into maturity) of an A_iA_j male (resp. female). An A_iA_j zygote is issued either from an $A_iA_r \times A_jA_s$ or an $A_jA_s \times A_iA_r$ mating, with arbitrary r,s. This gives the frequencies of the genotypes A_iA_j in the next zygote generation:

$$x'_{ij} = \frac{1}{\bar{F}} \sum_{r,s=1}^{n} \frac{1}{2}[h(ir,js)m(ir)f(js) + h(js,ir)m(js)f(ir)]x_{ir}x_{js} \, . \tag{26.3}$$

With

$$F(ir,js) = h(ir,js)m(ir)f(js) \tag{26.4}$$

and

$$f(ir,js) = \frac{1}{2}[F(ir,js) + F(js,ir)] \quad , \tag{26.5}$$

(26.3) can be interpreted as a *fertility selection equation*

$$x'_{ij} = \bar{F}^{-1} \sum_{r,s=1}^{n} f(ir,js)x_{ir}x_{js} \quad . \tag{26.6}$$

Here again the normalization factor

$$\bar{F} = \sum_{i,j,k,l} f(ij,kl)x_{ij}x_{kl} \tag{26.7}$$

is the *mean fertility* of the population.

The corresponding *continuous time fertility equation* is obtained as usual by replacing $x_i' - x_i$ by $\dot{x}_i$ and multiplying by the common factor $\bar{F}$:

$$\dot{x}_{ij} = \sum_{r,s=1}^{n} f(ir,js)x_{ir}x_{js} - x_{ij}\bar{F} \quad . \tag{26.8}$$

The state space of these equations is the simplex S_m with $m = \frac{n(n+1)}{2}$, which is forward invariant for both (26.6) and (26.8).

26.2 Two alleles

The two-allelic fertility equation (26.1) transforms for continuous time into

$$\begin{aligned}\dot{x} &= f_{11}x^2 + f_{12}xy + \frac{1}{4}f_{22}y^2 - x\bar{F}\\ \dot{y} &= \frac{1}{2}f_{22}y^2 + f_{12}xy + f_{23}yz + 2f_{13}xz - y\bar{F}\\ \dot{z} &= f_{33}z^2 + f_{32}yz + \frac{1}{4}f_{22}y^2 - z\bar{F}\end{aligned} \tag{26.9}$$

with symmetrized fertility coefficients $f_{ij} = \frac{1}{2}(F_{ij} + F_{ji})$ and

$$\bar{F}(x,y,z) = f_{11}x^2 + 2f_{12}xy + f_{22}y^2 + 2f_{13}xz + 2f_{23}yz + f_{33}z^2 \quad .$$

Since $x + y + z = 1$, this is a two-dimensional differential equation. The best way to analyse it is to use the variables

$$u = \frac{x}{y} \qquad v = \frac{z}{y} \quad . \tag{26.10}$$

Then (26.9) transforms - after dividing both equations by y - into

$$\begin{aligned}\dot{u} &= (f_{11} - f_{12})u^2 + (f_{12} - \frac{1}{2}f_{22})u + \frac{1}{4}f_{22} - f_{32}uv - 2f_{13}u^2v\\ \dot{v} &= (f_{33} - f_{32})v^2 + (f_{32} - \frac{1}{2}f_{22})v + \frac{1}{4}f_{22} - f_{12}uv - 2f_{13}uv^2 \, .\end{aligned} \tag{26.11}$$

This system is *competitive*: the off-diagonal terms of the Jacobian

$$\frac{\partial \dot{u}}{\partial v} = -f_{32}u - 2f_{13}u^2 \qquad \frac{\partial \dot{v}}{\partial u} = -f_{12}v - 2f_{13}v^2$$

are negative on the whole state space int $\mathbf{R}^2_+$. This excludes by Theorem 18.7 the possibility of periodic orbits. *The two-allelic fertility equation* (26.9) *is gradient-like.*

In fact (26.11) is even a gradient system. Indeed, let us write it in the concise form

$$\dot{u} = a(u) - vb(u)$$
$$\dot{v} = c(v) - ud(v) \tag{26.12}$$

with $b(u), d(v) > 0$ for $u,v > 0$. We define a Riemannian metric on int $\mathbf{R}^2_+$ by

$$<(\xi_1,\eta_1), (\xi_2,\eta_2)>_{(u,v)} = \frac{1}{b(u)}\xi_1\xi_2 + \frac{1}{d(v)}\eta_1\eta_2 \ . \tag{26.13}$$

Then

$$<(\dot{u},\dot{v}),(\xi,\eta)>_{(u,v)} = (\frac{a(u)}{b(u)} - v)\,\xi + (\frac{c(v)}{d(v)} - u)\eta = \frac{\partial V}{\partial u}\xi + \frac{\partial V}{\partial v}\eta$$

with

$$V(u,v) = \int \frac{a(u)}{b(u)}du - uv + \int \frac{c(v)}{d(v)}dv \ . \tag{26.14}$$

Thus (26.12) *is the gradient of the potential* V *with respect to the Riemannian metric* (26.13). Unfortunately, the explicit expression for V is rather complicated, and does not seem to allow a meaningful biological interpretation.

Exercise 1: Compute $V(u,v)$ explicitly.

Exercise 2: Determine the behaviour of the fertility equations (26.1) and (26.9), if there are no fertility differences (i.e. all $F_{ij} = 1$).

Exercise 3: For which fertility coefficients f_{ij} is the Hardy-Weinberg manifold $y^2 = 4xz$ invariant under (26.1) respectively (26.9)?

Exercise 4: Study the *symmetric fertility equation* (26.9) with $f_{11} = f_{33} = a$ and $f_{12} = f_{21} = f_{23} = f_{32} = b$, $f_{13} = f_{31} = c$, $f_{22} = 4d$.

(a) Show that $x = z$ is an invariant line which contains up to three fixed points. Determine their stability properties.

(b) Show that a pair of "asymmetric" fixed points (with $x \neq z > 0$) exists iff $0 < [2d(a+c) - ab]^{-1}[2(a-b)\sqrt{ad}\,] < 1$.

(c) Show that at an asymmetric fixed point, the Jacobian has negative trace and determinant $a(b-a)(x-z)^2$.

(d) If $a > b$ then the corners corresponding to pure homozygote populations are stable, and the asymmetric fixed points are saddles (whenever they exist).

(e) If $a < b$ then the corners are saddle points and the asymmetric fixed points (if they exist) are stable. Show that $\frac{1}{uv}$ is a Dulac function for (26.11) in this case.

(f) Sketch all possible (robust) phase portraits.

Exercise 5: Show that the symmetric difference equation (26.1) gives rise to a period doubling bifurcation (see **6.5**) on the line of symmetry $x = z$. Find parameter values a,b,c,d for which there is a stable orbit of period two, while none of the fixed points in S_3 are stable.

Exercise 6: Investigate the general two-allelic equation (26.9).

a) Show that there may be up to 5 interior fixed points.

b) If one of the homozygotes is stable and the other unstable, then there are at most 4 interior fixed points. In generic situations their number is even. (Hint: adapt the index theorem 19.4.1 to this situation).

Exercise 7: Consider the following model of *segregation distortion.* Assume that an $A_1A_1 \times A_1A_2$ mating produces offspring of genotype A_1A_1 and A_1A_2 in proportions $\lambda:(1-\lambda)$ ($\lambda = \frac{1}{2}$ is the usual case of Mendelian segregation), while the offspring of an $A_1A_2 \times A_2A_2$ mating is A_1A_2 resp. A_2A_2 with proportions $(1-\mu):\mu$ and the offspring of an $A_1A_2 \times A_1A_2$ mating splits up into A_1A_1, A_1A_2 and A_2A_2 as $\lambda(1-\mu):2(1-\lambda)(1-\mu):\mu(1-\lambda)$. Derive the recurrence equations for the genotype frequencies x,y,z. Find a projective change of coordinates which transforms this equation into the fertility equation (26.1). Analyse this equation (or its continuous time counterpart) in the three different cases where λ and μ are smaller resp. larger than $\frac{1}{2}$.

26.3 Multiplicative fertility

For more than two alleles the fertility equations (26.6) and (26.8) have not yet been analysed in their general form. We will restrict our attention to the special case when the fertility of a mating type can be split up into the contributions of the male and the female part. So let us assume that in (26.4)

$$F(ij,kl) = m_{ij} f_{kl} \quad . \tag{26.15}$$

We can then assign an average fertility to each allele A_i, namely

$$M(i) = \sum_{j=1}^{n} m_{ij} x_{ij} \tag{26.16}$$

in the male population, and

$$F(i) = \sum_{j=1}^{n} f_{ij} x_{ij} \tag{26.17}$$

in the female population. The fertility equations (26.6) and (26.8) take the form

$$x'_{ij} = \frac{M(i)\,F(j) + M(j)\,F(i)}{2\bar{F}} \tag{26.18}$$

and

$$\dot{x}_{ij} = \frac{1}{2}[M(i)\,F(j) + M(j)\,F(i)] - x_{ij}\bar{F} \tag{26.19}$$

with

$$\bar{F} = \sum_{i,j=1}^{n} M(i)\,F(j) = MF \tag{26.20}$$

where

$$M = \sum_{i=1}^{n} M(i) \quad \text{and} \quad F = \sum_{i=1}^{n} F(i) \quad . \tag{26.21}$$

If male and female contributions are equal, i.e. $m_{ij} = f_{ij}$, then we have $M(i) = F(i)$, $M = F$, and (26.18) and (26.19) reduce to

$$x'_{ij} = \frac{M(i)\,M(j)}{M^2} \tag{26.22}$$

$$\dot{x}_{ij} = M(i)\,M(j) - x_{ij}M^2 \quad . \tag{26.23}$$

In the discrete time model, genotype frequencies are in Hardy-Weinberg equilibrium after one generation. Indeed, the gene frequencies $x_i = \sum_{j=1}^{n} x_{ij}$ satisfy

$$x'_i = \sum_{j=1}^{n} x'_{ij} = \frac{M(i)}{M} \sum_{j=1}^{n} \frac{M(j)}{M} = \frac{M(i)}{M} \quad , \tag{26.24}$$

so that

$$x'_{ij} = x'_i\, x'_j \quad .$$

As soon as Hardy-Weinberg holds, (26.24) simplifies to

$$x'_i = x_i \frac{\sum_{j=1}^{n} m_{ij} x_j}{\sum_{k,l} m_{kl} x_k x_l} \quad . \tag{26.25}$$

The dynamics of the fertility equation (26.6) with sex-independent multiplicative fertilities is easy to describe: *after one generation the state of the population reaches the Hardy-Weinberg manifold* $x_{ij} = x_i x_j$ *and follows subsequently the classical selection equation* (3.5) *with viability parameters* m_{ij}. *In particular the "mean fitness"* $M = \sum m_{ij} x_{ij}$ *increases monotonically after the first step.*

Exercise 1: Show that the differential equation (26.23) has all fixed points on the Hardy-Weinberg manifold, but that this manifold is in general not invariant. However, (26.23) is still equivalent to (23.2). (Hint: Use the variables $X_i = \frac{M(i)}{M}$ instead of x_i).

In the general case of different male and female contributions, (26.18) cannot be reduced to the classical equation on S_n. But still, the equations can be considerably simplified by introducing the variables

$$X_i = \frac{M(i)}{M} \quad \text{and} \quad Y_i = \frac{F(i)}{F} \quad . \tag{26.26}$$

Indeed X_i' is proportional to

$$\sum_j m_{ij} x_{ij}' = \sum_j m_{ij} \frac{M(i)\,F(j) + M(j)\,F(i)}{2MF} = \frac{1}{2}[X_i \sum_j m_{ij} Y_j + Y_i \sum_j m_{ij} X_j] \; .$$

So (26.18) is equivalent to the following difference equation on $S_n \times S_n$:

$$\begin{aligned} X_i' &= \frac{1}{2\bar{m}}(X_i \sum_{j=1}^{n} m_{ij} Y_j + Y_i \sum_{j=1}^{n} m_{ij} X_j) \\ Y_i' &= \frac{1}{2\bar{f}}(X_i \sum_{j=1}^{n} f_{ij} Y_j + Y_i \sum_{j=1}^{n} f_{ij} X_j) \end{aligned} \tag{26.27}$$

with

$$\bar{m} = \sum_{i,j=1}^{n} m_{ij} X_i Y_j \quad \text{and} \quad \bar{f} = \sum_{i,j=1}^{n} f_{ij} X_i Y_j \quad .$$

In fact we do not need these computations to see that the fertility equation reduces to (26.27) for multiplicative fertility rates. Comparing with (26.4), we see that the case of multiplicative fertilities (26.15) is biologically equivalent to a pure viability selection model, without any fertility differences but with sex-dependent viability coefficients. Proceeding as in the derivation of the classical selection model (3.1), considering the gene frequencies X_i and Y_i of A_i in the adult gene pool one arrives immediately at (26.27). Since the x_{ij} in the fertility equation count genotype frequencies at the *zygote* stage, the *adult* gene frequencies X_i and Y_i

are related to them by (26.26), (26.16), (26.17) and (26.21).

Unfortunately not much is known on the *two-sex selection equation* (if $n \geq 3$), nor on its continuous counterpart

$$\dot{X}_i = \frac{1}{2}[X_i \sum_j m_{ij} Y_j + Y_i \sum_j m_{ij} X_j] - X_i \bar{m}$$
$$\dot{Y}_i = \frac{1}{2}[Y_i \sum_j f_{ij} X_j + X_i \sum_j f_{ij} Y_j] - Y_i \bar{f} \quad . \tag{26.28}$$

A particularly important open problem is the extent to which the Fundamental Theorem can be generalized to (26.27) and (26.28).

Exercise 2: Analyse (26.18), (26.19) or the two-sex equations (26.27) and (26.28) for $n = 2$ and compare the results with the one-sex case (cf. **3.5**):

(a) The homozygote A_1A_1 is stable iff $\dfrac{m_{12}}{m_{11}} + \dfrac{f_{12}}{f_{11}} \leq 2$.

(b) Show that there are at most 3 interior fixed points.

(c) Sketch all possible phase portraits of the differential equation.

Exercise 3: In the two-sex equations, with $m_{ij} = f_{ij}$, the subspace $\{(\mathbf{x},\mathbf{y}) \in S_n \times S_n : x_i = y_i \text{ for all } i\}$ is invariant and globally attracting.

Exercise 4: $f_{ij} = 1$ means that selection acts only in one sex, $f_{ij} = am_{ij} + b$ ($a > 0$) means that selection acts in the same way in both sexes but on a different scale. Show that in these cases $\mathbf{x} = \mathbf{y}$ holds at all equilibria and that all eigenvalues of the linearization are real. Equation (26.28) seems a good candidate for a gradient system.

Exercise 5: Simplify the differential equation (26.19) using (26.26) as new variables. Does this lead to (26.28)?

26.4 Additive fertility

The assumption of multiplicative fertilities works out much better for the difference equation than for the differential equation. It turns out that an additive splitting is a more appropriate simplification for the continuous time case. So let us assume that in (26.4)

$$F(ij,kl) = m_{ij} + f_{kl} \quad . \tag{26.29}$$

An important special case of (26.29) is when one of the two sexes has no influence on the fertility at all. (26.29) implies that the symmetrized fertility coefficients are of the form

$$f(ij,kl) = F_{ij} + F_{kl} \tag{26.30}$$

with

$$F_{ij} = \frac{m_{ij}+f_{ij}}{2} \quad . \tag{26.31}$$

The differential equation (26.8) simplifies then to

$$\begin{aligned} \dot{x}_{ij} &= \sum_{r,s}(F_{ir} + F_{js})x_{ir}x_{js} - x_{ij}\bar{F} \\ &= x_i F(j) + x_j F(i) - 2x_{ij}\bar{F} \quad . \end{aligned} \tag{26.32}$$

Here x_i is again the frequency (26.2) of the allele A_i, while

$$F(i) = \sum_{k=1}^{n} F_{ik}x_{ik} \tag{26.33}$$

is its average fertility in the population, and

$$\bar{F} = 2F = 2\sum_{i=1}^{n} F(i) = 2\sum_{i,j=1}^{n} F_{ij}x_{ij} \tag{26.34}$$

is the mean fertility of the whole population. For the gene frequency x_i we obtain

$$\dot{x}_i = F(i) - x_i F \quad . \tag{26.35}$$

Then

$$\begin{aligned} (x_{ij} - x_i x_j)^{\cdot} &= x_i F(j) + x_j F(i) - 2x_{ij}F - x_j F(i) + x_i x_j F - x_i F(j) + x_i x_j F \\ &= -(x_{ij} - x_i x_j)2F \quad . \end{aligned}$$

Since $F > 0$, $x_{ij} - x_i x_j$ converges to zero and in the limit $t \to +\infty$ the *Hardy-Weinberg relation*

$$x_{ij} = x_i x_j \tag{26.36}$$

holds. On the Hardy-Weinberg manifold we obtain

$$\dot{x}_i = F(i) - x_i F = x_i[\sum_k F_{ik}x_k - \sum_{r,s} F_{rs}x_r x_s] \quad . \tag{26.37}$$

This is the continuous time selection equation (23.2) with selection parameters F_{ij}. Hence *the mean fertility $\bar{F}$ increases monotonically,* according to the Fundamental Theorem.

Exercise: Show that the difference equation (26.6) with additive fertilities (26.30) is equivalent to a two-sex selection equation with no selection in one sex. The Hardy-Weinberg manifold is not invariant, however.

26.5 The fertility-mortality equation

The discrete time model in (26.1) was derived quite carefully, but the differential equation (26.8) was obtained in a rather high handed way. A biologically more satisfactory way would be to consider a continuous time model for overlapping generations with $x_{ij}(t)$ as frequencies of the A_iA_j genotypes at time t. During a small time interval Δt, $x_{ij}(t)$ will increase due to births by $\sum f(ir,js)x_{ir}x_{js}\Delta t$, with $f(ir,js)$ measuring again the fertility of an $A_iA_r \times A_jA_s$ mating, and it will decrease due to deaths by $d_{ij}x_{ij}\Delta t$, with d_{ij} as the death rate of the genotype A_iA_j. This leads to

$$\dot{x}_{ij} = \sum_{r,s=1}^{n} f(ir,js)x_{ir}x_{js} - d_{ij}x_{ij} - x_{ij}\bar{F} \quad . \tag{26.38}$$

Here

$$\bar{F} = \sum_{i,j,k,l} f(ij,kl)x_{ij}x_{kl} - \sum_{i,j} d_{ij}x_{ij}$$

ensures again that (26.38) maintains the relation $\sum_{i,j} x_{ij} = 1$. Thus the natural continuous time counterpart corresponding to the discrete time model (26.6) involves both different fertility and mortality rates, the latter corresponding to the viabilities of the genotypes.

Exercise 1: Show that adding a constant to the death rates d_{ij} does not change the equation (26.38). In particular, if all death rates are equal, $d_{ij} = d$, then (26.38) reduces to the pure fertility equation (26.8).

Exercise 2: Analyse the pure mortality equation with all $f(ij,kl) = 1$.

(a) Show that the Hardy-Weinberg manifold is invariant iff the death rates are additive, i.e. $d_{ij} = d_i + d_j$.

(b) If there exists an interior fixed point with Hardy-Weinberg proportions then all death rates are equal: $d_{ij} = d$.

(c) Analyse the case of two alleles and compare with the classical selection equation (23.2).

Exercise 3: If (26.38) leaves the Hardy-Weinberg manifold invariant, then the fertilities are additive (cf. (26.30)) and the death rates are additive.

Assume now that both fertility and mortality rates are additive, i.e. that

$$f(ij,kl) = f_{ij} + f_{kl} \quad \text{and} \quad d_{ij} = d_i + d_j \quad . \tag{26.39}$$

Then a similar computation as in **26.4** shows that the Hardy-Weinberg relations $x_{ij} = x_i x_j$ are invariant, and that the flow on the Hardy-Weinberg manifold is given by

$$\dot{x}_i = x_i[\sum_{k=1}^{n} f_{ik} x_k - d_i + \bar{d} - \bar{f}] \tag{26.40}$$

with

$$\bar{d} = \sum_{k=1}^{n} d_k x_k \quad \text{and} \quad \bar{f} = \sum_{i,j} f_{ij} x_i x_j \quad .$$

(26.40) is equivalent to the selection equation (23.2) with Malthusian parameters $m_{ij} = f_{ij} - d_i - d_j$, f_{ij} corresponding to the birth rate and $d_i + d_j$ to the death rate of A_iA_j.

Now remember that the derivation of the continuous selection equation (23.2) relied on the a priori assumption of a Hardy-Weinberg equilibrium. Strictly speaking this assumption was not at all justified. The more satisfactory approach would be to start with genotype frequencies and *prove* that Hardy-Weinberg holds. The above derivation (in particular, exercise 26.5.3) shows exactly when this approach works:

a) the fertilities of a mating have to split up into additive contributions of each genotype; and (more seriously)

b) the genes A_i and A_j have to contribute additively to the death rate of the genotype A_iA_j.

The moral is that the classical selection equation (23.2) is "correct" only under the additional assumption (b). Of course, this does not alter its value as the basic continuous time model in population genetics, in particular as there is no understanding of the dynamic behaviour of the fertility equation (26.8) or the fertility-mortality equation (26.38) for 3 or more alleles.

In contrast to the result in **26.2**, the fertility-mortality equation gives rise to periodic orbits even for two alleles:

Exercise 4: Show that after adding to the right hand side of (26.9) the mortality terms $-d_1 x, -d_2 y$ and $-d_3 z$, a Hopf bifurcation can occur. (Hint: Try for example $f_{11} = 14.55$, $f_{12} = 0.25$, $f_{22} = 14.5$, $f_{13} = 0.02$,

$f_{23} = 3.1$, $f_{33} = 2.89$, $d_1 = 0.8$, $d_2 = 0.9$, $d_3 = 0.05$. The corners (1,0,0) and (0,0,1) are asymptotically stable. There are two saddles in S_3 and a center $(\frac{1}{8}, \frac{1}{4}, \frac{5}{8})$. Use f_{22} as bifurcation parameter).

26.6 Notes

The general fertility equation for separated generations was set up by Roux (1977). Bodmer (1965) had analysed the case of multiplicative fertilities and its relation to the two-sex equation, and studied the two-allelic model: see also Karlin (1972) and Ewens (1979) for reviews. An excellent survey on discrete time selection models was given by Pollak (1979). The symmetric two-allelic case (exercises 26.2.4-5) was analysed by Hadeler and Libermann (1975) and Butler et al. (1982). Exercise 26.2.7 is from Karlin (1972). A detailed analysis of the two-sex equation, in particular the case $m_{ij} + f_{ij} = 1$ which arises in sex ratio models, is due to Karlin (1984) and Karlin and Lessard (1986). The continuous time model (26.38) was derived by Nagylaki and Crow (1974), see also Nagylaki (1977), Ewens (1979), and Hoppensteadt (1975) for a related approach. Hadeler and Glas (1983) showed that the fertility equation (26.9) is gradient-like whereas Koth and Kemler (1986) found stable limit cycles in the fertility-mortality equation (26.38). This is also a consequence of Akin (1987), see Hofbauer (1985a).

PART VII: ON SEX AND GAMES: STRATEGIC AND GENETIC EVOLUTION

27. Evolutionary Dynamics for Bimatrix Games

27.1. Dynamics for bimatrix games

In this section we resume the study of asymmetric conflicts, described by bimatrix games. Let us recall from chapter 17 that we consider two different populations playing the strategies $E_1,\dots,E_n$ and $F_1,\dots,F_m$ with frequencies $x_1,\dots,x_n$ and $y_1,\dots,y_m$ respectively. After a contest E_i versus F_j, the payoff for the first player is a_{ij} and for the second player b_{ji}. As the standard evolutionary dynamics for this type of games we introduced the following differential equation on $S_n \times S_m$:

$$\begin{aligned} \dot{x}_i &= x_i[(A\mathbf{y})_i - \mathbf{x}\cdot A\mathbf{y}] \\ \dot{y}_j &= y_j[(B\mathbf{x})_j - \mathbf{y}\cdot B\mathbf{x}] \quad . \end{aligned} \tag{27.1}$$

A slightly different dynamics includes normalization by mean payoff:

$$\begin{aligned} \dot{x}_i &= x_i\frac{(A\mathbf{y})_i - \mathbf{x}\cdot A\mathbf{y}}{\mathbf{x}\cdot A\mathbf{y}} \\ \dot{y}_j &= y_j\frac{(B\mathbf{x})_j - \mathbf{y}\cdot B\mathbf{x}}{\mathbf{y}\cdot B\mathbf{x}} \quad . \end{aligned} \tag{27.2}$$

These equations are motivated by the discrete time model, which is based on the assumption that the frequency of the E_i players in the next generation, x_i', is proportional to $x_i(A\mathbf{x})_i$, since $(A\mathbf{x})_i$ is the mean payoff for E_i. Since $\sum x_i' = 1$, the multiplication rate from one generation to the next has to be normalized by the mean payoff for the respective population. This gives

$$x_i' = x_i\frac{(A\mathbf{y})_i}{\mathbf{x}\cdot A\mathbf{y}} \qquad y_j' = y_j\frac{(B\mathbf{x})_j}{\mathbf{y}\cdot B\mathbf{x}} \quad . \tag{27.3}$$

The above differential equation (27.2) follows then from the straightforward approximation $x_i' - x_i \sim \dot{x}_i$. Since in general $\mathbf{x}\cdot A\mathbf{y} \neq \mathbf{y}\cdot B\mathbf{x}$ except for partnership games ($A^t = B$), (27.2) cannot be reduced to (27.1) by a change of velocity. In fact we will observe soon that (27.1) and (27.2) show quite different qualitative behaviour.

Equations (27.3) and (27.2) are well defined on $S_n \times S_m$ provided the payoffs a_{ij} and b_{ji} are positive numbers, a restriction which is not needed for (27.1). The biological reason for this is the different meaning of payoffs: In the discrete model (27.3) and hence also in its offshoot (27.2), the a_{ij} are interpreted as *multiplication rates* from one generation to the next (as e.g. the selection parameters w_{ij} in the discrete time selection model (3.1)). The a_{ij} in (27.1) however, measure only the changes in fitness caused by the contests (thus corresponding to the *Malthusian fitness* m_{ij} in (23.1)). In order to obtain a better understanding of the relation between (27.1) and the other two equations, we should therefore add some large constant C, measuring the common "background fitness" of the individuals, to the payoffs a_{ij} and b_{ji} in (27.2) and (27.3). It is then easy to see that (27.1) is the limiting case of (27.2) as $C \to \infty$, after a compression of the time-scale with factor C. This suggests that the dynamics of (27.1) should be somewhat simpler than that of the other two equations.

Exercise 1: Show that all three dynamics have the same fixed points.

Exercise 2: Compute the linearization of (27.2) and obtain a result for the position of the eigenvalues similar to the one shown for (27.1) in **17.4**. Prove that an isolated interior equilibrium is always unstable for the discrete dynamics (27.3).

Exercise 3: If no interior equilibrium exists then all interior orbits of (27.2) converge to the boundary of $S_n \times S_m$. (For (27.1) cf. exercise 17.4.2).

Exercise 4. A Nash equilibrium is a fixed point of any of the three dynamics, but the converse does not hold.

Exercise 5: If all Nash equilibria are regular, then the sum of their indices is $(-1)^{n+m}$. In particular, their number is odd. Every bimatrix game has at least one Nash equilibrium. (Hint: Proceed as in the proof of the index theorem 19.4.1. Show that the Nash equilibria correspond to the rest points of the modified equation

$$\begin{aligned} \dot{x}_i &= x_i[(A\mathbf{y})_i - \mathbf{x}\cdot A\mathbf{y} - n\epsilon] + \epsilon \\ \dot{y}_j &= y_j[(B\mathbf{x})_j - \mathbf{y}\cdot B\mathbf{x} - m\epsilon] + \epsilon) \quad . \end{aligned} \tag{27.4}$$

27.2. Partnership games and zero-sum games

(27.1) is not only simpler than the two other equations, it is also immune to rescalings of the payoffs by the addition of arbitrary constants to the columns of the payoff matrices A and B. More generally, a game (A',B') is said to be a *rescaling* of (A,B) if there exist constants c_j, d_i and $\alpha > 0$, $\beta > 0$ such that

$$a'_{ij} = \alpha a_{ij} + c_j \qquad b'_{ji} = \beta b_{ji} + d_i \quad . \tag{27.5}$$

We then write $(A,B) \sim (A',B')$.

Exercise 1: Show that rescaling a game does not change its equilibrium points.

Rescalings are of particular interest for partnership games, where the payoff is equally shared between the two players, and for zero-sum games, where the gain of one player is the loss of the other. Thus (A,B) is a *rescaled partnership game* if $(A,B) \sim (C,C^t)$ and a *rescaled zero-sum game* if $(A,B) \sim (C,-C^t)$ for some suitable $n \times m$-matrix C. More precisely, we shall say that (A,B) is a *c-partnership game* (with $c > 0$) or a *c-zero-sum game* (with $c < 0$) if there exist suitable c_{ij}, c_j and d_i such that

$$a_{ij} = c_{ij} + c_j \qquad b_{ji} = c c_{ij} + d_i \tag{27.6}$$

for all i and j.

We shall first show that c-partnership games correspond exactly to Shahshahani-type gradients. The tangent space at a point $(\mathbf{x},\mathbf{y}) \in \operatorname{int} S_n \times S_m$ consists of vectors $(\boldsymbol{\xi},\boldsymbol{\eta})$ with $\boldsymbol{\xi} \in \mathbf{R}^n_0$ and $\boldsymbol{\eta} \in \mathbf{R}^m_0$ (i.e. $\sum \xi_i = \sum \eta_j = 0$). As inner product in the tangent space, we define

$$<(\boldsymbol{\xi},\boldsymbol{\eta}),(\boldsymbol{\xi}',\boldsymbol{\eta}')>_{(\mathbf{x},\mathbf{y})} = \sum \frac{\xi_i \xi'_i}{x_i} + \frac{1}{c} \sum \frac{\eta_j \eta'_j}{y_j} \quad . \tag{27.7}$$

The corresponding gradient (cf. **25.2**) is said to be a *c-gradient*.

Theorem 1. *The following conditions are equivalent:*
(RP1) (A,B) *is a c-partnership game;*
(RP2) (27.1) *is a c-gradient;*
(RP3) $c\boldsymbol{\xi} \cdot A\boldsymbol{\eta} = \boldsymbol{\eta} \cdot B\boldsymbol{\xi}$ *for all* $\boldsymbol{\xi} \in \mathbf{R}^n_0, \boldsymbol{\eta} \in \mathbf{R}^m_0$;
(RP4) *For all* $i,k \in \{1,\dots,n\}$ *and* $j,l \in \{1,\dots,m\}$

$$c(a_{ij} - a_{il} - a_{kj} + a_{kl}) = b_{ji} - b_{li} - b_{jk} + b_{lk} \quad ; \tag{27.8}$$

(RP5) *There exist* u_i, v_j *such that* $Q = cA - B^t$ *satisfies* $q_{ij} = u_i + v_j$ *for all* i *and* j.
The potential of (27.1) *is then given by* $\mathbf{x} \cdot C\mathbf{y}$. *If* (A,B) *is a partnership game, then this is just the average payoff.*

Proof: (RP1) ⇒ (RP2). For every $(\xi,\eta) \in \mathbf{R}_0^n \times \mathbf{R}_0^m$ we have, according to (27.7):

$$\langle(\dot{\mathbf{x}},\dot{\mathbf{y}}),(\xi,\eta)\rangle_{(\mathbf{x},\mathbf{y})} = \sum \xi_i[(A\mathbf{y})_i - \mathbf{x}\cdot A\mathbf{y}] + \frac{1}{c}\sum \eta_j[(B\mathbf{x})_j - \mathbf{y}\cdot B\mathbf{x}]$$

$$= \xi\cdot C\mathbf{y} + \eta\cdot C^t\mathbf{x} = \xi\cdot C\mathbf{y} + \mathbf{x}\cdot C\eta \quad .$$

This last expression is just $D_{(\mathbf{x},\mathbf{y})}V(\xi,\eta)$, where V is the function $(\mathbf{x},\mathbf{y}) \to \mathbf{x}\cdot C\mathbf{y}$.
(RP2) ⇒ (RP3). If

$$\langle(\dot{\mathbf{x}},\dot{\mathbf{y}}),(\xi,\eta)\rangle_{(\mathbf{x},\mathbf{y})} = D_{(\mathbf{x},\mathbf{y})}V(\xi,\eta)$$

for some function V defined in a neighbourhood of $\operatorname{int} S_n \times S_m$, then

$$\sum \xi_i[(A\mathbf{y})_i - \mathbf{x}\cdot A\mathbf{y}] + \frac{1}{c}\sum \eta_j[(B\mathbf{x})_j - \mathbf{y}\cdot B\mathbf{x}] = \sum \frac{\partial V}{\partial x_i}\xi_i + \sum \frac{\partial V}{\partial y_j}\eta_j$$

for all $\xi \in \mathbf{R}_0^n, \eta \in \mathbf{R}_0^m$. Hence

$$\sum \xi_i (A\mathbf{y})_i = \sum \frac{\partial V}{\partial x_i}\xi_i$$

for all $\xi \in \mathbf{R}_0^n$, which implies

$$\frac{\partial V}{\partial x_i}(\mathbf{x},\mathbf{y}) = (A\mathbf{y})_i + g(\mathbf{x},\mathbf{y})$$

for a suitable function g which does not depend on i, and similarly

$$\frac{\partial V}{\partial y_j}(\mathbf{x},\mathbf{y}) = \frac{1}{c}(B\mathbf{x})_j + h(\mathbf{x},\mathbf{y}) \quad .$$

Differentiating again, we obtain

$$a_{ij} = \frac{\partial^2 V}{\partial x_i \partial y_j} - \frac{\partial g}{\partial y_j} \quad \text{and} \quad \frac{1}{c}b_{ji} = \frac{\partial^2 V}{\partial y_j \partial x_i} - \frac{\partial h}{\partial x_i} \quad .$$

This implies that $Q = cA - B^t$ satisfies $\xi\cdot Q\eta = 0$ for all $\xi \in \mathbf{R}_0^n$ and $\eta \in \mathbf{R}_0^m$.

(RP3) ⇒ (RP4). This follows by choosing $\xi = \mathbf{e}_i - \mathbf{e}_k$ (where $\mathbf{e}_i,\mathbf{e}_k$ are from the standard basis in $\mathbf{R}^n$) and $\eta = \mathbf{f}_j - \mathbf{f}_l$ (with $\mathbf{f}_j, \mathbf{f}_l$ in $\mathbf{R}^m$).

(RP4) ⇒ (RP5). By (27.8) we have for $Q = cA - B^t$

$$q_{ij} - q_{il} - q_{kj} + q_{kl} = 0 \quad .$$

With $k = n, l = m$ this implies

$$q_{ij} = q_{im} + q_{nj} - q_{nm} = u_i + v_j$$

where $u_i = q_{im}$ and $v_j = q_{nj} - q_{nm}$.

(RP5) ⇒ (RP1). Since $ca_{ij} - b_{ji} = u_i + v_j$, we have only to set $c_{ij} = \frac{1}{c}(b_{ji} + u_i)$ to obtain (27.6).

Exercise 2: Show that a strict maximum of $\mathbf{x} \cdot C\mathbf{y}$ on $S_n \times S_m$ corresponds to a pure strategy. Deduce that every rescaled partnership game has a pure Nash equilibrium.

Exercise 3: If (A,B) is a c-partnership game, then (27.2) is a gradient. (Hint: the definition of the inner product has to be changed, by replacing, in the right hand side of (27.7), x_i by $x_i(\mathbf{x} \cdot A\mathbf{y})^{-1}$ and y_j by $y_j(\mathbf{y} \cdot B\mathbf{x})^{-1}$).

Theorem 2. *If* (27.6) *is valid for some* $c \in \mathbf{R}$, *and if* (A,B) *has an interior Nash equilibrium* $(\mathbf{p},\mathbf{q})$, *then the function*

$$H(\mathbf{x},\mathbf{y}) = c\sum p_i \log x_i - \sum q_j \log y_j \tag{27.9}$$

is a constant of motion for (27.1).

Proof: $\dot{H}(\mathbf{x},\mathbf{y}) = c\sum p_i \frac{\dot{x}_i}{x_i} - \sum q_j \frac{\dot{y}_j}{y_j}$

$$= c(\mathbf{p} - \mathbf{x}) \cdot A\mathbf{y} - (\mathbf{q} - \mathbf{y}) \cdot B\mathbf{x}$$

$$= c(\mathbf{p} - \mathbf{x}) \cdot A(\mathbf{y} - \mathbf{q}) - (\mathbf{q} - \mathbf{y}) \cdot B(\mathbf{x} - \mathbf{p}) = 0 \quad .$$

In particular, an interior equilibrium $(\mathbf{p},\mathbf{q})$ for a c-zero-sum game is always stable (but not asymptotically stable).

Exercise 4: Prove that the time average of every interior orbit of (27.1) converges to $(\mathbf{p},\mathbf{q})$ in a rescaled zero sum game.

Exercise 5: Show that the Battle of the Sexes from **17.2** is a rescaled zero-sum game.

Exercise 6: Show that all eigenvalues at an interior equilibrium of a zero-sum game are purely imaginary.

Exercise 7: Show that a game (A,B) is a c-zero-sum game iff (27.8) holds for all i,j,k,l. (Here, $c < 0$).

Exercise 8: Analyze the behaviour of (27.1) for a rescaled zero-sum game without interior equilibrium: Show that (27.9) is a monotonically increasing Ljapunov function, if $(\mathbf{p},\mathbf{q})$ is a Nash equilibrium on the boundary.

Condition (27.8) can be interpreted in terms of phenotypes playing mixed strategies. Let us assume that there are N phenotypes in the first population, corresponding to the strategies $\mathbf{p}^i \in S_n$ and occurring with frequencies $x_i (i = 1,...,N)$; similarly, that there are M phenotypes in the second population, corresponding to strategies $\mathbf{q}^j \in S_m$ and occurring with frequencies $y_j (j = 1,...,M)$. The strategy mixtures in the two populations are given by $\mathbf{p} = \sum x_i \mathbf{p}^i$ and $\mathbf{q} = \sum y_j \mathbf{q}^j$. Since the payoff for the phenotype $\mathbf{p}^i$ is given by $\mathbf{p}^i \cdot A\mathbf{q}$ and that for $\mathbf{q}^j$ by $\mathbf{q}^j \cdot B\mathbf{p}$, the game dynamical equations for the *mixed strategist game* are given by

$$\dot{x}_i = x_i[(\bar{A}\mathbf{y})_i - \mathbf{x} \cdot \bar{A}\mathbf{y}] \qquad i = 1,...,N$$
$$\dot{y}_j = y_j[(\bar{B}\mathbf{x})_j - \mathbf{y} \cdot \bar{B}\mathbf{x}] \qquad j = 1,...,M$$

where $\bar{a}_{ij} = \mathbf{p}^i \cdot A\mathbf{q}^j$ and $\bar{b}_{ji} = \mathbf{q}^j \cdot B\mathbf{p}^i$. Thus $\bar{A} = PAQ^t$ and $\bar{B} = QBP^t$, where P is the stochastic $N \times n$ matrix with rows $\mathbf{p}^i$ and Q is the stochastic $M \times m$ matrix with rows $\mathbf{q}^j$. (A nonnegative matrix A is *stochastic* if each row sums up to 1.)

Exercise 9: If (A',B') is a rescaling of (A,B), then $(\bar{A}',\bar{B}')$ is a rescaling of $(\bar{A},\bar{B})$. In particular, if (A,B) is a c-partnership game (or a c-zero sum game), then so are all mixed strategist games (PAQ^t,QBP^t), for all stochastic $N \times n$-matrices P and $M \times m$-matrices Q.

Exercise 10: (A,B) is a c-partnership game (or a c-zero-sum game) iff the mixed strategists games (PAQ^t,QBP^t) are c-partnership games (resp. c-zero sum games), for all $2 \times n$ matrices P having as rows the basis vectors $\mathbf{e}_i$ and $\mathbf{e}_k$ from $\mathbf{R}^n$, and all $2 \times m$-matrices Q having as rows the basis vectors $\mathbf{f}_j$ and $\mathbf{f}_l$ from $\mathbf{R}^m$. (Hint: In this case

$$\bar{A} = \begin{bmatrix} a_{ij} & a_{il} \\ a_{kj} & a_{kl} \end{bmatrix} \quad \text{and} \quad \bar{B} = \begin{bmatrix} b_{ji} & b_{jk} \\ b_{li} & b_{lk} \end{bmatrix}$$

and $(\bar{A},\bar{B})$ is a c-partnership game (resp. a c-zero sum game) iff (27.8) is valid for some $c > 0$ (resp. < 0).

In particular, the following conditions are equivalent to (RP1)-(RP5):

(RP6) *The mixed strategist games* (PAQ^t, QBP^t) *are* c*-partnership games, for all stochastic* $N \times n$ *matrices* P *and all stochastic* $M \times m$ *matrices* Q.

(RP7) *If both populations consist of two phenotypes corresponding to pure strategies, the mixed strategist game is a* c*-partnership game.*

The dynamic behaviour of (27.2) for general rescaled zero-sum games is not yet known. The results in **27.4** suggest that a Nash equilibrium is asymptotically stable. A partial result in this direction is obtained in the following exercise.

Exercise 11: Show for $\beta a_{ij} + \alpha b_{ji} =$ const., that (27.9) is a monotonically increasing Ljapunov function for (27.2).

Exercise 12: Show that generically a 2×2 game is either a rescaled zero-sum or a rescaled partnership game.

Exercise 13: As a typical example for a game which can be rescaled neither to a zero-sum nor to a partnership game, consider the 3×3 game

$$A = \begin{pmatrix} 1 & 0 & 0 \\ 0 & 1 & 0 \\ 0 & 0 & 1 \end{pmatrix} \qquad B = \begin{pmatrix} a & b & c \\ c & a & b \\ b & c & a \end{pmatrix}$$

Under what conditions is (A,B) a rescaled zero-sum game? When does (27.1) have an attractor on the boundary? When is (27.2) permanent? When does a Hopf bifurcation occur?

27.3 An example: the origin of the male-female phenomenon

In sexually reproducing species, *anisogamy* is a very frequent phenomenon. Some individuals produce many small gametes (sperm), the others produce fewer and larger ones (the eggs), and zygotes occur by the fusion of an egg with a sperm. The sperm-producers are by definition the *males*, and the egg-producers the *females* of the species.

We shall model the origin of anisogamy by a partnership game, supposing that the species consist of two mating types to begin with. Fusion can only occur between gametes of different types. These mating types are the two populations of the game. The individuals are assumed to produce a total gametic mass M. A strategy consists in dividing this mass into individual gametes. The strategy is pure if all gametes are of the same size. The common payoff is the number of viable zygotes. There is a trade-off between producing many zygotes and ensuring the survival of the zygotes, that is between small gametes in large numbers or few gametes with lots of nutritive material.

With $W(m)$ we denote the survival probability of a zygote of initial mass m. It is reasonable to assume that it is monotonically increasing to some constant ≤ 1, and zero for all m in some interval $[0,d]$. If one mating type produces i and the other mating type j gametes of equal size, then their common payoff is the product of the number of zygotes with their survival probability, and hence proportional to $ij\,W(\frac{M}{i}+\frac{M}{j})$. In other words, if $c_1 < c_2 < \cdots < c_n$ are the gamete sizes available, then the payoff matrices are given, up to a constant factor, by A and $B = A^t$, with $a_{ij} = (c_i c_j)^{-1} W(c_i + c_j)$.

The resulting dynamics is a gradient leading to the establishment of pure strategies: mixed strategies (i.e. hermaphrodites) will be selected against. The stable equilibria are the maxima of $\mathbf{x} \cdot A\mathbf{y}$, and hence correspond to those pairs (c_i,c_j) satisfying $a_{kj} \leq a_{ij}$ for all k and $a_{il} \leq a_{ij}$ for all l.

In order to analyze these pairs, it is convenient to look at the corresponding continuous problem obtained by assuming that all u in some compact interval $[k,M]$ are available as gametic masses (the minimal gametic mass k consists only of the DNA molecule). We must then look for the maxima of $(uv)^{-1} W(u+v)$ on the square $k \leq u,\ v \leq M$. With $u+v=t$ and $u-v=s$, we have to analyze $(t^2-s^2)^{-1} W(t)$. Its critical points $(\bar{t},\bar{s})$ are given by

$$\bar{s} = 0 \qquad \bar{t}\frac{dW}{dt}(\bar{t}) = 2\,W(\bar{t}) \quad .$$

The function $t \rightarrow t^{-2} W(t)$ is positive for $t \geq 0$, its value is 0 for $t \leq d$ and tends to 0 for $t \rightarrow +\infty$. Thus there exists at least one local maximum. It is unique under mild restrictions. We take $\hat{t}$ to be the smallest value where a maximum is attained. The point $(\hat{t},0)$ is obviously a saddle for $(t^2-s^2)^{-1} W(t)$.

It follows that the maxima of $(uv)^{-1} W(u+v)$ are on the boundary of the square $[k,M] \times [k,M]$. If $\hat{t} \geq M$, then the unique maximum is at the corner (M,M) of the square. If $\hat{t} \in (k,M)$, then the maxima occur in symmetric pairs on the boundary, and no longer at the corner (M,M).

Let us assume now that in the course of evolution, the total gametic mass is slowly increasing. If M is less than $\hat{t}$, both mating types produce the largest gametes possible. But if M crosses the threshold value $\hat{t}$, a disruption of the gametic sizes occurs. One type produces larger gametes than the other. For M slightly larger than $\hat{t}$, the "females" produce gametes of largest possible mass M and the "males" adjust such that the payoff is optimal. For still larger values of M, the males produce gametes with the smallest possible mass k and the females adjust.

In the fitness landscape, the state moves (with increasing M) at first along the diagonal valley up to the "pass" $u = v = \hat{t}/2$, and then further uphill, away from the diagonal.

27.4. Conservation of volume

In higher dimensions, (27.1) need not admit a constant of motion, but it will still preserve some (properly modified) volume.

First we divide the vectorfields (27.1) and (27.2) by the positive function $P = \prod_{i=1}^{n} x_i \prod_{j=1}^{m} y_j$. We call these modified vectorfields (I) and (II) and compute their divergence. The divergence of (I) in $\operatorname{int} \mathbf{R}^n \times \mathbf{R}^m$ is given by

$$\operatorname{div}(\mathrm{I}) = \sum_{i=1}^{n} \frac{\partial \dot{x}_i}{\partial x_i} + \sum_{j=1}^{m} \frac{\partial \dot{y}_j}{\partial y_j} =$$

$$= \frac{1}{P}\left\{-\sum_{i=1}^{n} x_i(A\mathbf{y})_i - \sum_{j=1}^{m} y_j(B\mathbf{x})_j\right\} = -\frac{1}{P}(\mathbf{x}\cdot A\mathbf{y} + \mathbf{y}\cdot B\mathbf{x})\,. \qquad (27.10)$$

In order to obtain the divergence div_0 within the state space $\operatorname{int} S_n \times S_m$, we have to subtract the eigenvalues of the Jacobian which are orthogonal to S_n and S_m. Since

$$\left(\sum x_i\right)^{\cdot} = \left(1 - \sum x_i\right)\mathbf{x}\cdot A\mathbf{y} \quad ,$$

for (27.1) these two superfluous eigenvalues are $-\mathbf{x}\cdot A\mathbf{y}$ and $-\mathbf{y}\cdot B\mathbf{x}$. Taking into account the factor $\frac{1}{P}$, we see from (27.10) that the standard dynamics (27.1) is divergence free:

$$\operatorname{div}_0(\mathrm{I}) \equiv 0 \ . \qquad (27.11)$$

By Liouville's formula (19.23), (I) *preserves volume. Up to a change in velocity the flow* (27.1) *is incompressible.*

Exercise 1: Show that (27.1) cannot have attractors in int $S_n \times S_m$. In particular, an interior equilibrium cannot be asymptotically stable for (27.1).

By a similar calculation we obtain for the modification (II) of (27.2):

$$\operatorname{div}(\mathrm{II}) = -\frac{1}{P}\left\{\sum_{i=1}^{n} x_i \frac{(A\mathbf{y})_i^2}{(\mathbf{x}\cdot A\mathbf{y})^2} + \sum_{j=1}^{m} y_j \frac{(B\mathbf{x})_j^2}{(\mathbf{y}\cdot B\mathbf{x})^2}\right\} .$$

The eigenvalues orthogonal to S_n and S_m are seen to be both $= -1$. Thus

$$\operatorname{div}_0(\mathrm{II}) = \operatorname{div}(\mathrm{II}) + \frac{2}{P} \tag{27.12}$$

$$= -\frac{1}{P}\left\{\sum_{i=1}^{n} x_i \left[\frac{(A\mathbf{y})_i - \mathbf{x}\cdot A\mathbf{y}}{\mathbf{x}\cdot A\mathbf{y}}\right]^2 + \sum_{j=1}^{m} y_j \left[\frac{(B\mathbf{x})_j - \mathbf{y}\cdot B\mathbf{x}}{\mathbf{y}\cdot B\mathbf{x}}\right]^2\right\}$$

Thus $\operatorname{div}_0(\mathrm{II}) \le 0$ on int $S_n \times S_m$ and by (19.23), (27.2) is *volume contracting*. The rate of contraction corresponds to the variance of the vector field and tends to zero near an interior equilibrium point. This reflects the fact that (27.1) and (27.2) have essentially the same linearization near fixed points.

Exercise 2: Show that the equations (27.4) are volume contracting too.

Let us apply this result now to complete the discussion of two-strategy games ($n = m = 2$). What is still lacking is the behaviour of (27.2) for rescaled zero-sum games.

Consider the general Battle of the Sexes

$$A = \begin{bmatrix} \alpha & b+\beta \\ a+\alpha & \beta \end{bmatrix} \qquad B = \begin{bmatrix} c+\gamma & \delta \\ \gamma & d+\delta \end{bmatrix} \tag{27.13}$$

with $a,b,c,d > 0$ and $\alpha,\beta,\gamma,\delta \ge 0$.

Exercise 3: Show that if $n = m = 2$ this is the general form of a rescaled zero-sum game with a unique interior fixed point.

For (27.1) we may discard the constants $\alpha,\beta,\gamma,\delta$ and compute the interior Nash equilibrium to

$$\bar{p} = \frac{d}{c+d} \qquad \bar{q} = \frac{b}{a+b} \quad . \tag{27.14}$$

For (27.2), formula (27.12) shows that P^{-1} is a Dulac function and that periodic orbits are impossible (see **18.1**). Since the area is decreasing we conclude that *the interior equilibrium* (27.14) *is globally asymptotically stable*. Although the mixed strategy $(\bar{p},\bar{q})$ is not an ESS (recall exercise 17.1), it is an attractor for (27.2).

Exercise 4. Give another proof of this by finding a global Ljapunov function.

Exercise 5: Prove that the boundary of $S_2 \times S_2$ is a repellor. (Hint: Construct an average Ljapunov function as shown in **19.6**).

Exercise 6: Analyze the behaviour of the difference equation (27.3) for (27.13).

27.5. Nash-Pareto pairs

We have seen in the last section that in the Battle of the Sexes endowed with dynamics (27.2), the mixed Nash equilibrium is asymptotically stable. This indicates that mixed strategies also are of some relevance in evolutionary terms, although they cannot be evolutionarily stable. This raises the problem of relaxing the notion of ESS, which is a rather narrow one for asymmetric contests, so that it may include mixed strategies.

Suppose that the two populations are in state $(\mathbf{p},\mathbf{q}) \in S_n \times S_m$. This state will certainly *not* be stable in any evolutionary sense when there exists a state $(\mathbf{x},\mathbf{y})$ near $(\mathbf{p},\mathbf{q})$ such that *both* populations can increase their mean payoff by switching to it. More formally, we say that a pair of strategies $(\mathbf{p},\mathbf{q})$ is a *Nash-Pareto pair* for an asymmetric game with payoff matrices A and B if the following two conditions hold:

(i) *Equilibrium condition:*
$\mathbf{p}\cdot A\,\mathbf{q} \geq \mathbf{x}\cdot A\mathbf{q}$ and $\mathbf{q}\cdot B\,\mathbf{p} \geq \mathbf{y}\cdot B\mathbf{p}$ for all $(\mathbf{x},\mathbf{y}) \in S_n \times S_m$.

(ii) *Stability condition:* For all states $(\mathbf{x},\mathbf{y}) \in S_n \times S_m$ for which equality holds in (i) we have

$$\text{if } \mathbf{x}\cdot A\,\mathbf{y} > \mathbf{p}\cdot A\,\mathbf{y} \quad \text{then} \quad \mathbf{y}\cdot B\,\mathbf{x} < \mathbf{q}\cdot B\mathbf{x}$$

and

$$\text{if } \mathbf{y}\cdot B\mathbf{x} > \mathbf{q}\cdot B\,\mathbf{x} \quad \text{then} \quad \mathbf{x}\cdot A\,\mathbf{y} < \mathbf{p}\cdot A\,\mathbf{y} \quad .$$

The first condition just says that $(\mathbf{p},\mathbf{q})$ is a Nash equilibrium pair. The second condition says that it is impossible for *both* players to take simultaneously advantage from a deviation from the equilibrium $(\mathbf{p},\mathbf{q})$: at least one of them gets penalized. This corresponds to the concept of

Pareto optimality or *efficiency* in game theory. Note that in condition (ii) it is essential to allow both populations to deviate simultaneously. If $\mathbf{p}$ is replaced by $\mathbf{x}$ and $\mathbf{y}$ remains equal to $\mathbf{q}$ then condition (ii) is void.

Exercise 1: Show that the Nash equilibrium of the Battle of Sexes game (27.13) is a Nash-Pareto pair. Find other examples of Nash-Pareto pairs. Show that for $A = B = 0$ every state is Nash-Pareto.

We now characterize Nash-Pareto pairs.

Theorem 1. *$(\mathbf{p},\mathbf{q})$ is a Nash-Pareto pair of the bimatrix game (A,B) iff there exists a constant $c > 0$ such that*

$$(\mathbf{x}-\mathbf{p})\cdot A\mathbf{y} + c(\mathbf{y}-\mathbf{q})\cdot B\mathbf{x} \le 0 \tag{27.15}$$

holds for all states $(\mathbf{x},\mathbf{y})$ near $(\mathbf{p},\mathbf{q})$. If $(\mathbf{p},\mathbf{q})$ is a Nash-Pareto pair then equality holds in (27.15) whenever $\operatorname{supp}(\mathbf{x}) \subseteq \operatorname{supp}(\mathbf{p})$ and $\operatorname{supp}(\mathbf{y}) \subseteq \operatorname{supp}(\mathbf{q})$. In particular, if $(\mathbf{p},\mathbf{q})$ is totally mixed, then

$$(\mathbf{x}-\mathbf{p})\cdot A\mathbf{y} + c(\mathbf{y}-\mathbf{q})\cdot B\mathbf{x} = 0 \tag{27.16}$$

holds for all $(\mathbf{x},\mathbf{y}) \in S_n \times S_m$.

This shows that the harmlessly looking conditions (i) and (ii) (which are purely qualitative assertions, involving only inequalities) imply an identity between the payoff matrices.

Proof: It is straightforward to show that (27.15) implies the Nash-Pareto conditions. The converse direction relies on a convexity argument. Let $I = \{i : (A\mathbf{q})_i = \mathbf{p}\cdot A\mathbf{q}\}$ and $J = \{j : (B\mathbf{p})_j = \mathbf{q}\cdot B\mathbf{p}\}$ be the *extended carriers.* Of course we have $I \supseteq \operatorname{supp}(\mathbf{p})$ and $J \supseteq \operatorname{supp}(\mathbf{q})$. We first restrict the game to the strategies in I and J. Then we have equality in (i) and the second condition (ii) becomes effective. Consider the bilinear form $F : \mathbf{R}_0^I \times \mathbf{R}_0^J \to \mathbf{R}^2$ given by

$$F(\boldsymbol{\xi},\boldsymbol{\eta}) = (\boldsymbol{\xi}\cdot A\boldsymbol{\eta}, \boldsymbol{\eta}\cdot B\boldsymbol{\xi}) \ . \tag{27.17}$$

With $\boldsymbol{\xi} = \mathbf{x}-\mathbf{p}$ and $\boldsymbol{\eta} = \mathbf{y}-\mathbf{q}$, (ii) means that F does not take values in the positive quadrant $\mathbf{R}_+^2\backslash\{0\}$. Now $\boldsymbol{\xi} = \mathbf{x}-\mathbf{p}$ lies in

$$V = \{\boldsymbol{\xi} \in \mathbf{R}_0^I : \xi_i \ge 0 \text{ for all } i \notin \operatorname{supp}(\mathbf{p})\}$$

and $\boldsymbol{\eta} = \mathbf{y}-\mathbf{q}$ in

$$W = \{\boldsymbol{\eta} \in \mathbf{R}_0^J : \eta_j \ge 0 \text{ for all } j \notin \operatorname{supp}(\mathbf{q})\} \ .$$

So (ii) is equivalent to the fact that the two sets $F(V \times W)$ and $\mathbf{R}_+^2\backslash\{0\}$ are disjoint. Since both are convex they can be separated by a line in

$\mathbf{R}^2$, i.e. by a linear functional $(c_1,c_2) \neq \mathbf{0}$ on $\mathbf{R}^2$:

$$c_1\xi \cdot A\eta + c_2\eta \cdot B\xi < c_1x_1 + c_2x_2$$

for all $\xi \in V, \eta \in W$, $(x_1,x_2) \in \mathbf{R}^2_+\backslash\{0\}$. This implies

$$c_1\xi \cdot A\eta + c_2\eta \cdot B\xi \leq 0 \text{ and } c_1,c_2 > 0 \ . \tag{27.18}$$

This is (27.15) for $(\mathbf{x},\mathbf{y}) \in S_n(I) \times S_m(J)$, with $c = c_2/c_1$. If $\text{supp}\,(\mathbf{x}) \subseteq \text{supp}\,(\mathbf{p})$ then (27.15) must also hold for ξ replaced by $-\xi$. This shows (27.16).

Now consider arbitrary $\mathbf{x} \in S_n$ and $\mathbf{y} \in S_m$ near $(\mathbf{p},\mathbf{q})$. Let

$$\delta_1 = \sum_{i \notin I} x_i \quad\text{and}\quad \delta_2 = \sum_{j \notin J} y_j \ .$$

Then $\mathbf{x},\mathbf{y}$ can be written as

$$\begin{aligned} \mathbf{x} &= (1-\delta_1)(\mathbf{p}+\xi) + \delta_1\bar{\mathbf{x}} \\ \mathbf{y} &= (1-\delta_2)(\mathbf{q}+\eta) + \delta_2\bar{\mathbf{y}} \end{aligned} \tag{27.19}$$

with $(\xi,\eta) \in V \times W$ as above and the supports of $\bar{\mathbf{x}},\bar{\mathbf{y}}$ being disjoint from I,J respectively. Then, with c as above

$$\begin{aligned} &(\mathbf{x}-\mathbf{p}) \cdot A\mathbf{y} + c(\mathbf{y}-\mathbf{q}) \cdot B\mathbf{x} = \\ &\quad = (1-\delta_1)(1-\delta_2)[\xi \cdot A(\mathbf{q}+\eta) + c\eta \cdot B(\mathbf{p}+\xi)] + \\ &\quad + \delta_1(1-\delta_2)[(\bar{\mathbf{x}}-\mathbf{p}) \cdot A(\mathbf{q}+\eta) + c\eta \cdot B\bar{\mathbf{x}}] + \\ &\quad + \delta_2(1-\delta_1)[\xi \cdot A\bar{\mathbf{y}} + c(\bar{\mathbf{y}}-\mathbf{q}) \cdot B(\mathbf{p}+\xi)] + \\ &\quad + \delta_1\delta_2[(\bar{\mathbf{x}}-\mathbf{p}) \cdot A\bar{\mathbf{y}} + c(\bar{\mathbf{y}}-\mathbf{q}) \cdot B\bar{\mathbf{x}}] \ . \end{aligned} \tag{27.20}$$

The first line of the right hand side is nonpositive by (27.18) and the fact that $\xi \cdot A\mathbf{q} = \eta \cdot B\mathbf{p} = 0$. For the second and third line observe that due to the definition of I and J we can find a number $\alpha > 0$ such that

$$(A\mathbf{q})_i < \mathbf{p} \cdot A\mathbf{q} - \alpha \text{ and } (B\mathbf{p})_j < \mathbf{q} \cdot B\mathbf{p} - \alpha \tag{27.21}$$

for $i \notin I$ and $j \notin J$. So $(\bar{\mathbf{x}}-\mathbf{p}) \cdot A\mathbf{q} < -\alpha$ and $(\bar{\mathbf{y}}-\mathbf{q}) \cdot B\mathbf{p} < -\alpha$. The remaining terms in the second and third line are of order $O(\|\xi\| + \|\eta\|)$ and can therefore be made smaller than α as well as the terms in the fourth line which are of even smaller order $O(\delta_1\delta_2)$. So (27.15) holds also for $(\mathbf{x},\mathbf{y}) \in S_n \times S_m$ with $\delta_1,\delta_2,\xi,\eta$ sufficiently small. The inequality is even strict if at least one of the δ_i is positive.

Consider now the case of a *totally mixed Nash-Pareto point* $(\mathbf{p},\mathbf{q})$ in more detail: Then (27.16) can be written as

$$(\mathbf{x}-\mathbf{p})\cdot A(\mathbf{y}-\mathbf{q}) + c(\mathbf{y}-\mathbf{q})\cdot B(\mathbf{x}-\mathbf{p}) \equiv 0$$

or

$$\xi\cdot A\eta + c\eta\cdot B\xi \equiv 0 \text{ for } (\xi,\eta)\in \mathbf{R}_0^n \times \mathbf{R}_0^m \quad . \tag{27.22}$$

This means that the bilinear form $A + cB^t$ vanishes on $\mathbf{R}_0^n \times \mathbf{R}_0^m$. In order to obtain an explicit condition, choose $\xi = \mathbf{e}_i - \mathbf{e}_k$ and $\eta = \mathbf{f}_j - \mathbf{f}_l$ ($\mathbf{e}_i$,$\mathbf{f}_j$ denote the unit vectors, i.e. the corners of S_n,S_m). Then (27.22) transforms into

$$a_{ij} - a_{il} - a_{kj} + a_{kl} = -c(b_{ij} - b_{il} - b_{kj} + b_{kl}) \tag{27.23}$$

for all i,j,k,l. As shown in (27.8) and exercise 27.2.7, this means precisely that (A,B) is a rescaling of a zero-sum game.

Thus *totally mixed Nash-Pareto points are just the Nash equilibria of rescaled zero-sum games* (which form a rare set of games if $n,m \geq 3$).

Theorem 2. *If* $(\mathbf{p},\mathbf{q})$ *is a Nash-Pareto pair, then it is stable for the standard dynamics* (27.1).

This is an immediate consequence of (27.15), since it means that the function

$$H = \sum_{i=1}^{n} p_i \log x_i + c\sum_{j=1}^{m} q_j \log y_j$$

satisfies $\dot{H} \geq 0$ (with $\dot{H} = 0$ if $\mathbf{x}$ and $\mathbf{y}$ have the same support as $\mathbf{p}$ and $\mathbf{q}$ respectively).

It is instructive to compare this with the notion of evolutionary stability for symmetric conflicts. As we have seen in **16.4**, $\mathbf{p} \in S_n$ is evolutionarily stable iff $P(\mathbf{x}) = \prod x_i^{p_i}$ is a strict local Ljapunov function: $\dot{P}(\mathbf{x}) > 0$ for $\mathbf{x}$ near $\mathbf{p}$. For asymmetric conflicts the above shows that $(\mathbf{p},\mathbf{q})$ is a Nash-Pareto pair iff the power product $P(\mathbf{x},\mathbf{y}) = \prod x_i^{p_i} \prod y_j^{cq_j}$ is a Ljapunov function for the dynamics (27.1): $\dot{P}(\mathbf{x},\mathbf{y}) \geq 0$ for $(\mathbf{x},\mathbf{y})$ near $(\mathbf{p},\mathbf{q})$.

Exercise 2: Show that if $P(\mathbf{x},\mathbf{y})$ is a strict Ljapunov function for (27.1), i.e. if $\dot{P}(\mathbf{x},\mathbf{y}) > 0$ for all $(\mathbf{x},\mathbf{y})$ near $(\mathbf{p},\mathbf{q})$, then both $\mathbf{p}$ and $\mathbf{q}$ are pure strategies.

In symmetric games, an evolutionarily stable state is in general robust against small perturbations in the payoffs. In asymmetric conflicts the situation is different:

Let $(\mathbf{p},\mathbf{q})$ be a Nash-Pareto pair for a bimatrix game (A,B). It is said to be *robust*, if every game $(\tilde{A},\tilde{B})$ in a suitable neighbourhood of the given (A,B) has a Nash-Pareto pair $(\tilde{\mathbf{p}},\tilde{\mathbf{q}})$ near $(\mathbf{p},\mathbf{q})$.

Exercise 3: A robust Nash-Pareto pair is a mixture of *at most two* strategies in each population. (Hint: if n or $m \geq 3$, then a slight perturbation can destroy the identity (27.16)).

So if $(\mathbf{p},\mathbf{q})$ is totally mixed and robust, then $n,m \leq 2$. Thus there are only two possibilities for a robust Nash-Pareto pair: (a) either it is a pure strategy (and then evolutionarily stable) or (b) it is the Nash equilibrium of a subgame of type (27.13).

27.6 Game dynamics and Nash-Pareto pairs

In **16.2** we interpreted the ESS conditions in game dynamical terms. We shall now do this for Nash-Pareto pairs too, and ask in which sense they are immune against mutant strategies.

Suppose that the two populations are homogeneous, and consist of $\mathbf{p}$- and $\mathbf{q}$-players, respectively. Let small numbers of $\mathbf{x}$-resp. $\mathbf{y}$-players invade. We shall denote the frequency of $\mathbf{x}$-players by x, that of $\mathbf{y}$-players by y. The dynamics (27.1) of this game reads

$$\begin{aligned} \dot{x} &= x(1-x)(b-(a+b)y) \\ \dot{y} &= y(1-y)(d-(c+d)x) \end{aligned} \tag{27.24}$$

on $[0,1] \times [0,1]$, with $a = (\mathbf{p}-\mathbf{x}) \cdot A\mathbf{y}$, $b = (\mathbf{x}-\mathbf{p}) \cdot A\mathbf{q}$, $c = (\mathbf{q}-\mathbf{y}) \cdot B\mathbf{x}$, $d = (\mathbf{y}-\mathbf{q}) \cdot B\mathbf{q}$. The pair $(\mathbf{p},\mathbf{q})$ is a Nash equilibrium iff $b \leq 0$ and $d \leq 0$ for all $\mathbf{x}$ and $\mathbf{y}$. It is stable against $(\mathbf{x},\mathbf{y})$, in the sense of that the corner $(\mathbf{p},\mathbf{q})$ is an attractor (see Fig. 27.1) iff $b < 0$ and $d < 0$, i.e.

$$\mathbf{p} \cdot A\,\mathbf{q} > \mathbf{x} \cdot A\,\mathbf{q} \text{ and } \mathbf{q} \cdot B\,\mathbf{p} > \mathbf{y} \cdot B\,\mathbf{p} \text{ for all } \mathbf{x} \neq \mathbf{p} \text{ and } \mathbf{y} \neq \mathbf{q}. \tag{27.25}$$

Thus we recovered exactly the ESS-definition (17.1-2) for asymmetric games. It reduces to the notion of strict Nash equilibrium. There is no "second condition" in this case, since if $b = 0$, i.e.

$$\mathbf{p} \cdot A\,\mathbf{q} = \mathbf{x} \cdot A\,\mathbf{q} \text{ for some } \mathbf{x} \neq \mathbf{p} \quad , \tag{27.26}$$

a fixed point line arises and $\mathbf{p}$ can be invaded by $\mathbf{x}$-mutants due to stochastic fluctuations (see Figs. 27.2,3,4,5). Thus in the strict sense,

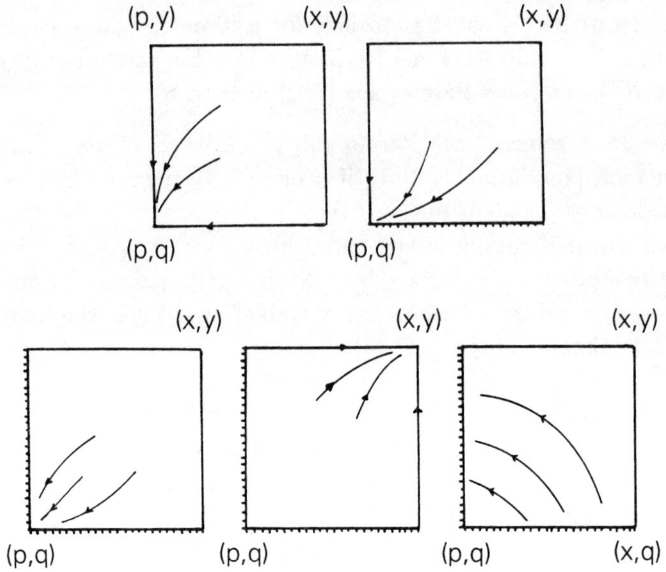

Figures 27.1,2,3,4,5

(27.26) is incompatible with evolutionary stability. On the other hand the introduction of *any* **y**-mutant satisfying $a > 0$ and $d < 0$, i.e. $\mathbf{p} \cdot A\,\mathbf{y} > \mathbf{x} \cdot A\,\mathbf{y}$ and $\mathbf{q} \cdot B\,\mathbf{p} > \mathbf{y} \cdot B\,\mathbf{p}$, into the second population would drive back *both* populations to $(\mathbf{p},\mathbf{q})$. So the following concept of a "weak ESS" could still guarantee some evolutionary stability: $b \leq 0$, with $a > 0$ if $b = 0$; and $d \leq 0$, with $c > 0$ if $d = 0$, i.e.

$$\begin{gathered}
\mathbf{p} \cdot A\,\mathbf{q} \geq \mathbf{x} \cdot A\,\mathbf{q} \quad \text{for all } \mathbf{x} \\
\text{if } \ \mathbf{p} \cdot A\,\mathbf{q} = \mathbf{x} \cdot A\,\mathbf{q}, \ \text{ then } \ \mathbf{p} \cdot A\,\mathbf{y} > \mathbf{x} \cdot A\,\mathbf{y} \ \text{ for all } \mathbf{y} \\
\mathbf{q} \cdot B\mathbf{p} \geq \mathbf{y} \cdot B\mathbf{p} \quad \text{for all } \mathbf{y} \\
\text{if } \ \mathbf{q} \cdot B\,\mathbf{p} = \mathbf{y} \cdot B\,\mathbf{p}, \ \text{ then } \mathbf{q} \cdot B\mathbf{x} > \mathbf{y} \cdot B\mathbf{x} \ \text{ for all } \mathbf{x}.
\end{gathered} \tag{27.27}$$

Exercise 1: Show that (27.27) implies that the portrait of the reduced game endowed with any of the three dynamics (27.1), (27.2) or (27.3) looks as shown in Fig. 27.2,3. Show that (27.27) again implies that both **p** and **q** are pure strategies.

Now for mixed strategies $\mathbf{p}$,$\mathbf{q}$, the equalities $b = 0$ and $d = 0$, i.e., $\mathbf{p} \cdot A\,\mathbf{q} = \mathbf{x} \cdot A\,\mathbf{q}$ and $\mathbf{q} \cdot B\,\mathbf{p} = \mathbf{y} \cdot B\,\mathbf{p}$, hold for many $\mathbf{x}$,$\mathbf{y}$, and thus by (27.26) fixed point edges are inevitable. What we have to forbid in any case is that both $a < 0$ and $c < 0$, i.e. $\mathbf{x} \cdot A\,\mathbf{y} > \mathbf{p} \cdot A\,\mathbf{y}$ and $\mathbf{y} \cdot B\,\mathbf{x} > \mathbf{q} \cdot B\,\mathbf{x}$, holds for some ($\mathbf{x}$,$\mathbf{y}$), for then ($\mathbf{x}$,$\mathbf{y}$) would certainly invade ($\mathbf{p}$,$\mathbf{q}$). (See Fig. 27.4.)

Due to the asymmetry of the game, however, there remains one intermediate possibility, which has no analogue in the symmetric case. This is exactly our concept of Nash-Pareto pair: $b \leq 0$ and $d \leq 0$; if $b = 0$ and $d = 0$, then $ac < 0$.

Exercise 2: Show that for a Nash-Pareto pair ($\mathbf{p}$,$\mathbf{q}$) and for ($\mathbf{x}$,$\mathbf{y}$) having support contained in that of ($\mathbf{p}$,$\mathbf{q}$), the phase portrait of the reduced game, with any dynamics, looks like Fig. 27.5. All orbits go from some fixed point on one line to some fixed point on the other line.

Hence evolution will not further increase the "distance" of the deviation from a Nash-Pareto pair ($\mathbf{p}$,$\mathbf{q}$), caused by some mutation. It depends then on the actual dynamics and on the strength of stochastic effects whether this distance will decay to zero and the mutant will be eliminated again. This may be conjectured at least for dynamics (27.2). In any case, the state ($\mathbf{p}$,$\mathbf{q}$), if it is a Nash-Pareto point, will be physically relevant as the time-average of the orbits.

27.7 Notes

The model (27.1) was introduced by Taylor (1979), see also Schuster et al. (1981b). The model (27.2) was proposed by Maynard Smith (1982, Appendix J). Cressman et al. (1986) study yet another dynamics for bimatrix games. We owe the idea of **27.4** to Akin (personal communication), who proved that (27.1) is always volume preserving (see also Eshel and Akin (1983)). The rest of the chapter follows Hofbauer (1988d) and Koth and Sigmund (1987), with some new material added. The first to view the evolution of anisogamy as a packaging problem were Parker et al (1972). These ideas were further elaborated by Parker (1978), Charlesworth (1978) and Maynard Smith (1978). The approach presented here assumes that the establishment of mating types occurred before the disruption of gametic sizes. This is supported by Charlesworth (1978) and Hoekstra (1982).

28. Game Dynamics for Mendelian Populations

28.1. Strategy and genetics

The game theoretical models considered so far operated on the phenotypic level. In particular, game dynamics rested on the implicit assumption of asexual reproduction. This simplified the analysis considerably; moreover, in the absence of detailed knowledge on the genetic background of a behavioural trait, the corresponding assumptions are bound to be arbitrary, and may obfuscate the essential aspects of the model. On the other hand, the neglect of Mendelian inheritance entails a serious loss of realism. For the Battle of the Sexes or the sex ratio game, an asexual model may seem paradoxical.

It is obvious that a genetic mechanism may prevent the establishment of an ESS. For example, if the corresponding phenotype is realized by the heterozygote, but not by the homozygotes, then it can never make up the whole population. But this objection can be levelled against every kind of adaptationist argument, not only against game theoretical thinking.

We shall presently see that in many cases, strategic and genetic models agree quite well.

28.2 Viability selection and gradients

Let us consider a game with N pure strategies R_1 and R_N, where the payoff F_i for strategy R_i depends only on the average frequencies p_k of the strategies R_k in the population.

Let us furthermore assume that the gene pool contains the alleles A_1 to A_n with frequencies x_1 to x_n, that the frequency of the gene pair (A_i,A_j) is given by $x_i x_j$ according to the Hardy-Weinberg relation, and that such a gene pair yields the phenotype $\mathbf{p}(ij) \in S_N$. The strategy mixture $\mathbf{p} \in S_N$ in the population is given by

$$\mathbf{p} = \mathbf{p}(\mathbf{x}) = \sum_{ij} x_i x_j \mathbf{p}(ij) \quad . \tag{28.1}$$

The frequency of the strategies "played" by allele A_i is

$$\mathbf{p}^i(\mathbf{x}) = \sum_j x_j \mathbf{p}(ij) = \frac{1}{2}\frac{\partial \mathbf{p}}{\partial x_i} \tag{28.2}$$

(the allele A_i belongs with probability x_j to an (A_i,A_j) - or an (A_j,A_i)-individual playing strategy $\mathbf{p}(ij)$). The (frequency dependent) payoff for (A_i,A_j) - individuals is

$$w_{ij} = \mathbf{p}(ij) \cdot \mathbf{F}(\mathbf{p}) \quad , \tag{28.3}$$

and the average payoff for allele A_i is accordingly

$$f_i(\mathbf{x}) = \mathbf{p}^i \cdot \mathbf{F}(\mathbf{p}) \quad . \tag{28.4}$$

The evolution of the gene pool is given by the selection equation (23.2), which now reads

$$\dot{x}_i = x_i[(\mathbf{p}^i - \mathbf{p}) \cdot \mathbf{F}(\mathbf{p})] = x_i[f_i(\mathbf{x}) - \bar{f}] = \hat{f}_i(\mathbf{x}) \tag{28.5}$$

and is a replicator equation on S_n. This equation is closely related to the game dynamical equation for the pure strategists,

$$\dot{y}_i = y_i[F_i(\mathbf{y}) - \bar{F}] = \hat{F}_i(\mathbf{y}) \tag{28.6}$$

on S_N. Indeed, one has the following

Theorem. *If the pure strategist equation* (28.6) *is a Shahshahani gradient with potential* $V(\mathbf{y})$, *then the frequency dependent selection equation* (28.5) *is a Shahshahani gradient with potential* $\frac{1}{2}V(\mathbf{p}(\mathbf{x}))$.

Proof: If $\hat{\mathbf{F}}$ has the Shahshahani potential V, then by **24.4** there exists a function $G : S_n \to \mathbf{R}$ such that

$$\mathbf{F}(\mathbf{y}) = \text{grad } V(\mathbf{y}) + G(\mathbf{y})\mathbf{1} \tag{28.7}$$

holds for all $\mathbf{y} \in \text{int}\, S_N$. Then

$$f_i(\mathbf{x}) = \mathbf{p}^i \cdot \mathbf{F}(\mathbf{p}) = \frac{1}{2} \cdot 2\mathbf{p}^i \cdot \text{grad } V(\mathbf{p}) + G(\mathbf{p}) \quad .$$

This implies by (28.2)

$$\mathbf{f}(\mathbf{x}) = \frac{1}{2} \text{ grad } V(\mathbf{p}(\mathbf{x})) + G(\mathbf{p}(\mathbf{x}))\mathbf{1} \quad . \tag{28.8}$$

Hence $\hat{\mathbf{f}}$ has the Shahshahani potential $\frac{1}{2} V(\mathbf{p}(\mathbf{x}))$.

In particular, if $N = 2$ (or, of course, if $n = 2$), then (28.5) is a Shahshahani gradient.

Exercise 1: Prove directly that if (28.6) satisfies the triangular integrability condition (24.17), then so does (28.5).

Exercise 2: If $\mathbf{F}$ is linear and (28.6) is a Shahshahani gradient, i.e. if $F(\mathbf{x}) = (B + S)\mathbf{x}$, where B is a symmetric $N \times N$-matrix and S a matrix with constant rows, then $\mathbf{x} \to \frac{1}{4}\mathbf{p} \cdot B\mathbf{p}$ is a Shahshahani potential for (28.5).

Exercise 3: If $N = 2$ and $\mathbf{F}$ is linear, then a Shahshahani potential is given by

$$V(\mathbf{x}) = \frac{1}{2}\alpha\,(\sum_{r,s} x_r x_s p_1(rs) - \frac{a_{22} - a_{12}}{\alpha})^2 \tag{28.9}$$

(provided $\alpha = a_{11} - a_{21} + a_{22} - a_{12} \neq 0$). If in particular the game admits a mixed ESS, i.e. if $a_{11} - a_{21}$ and $a_{22} - a_{12}$ are both negative, then the strategy mix in the population converges to the ESS "whenever possible". Describe the states in the gene pool for which this ESS is obtained.

Exercise 4: Discuss all possible phase portraits if furthermore $n = 2$. (Hint: Draw $p_1(\mathbf{x})$ as a function of x_1. The equilibria of (28.5) in (0,1) are the critical points of p_1 and the solutions of $p_1(x_1) = (a_{22} - a_{12})$ $(a_{11} - a_{21} + a_{22} - a_{12})^{-1}$).

Exercise 5: Find a common extension of Theorems 24.5 and 28.2.

Exercise 6: If F is linear and the genetics is dominance free in the sense that $\mathbf{p}(ij) = \frac{1}{2}(\mathbf{p}(ii) + \mathbf{p}(jj))$, then (28.5) reduces to the haploid model (12.5).

28.3 The discrete model for two strategies

The discrete counterpart of (28.5) is, as usual, considerably more difficult to analyze, and we shall only discuss the case of two strategies R_1 and R_2 and linear payoff given by the 2×2-matrix A. We denote by $p(ij)$ (resp. p) the first component of $\mathbf{p}(ij)$ (resp. $\mathbf{p}(\mathbf{x})$), i.e. the frequency of R_1 among the A_iA_j-phenotype (resp. the whole population). Then

$$a_1 = a_{11}p + a_{12}(1-p) \quad \text{and} \quad a_2 = a_{21}p + a_{22}(1-p) \tag{28.10}$$

are the payoffs for R_1 and R_2, while the average payoff for A_iA_j is

$$w_{ij}(\mathbf{x}) = p(ij)a_1 + (1 - p(ij))a_2 \quad . \tag{28.11}$$

Since the probability that allele A_i "plays" R_1 (i.e. that it belongs to an R_1-strategist) is

$$p_i = \sum_{j=1}^{n} p(ij)x_j \tag{28.12}$$

we obtain as average payoff for allele A_i

$$p_i a_1 + (1-p_i)a_2 = \sum_{j=1}^{n} x_j w_{ij}(\mathbf{x}) \quad . \tag{28.13}$$

If we denote by x_i' the frequency of A_i in the next generation, then the assumption that the increase $\frac{x_i'}{x_i}$ of allele A_i is proportional to its payoff leads to the difference equation

$$x_i' = x_i \frac{p_i a_1 + (1-p_i)a_2}{\bar{p}} \tag{28.14}$$

where

$$\bar{p} = p a_1 + (1-p) a_2 \tag{28.15}$$

is the average payoff, or equivalently to

$$x_i' = x_i \frac{\sum_j w_{ij}(\mathbf{x}) x_j}{\sum_{rs} w_{rs}(\mathbf{x}) x_r x_s} \quad . \tag{28.16}$$

This is the discrete equivalent of (28.5) for $N=2$.

If $\mathbf{x}$ is a fixed point of (28.14) then

$$(p_i - p)\,(a_1 - a_2) = 0 \tag{28.17}$$

whenever $x_i > 0$. Thus we may distinguish two types of rest points:

(a) *strategic equilibria* for which $a_1 = a_2$, i.e. where the payoffs for the two strategies are equal;

(b) *genetic equilibria* for which $p_i = p$ whenever $x_i > 0$. For those equilibria, all alleles which are present "play" the same strategy mixture. Genetic equilibria are exactly the fixed points of the classical selection equation (3.1) with matrix $p(ij)$.

In the following discussion, we shall only consider the generic case $a_{11} - a_{21} + a_{22} - a_{12} \neq 0$. From

$$a_1 - a_2 = (a_{11} - a_{21} + a_{22} - a_{12})p - (a_{22} - a_{12}) \tag{28.18}$$

we can easily deduce that $(a_1 - a_2)^2$ is a Ljapunov function for the continuous case (cf. (28.9)). This suggests the following

Theorem. *$(a_1 - a_2)^2$ is a strict Ljapunov function for* (28.14). *Furthermore, $a_1 - a_2$ does not change its sign.*

Proof: We shall first show that (28.14) is *locally adaptive*, in the sense that the frequency p of strategy R_1 increases if and only if its payoff a_1 is larger than a_2. Indeed, since $w_{ij}(\mathbf{x}) = w_{ji}(\mathbf{x})$, we can apply Theorem

3.3 to get

$$\sum_{i,j} x_i' \, x_j' \, w_{ij}(\mathbf{x}) \geq \sum_{i,j} x_i x_j w_{ij}(\mathbf{x}) \tag{28.19}$$

which means by (28.11) that

$$a_2 + (a_1 - a_2)\sum_{i,j} x_i' \, x_j' \, p(ij) \geq a_2 + (a_1 - a_2)\sum_{i,j} x_i x_j p(ij) \quad . \tag{28.20}$$

Since

$$p = \sum_{i,j} x_i x_j p(ij) \tag{28.21}$$

(28.20) yields

$$(a_1 - a_2)\,(p' - p) \geq 0 \tag{28.22}$$

which corresponds to local adaptivity.

Case 1: $\alpha = a_{11} - a_{21} + a_{22} - a_{12} > 0$. Since $a_1 - a_2$ has the same sign as $p' - p$, (28.18) implies that $a_1' - a_2' \geq a_1 - a_2$ if $a_1 - a_2 \geq 0$, and $a_1' - a_2' \leq a_1 - a_2$ if $a_1 - a_2 \leq 0$. Hence $(a_1 - a_2)^2$ is an increasing Ljapunov function.

Case 2: $\alpha = a_{11} - a_{21} + a_{22} - a_{12} < 0$. Let us assume (without restricting generality) that $a_1 - a_2 > 0$. Then $a_1' - a_2' \leq a_1 - a_2$ follows from local adaptivity as before, but we have also to show that

$$a_1' - a_2' \geq 0 \tag{28.23}$$

and this takes some more work. (28.21) and (28.14) imply

$$p' = \bar{p}^{-2}\sum_{i,j} p(ij) x_i x_j [p_i a_1 + (1 - p_i)a_2][p_j a_1 + (1 - p_j)a_2] \tag{28.24}$$

$$= \bar{p}^{-2}[(a_1 - a_2)^2 \sum p(ij) x_i x_j p_i p_j + \tag{28.25}$$

$$+ (a_1 - a_2)a_2 \sum p(ij) x_i x_j (p_i + p_j) + p a_2^2] \ .$$

(28.18), when applied to $a_1' - a_2'$, yields therefore

$$\bar{p}^2(a_1' - a_2') = \alpha p' \bar{p}^2 + (a_{12} - a_{22})[p^2(a_1 - a_2)^2 + 2p(a_1 - a_2)a_2 + a_2^2]$$

$$= (a_1 - a_2)R \tag{28.26}$$

where

$$R = \alpha[(a_1 - a_2)\sum p(ij) x_i x_j p_i p_j + a_2 \sum p(ij) x_i x_j (p_i + p_j)] + a_2^2 +$$

$$+ (a_{12} - a_{22})[p^2(a_1 - a_2) + 2pa_2]$$
$$= \alpha[a_1 \sum p(ij) x_i x_j p_i p_j - a_2 \sum p(ij) x_i x_j (-p_i - p_j + p_i p_j)] +$$
$$+ (a_1 - a_2 - \alpha p)[p^2(a_1 - a_2) + 2pa_2] + a_2^2$$
$$= \alpha[a_1 \sum p(ij) x_i x_j p_i p_j - \sum (1 - p(ij)) x_i x_j (1 - p_i)(1 - p_j)] -$$
$$- \alpha[p^3 a_1 + (1-p)^3 a_2] + \bar{p}^2 . \tag{28.27}$$

Since $0 \le p(ij) \le 1$, we have

$$\sum p(ij) x_i x_j p_i p_j \le (\sum p_i x_i)^2 \le p^2 \tag{28.28}$$

and the analogous estimate $(1-p)^2$ for the second sum in (28.27). Thus

$$R \ge \alpha[(p^2 - p^3)a_1 + ((1-p)^2 - (1-p)^3)a_2] + \bar{p}^2$$
$$= \alpha p(1-p)[pa_1 + (1-p)a_2] + \bar{p}^2$$
$$= \bar{p}[\bar{p} + \alpha p(1-p)] \tag{28.29}$$
$$= \bar{p}[a_{11} p + a_{22}(1-p)] \ge 0 . \tag{28.30}$$

Hence (28.26) proves (28.23). Thus $a_1 - a_2$ converges monotonically to 0, without changing its sign. Since every point with $a_1 = a_2$ is a fixed point, $(a_1 - a_2)^2$ is a strict Ljapunov function.

Exercise 1: Discuss the nongeneric case $a_{11} - a_{21} + a_{22} - a_{12} = 0$.

Exercise 2: Under what conditions will the strategy mix of the population converge to an ESS?

Exercise 3: Discuss the case of $n = 2$ alleles completely.

28.4 Genetics and ESS

Consider now again the genetic model (28.5) for an evolutionary game with linear payoff

$$\dot{x}_i = x_i(\mathbf{p}^i - \mathbf{p}) \cdot A\mathbf{p} . \tag{28.31}$$

For the frequencies $\mathbf{p}$ of the strategies we obtain

$$\dot{\mathbf{p}} = 2\sum \mathbf{p}^i \dot{x}_i = 2\sum x_i \mathbf{p}^i[(\mathbf{p}^i - \mathbf{p}) \cdot A\mathbf{p}] = 2C(\mathbf{x})A\mathbf{p} \tag{28.32}$$

where the component $c_{kl}(\mathbf{x})$ of the matrix $C(\mathbf{x})$ is given by

$$c_{kl}(\mathbf{x}) = \sum_i x_i p_k^i(\mathbf{x})(p_l^i(\mathbf{x}) - p_l(\mathbf{x}))$$

$$= \sum x_i (p_k^i(\mathbf{x}) - p_k(\mathbf{x}))\ (p_l^i(\mathbf{x}) - p_l(\mathbf{x})) \quad .$$

Hence $C(\mathbf{x})$ is the covariance matrix of the mean strategies $\mathbf{p}^i(\mathbf{x})$ of the allele A_i. Since

$$\xi \cdot C(\mathbf{x})\xi = \sum_{i=1}^{n} x_i[(\mathbf{p}^i - \mathbf{p}) \cdot \xi]^2 \geq 0 \tag{28.33}$$

the matrix $C(\mathbf{x})$ is positive semidefinite and its kernel consists of all vectors ξ orthogonal to all $\mathbf{p}^i - \mathbf{p}$, $i = 1,...,n$.

We can now show that if at some point $\mathbf{x} \in \operatorname{int} S_n$, the strategy distribution $\mathbf{p}$ does not change, i.e. $\dot{\mathbf{p}} = \mathbf{0}$, then also $\dot{\mathbf{x}} = \mathbf{0}$, so that $\mathbf{x}$ is an equilibrium on the genetic level as well.

Indeed, $\dot{\mathbf{p}} = \mathbf{0}$ in (28.32) means that $A\mathbf{p}$ lies in the kernel of the covariance matrix $C(\mathbf{x})$. By (28.33) $A\mathbf{p}$ is orthogonal to all $\mathbf{p}^i - \mathbf{p}$, and hence $\dot{x}_i = 0$ in (28.31).

Suppose now that $\hat{\mathbf{p}} \in \operatorname{int} S_N$ is an ESS for the payoff matrix A and that there exists at least one genetic state $\hat{\mathbf{x}}$ which displays this strategy, that is $\mathbf{p}(\hat{\mathbf{x}}) = \hat{\mathbf{p}}$. Let

$$S(\hat{\mathbf{p}}) = \{\mathbf{x} \in S_n : \mathbf{p}(\mathbf{x}) = \hat{\mathbf{p}}\} \tag{28.34}$$

be the set of all such genetic states. $S(\hat{\mathbf{p}})$ is the intersection of $N-1$ conic sections, and it consists of fixed points by the above remark.

Theorem. *Let $\hat{\mathbf{p}} \in \operatorname{int} S_N$ be an ESS for the payoff matrix A and assume that for all $\hat{\mathbf{x}} \in S(\hat{\mathbf{p}})$, the covariance matrix $C(\hat{\mathbf{x}})$ has full rank (the minimum of $N-1$ and $n-1$). Then for all $\mathbf{x}$ in a neighbourhood of $S(\hat{\mathbf{p}})$, $\mathbf{p}(\mathbf{x}(t))$ converges to $\hat{\mathbf{p}}$, and hence the set of fixed points $S(\hat{\mathbf{p}})$ is attracting.*

Proof: Let us consider the more interesting case $n > N$ first. By assumption the $N \times N$ matrix $C(\hat{\mathbf{x}})$ is then positive definite, when restricted to the invariant subspace $\mathbf{R}_0^N$. According to exercise 16.4.4, $C(\hat{\mathbf{x}})A$ is stable on $\mathbf{R}_0^N$. Thus all $N-1$ eigenvalues of the linearization

$$\dot{\mathbf{v}} = C(\hat{\mathbf{x}})A\mathbf{v} \tag{28.35}$$

of (28.32), with

$$\mathbf{v} = (D_{\hat{\mathbf{x}}}\mathbf{p})\ (\mathbf{x} - \hat{\mathbf{x}}) = 2\sum \hat{x}_i \mathbf{p}^i(\mathbf{x} - \hat{\mathbf{x}})$$

have negative real part. The remaining $n-N$ eigenvalues of (28.31), corresponding to the tangent space of the fixed manifold $S(\hat{\mathbf{p}})$, are zero.

Hence $S(\hat{\mathbf{p}})$ is attracting.

If $n \leq N$, $C(\hat{\mathbf{x}})$ has rank $n-1$ and so $S(\hat{\mathbf{p}})$ consists of one point $\hat{\mathbf{x}}$ only. By (28.35) and exercise 16.4.4, all $n-1$ eigenvalues at $\hat{\mathbf{x}}$ have negative real part and so $\hat{\mathbf{x}}$ is asymptotically stable.

Exercise 1: Consider the function $P(\mathbf{x}) = \prod x_i^{\hat{x}_i}$. Show that

$$\dot{P}/P = (\sum \hat{x}_i \mathbf{p}^i - \mathbf{p}) \cdot A\mathbf{p} \quad .$$

Show that P is a Ljapunov function under the assumption of the theorem if $n \leq N$, but in general not for $n > N$.

Exercise 2: If $C(\hat{\mathbf{x}}) = 0$ (i.e. $\hat{\mathbf{x}}$ is a genetic equilibrium) and $\mathbf{p}(\hat{\mathbf{x}}) = \hat{\mathbf{p}}$ is an ESS for the matrix A, then again $S(\hat{\mathbf{p}})$ is an attracting set. (Hint: P is a Ljapunov function in this case).

Theorem 28.4 and exercise 28.4.2 together show that $S(\hat{\mathbf{p}})$ is attracting if $C(\hat{\mathbf{x}})$ either has rank 0 or maximal rank. For the case of $N = 2$ strategies this covers all cases and gives (another) complete proof. For $N \geq 3$ open problems remain.

Exercise 3: Consider a semidominance of the genetic pattern:

$$\mathbf{p}(ij) = \lambda_{ij}\mathbf{p}(ii) + (1-\lambda_{ij})\mathbf{p}(jj) \quad 0 \leq \lambda_{ij} \leq 1 \quad .$$

Assume $\hat{\mathbf{p}}$ is an ESS and lies in the interior of the convex hull of $\{\mathbf{p}(ii) : 1 = 1,...,n\}$. Show that $S(\hat{\mathbf{p}}) \neq \emptyset$, that all orbits $\mathbf{x}$ near $S(\hat{\mathbf{p}})$ converge to $S(\hat{\mathbf{p}})$ and that there are no other interior fixed points. (Hint: Note that the kernel of the covariance matrix $C(\hat{\mathbf{x}})$ coincides with the vector space spanned by $\{\mathbf{p}(\mathbf{x}) - \hat{\mathbf{p}} : \mathbf{x} \in S_n\}$.

Exercise 4: Analyse a *genetic rock-scissors-paper game*: Use the payoff matrix A from (19.8) and assume $n = 3$ with $\mathbf{p}(ii) = \mathbf{e}_i$ and $\mathbf{p}(ij) = p\mathbf{e}_i + q\mathbf{e}_j + r\mathbf{e}_k$ $(p+q+r = 1,\ p \geq 0,\ q \geq 0,\ r \geq 0)$ if (i,j,k) is a cyclic permutation of (1, 2, 3).

Recall that $\hat{\mathbf{p}} = (\frac{1}{3}, \frac{1}{3}, \frac{1}{3})$ is an ESS for A iff $\epsilon < 0$.

a) Show that $\hat{\mathbf{x}} = (\frac{1}{3}, \frac{1}{3}, \frac{1}{3})$ is a fixed point and compute the linearization.

b) Show that $\hat{\mathbf{x}}$ is a sink iff $r < 1$ and $\epsilon < 0$.

c) Show that in the critical case $\epsilon = 0$, $\hat{\mathbf{x}}$ is asymptotically stable if $p < q$ (i.e. if the heterozygote prefers the "better" of the two strategies) and $\hat{\mathbf{x}}$ is unstable if $p > q$.

d) Deduce the occurrence of super- and subcritical Hopf bifurcations. Note that this shows that even in the case of semidominance ($r = 0$) for small $\epsilon < 0$ the state $\hat{\mathbf{x}}$ corresponding to the ESS $\hat{\mathbf{p}}$ need not be globally stable and can have a small basin of attraction (bounded by the unstable limit cycle).
e) Using exercise 22.1.6 determine the stability conditions for $\text{bd}S_3$ (it is a heteroclinic cycle for small r and (in reverse direction) for r close to 1).
f) For large $r(r > \frac{2}{3})$, additional fixed points arise in $\text{int}\, S_3$. Find phase portraits with 7 interior fixed points and 4 limit cycles.

28.5 Genetics for bimatrix games

Let us now turn to asymmetric conflicts and consider a game between two populations described by the payoff matrices A and B. For the first population we shall use the same notation as in **28.2:** the strategy mixture in the population is given by $\mathbf{p} \in S_N$, the gene pool state by $\mathbf{x} \in S_n$ and the relations (28.1) and (28.2) are still valid.

Similarly, the strategy mixture in the second population is given by $\mathbf{q} \in S_M$, the gene pool consists of the alleles B_1 to B_m and its state is given by $\mathbf{y} \in S_m$. The gene pair (B_r,B_s) corresponds to the phenotype $\mathbf{q}(rs) \in S_M$. Again, one has

$$\mathbf{q} = \mathbf{q}(\mathbf{y}) = \sum_{r,s} \mathbf{q}(rs)\, y_r y_s \tag{28.36}$$

and the allele B_s "plays" the strategy

$$\mathbf{q}^s(\mathbf{y}) = \sum_r y_r \mathbf{q}(sr) = \frac{1}{2}\frac{\partial \mathbf{q}}{\partial x_s} \quad . \tag{28.37}$$

We shall furthermore assume that the A-locus and the B-locus are unlinked: belonging, for example, to non-homologous chromosomes, or even to non-interbreeding populations.

The payoffs for the alleles A_i or B_j are given by

$$\mathbf{p}^i \cdot A\mathbf{q} \quad \text{resp.} \quad \mathbf{q}^j \cdot B\mathbf{p} \tag{28.38}$$

and the evolution in the gene pool is described by

$$\begin{aligned} \dot{x}_i &= x_i[(\mathbf{p}^i - \mathbf{p}) \cdot A\mathbf{q}] \\ \dot{y}_j &= y_j[\mathbf{q}^j - \mathbf{q}) \cdot B\mathbf{p}] \end{aligned} \tag{28.39}$$

on the state space $S_n \times S_m$.

Theorem. *If* (A,B) *is a rescaling of the partnership game* (C,C^t), *i.e. if* (27.6) *is valid, then* (28.39) *is a gradient of* $V(\mathbf{x},\mathbf{y}) = \frac{1}{2}\mathbf{p}\cdot C\mathbf{q}$.

Proof: Using the same metric as in the proof of Theorem 27.2.1, one has for $(\xi,\eta) \in \mathbf{R}_0^n \times \mathbf{R}_0^m$

$$<(\dot{\mathbf{x}},\dot{\mathbf{y}}),(\xi,\eta)>_{(\mathbf{x},\mathbf{y})} = \sum_i \xi_i \frac{1}{x_i} x_i[(\mathbf{p}^i - \mathbf{p})\cdot A\mathbf{q}] + \sum_j \eta_j \frac{1}{cy_j} y_j[(\mathbf{q}^j - \mathbf{q})\cdot B\mathbf{p}]$$

$$= \sum_i \xi_i(\mathbf{p}^i - \mathbf{p})\cdot C\mathbf{q} + \sum_j \eta_j(\mathbf{q}^j - \mathbf{q})\cdot C^t\mathbf{p}$$

$$= \sum_i \xi_i \mathbf{p}^i \cdot C\mathbf{q} + \sum_j \eta_j \mathbf{p}\cdot C\mathbf{q}^j \quad .$$

Since $\frac{\partial}{\partial x_i}\mathbf{p}\cdot C\mathbf{q} = 2\mathbf{p}^i\cdot C\mathbf{q}$ and $\frac{\partial}{\partial y_j}\mathbf{p}\cdot C\mathbf{q} = 2\mathbf{p}\cdot C\mathbf{q}^j$ we obtain

$$D_{(\mathbf{x},\mathbf{y})}V(\xi,\eta) = \sum_i \xi_i \mathbf{p}^i\cdot C\mathbf{q} + \sum_j \eta_j \mathbf{p}\cdot C\mathbf{q}^j$$

as had to be shown.

Let us now take a look at the case where the first population has only two pure strategies. For $N = 2$, it is enough to consider the first components of the vectors $\mathbf{p}(ij)$ and $\mathbf{p}(\mathbf{x})$ in S_2, which we shall denote by $p(ij)$ and $p = p(\mathbf{x})$, respectively. (28.39) yields

$$\dot{x}_i = x_i(p_i - p)[(A\mathbf{q})_1 - (A\mathbf{q})_2] \tag{28.40}$$

with

$$p_i = \sum_j p(ij)x_j = (P\mathbf{x})_i \quad \text{and} \quad p = \sum_{ij} p(ij)x_i x_j = \mathbf{x}\cdot P\mathbf{x} \tag{28.41}$$

(where P is the symmetric $n \times n$-matrix with elements $p(ij)$).

The gene pool at the A-locus is stationary if one of the following two conditions holds.

(a) *strategic equilibria:* the two strategies R_1 and R_2 yield the same payoff, i.e. $(A\mathbf{q})_1 = (A\mathbf{q})_2$. This depends only on the gene frequencies at the B-locus.

(b) *genetic equilibria:* $x_i(p_i - p) = 0$ for all i. This corresponds to the equilibria of the classical selection equation (23.2) with fitness parameters $p(ij)$.

A short computation yields

$$\dot{p} = \sum_i \frac{\partial p}{\partial x_i}\dot{x}_i = 2[(A\mathbf{q})_1 - (A\mathbf{q})_2][\sum_i (P\mathbf{x})^2 x_i - (\mathbf{x}\cdot P\mathbf{x})^2]$$

$$=2[(A\mathbf{q})_1-(A\mathbf{q})_2]\mathrm{Var}\,\tilde{P}$$

where $\mathrm{Var}\,\tilde{P}$ is the variance of the random variable taking the values $(P\mathbf{x})_i$ with probabilities x_i. This implies *local adaptivity:* if the payoff for one strategy is larger than for the other one, its frequency in the population will increase, except at the genetic equilibria (for which $\mathrm{Var}\,\tilde{P}=0$).

If $N=M=2$, we write

$$(A\mathbf{q})_1-(A\mathbf{q})_2=(a_{11}-a_{12}+a_{22}-a_{21})q-(a_{22}-a_{12}) \quad .$$

In the generic case, where $a=a_{11}-a_{12}+a_{22}-a_{21}\neq 0$, we set $\hat{q}=a^{-1}(a_{22}-a_{12})$ and obtain $(A\mathbf{q})_1-(A\mathbf{q})_2=a(q-\hat{q})$. Similarly $(B\mathbf{p})_1-(B\mathbf{p})_2=b(p-\hat{p})$ and thus (28.39) becomes

$$\begin{aligned}\dot{x}_i&=ax_i(p_i-p)(q-\hat{q})\\ \dot{y}_j&=by_j(q_j-q)(p-\hat{p}) \quad .\end{aligned} \tag{28.42}$$

If we define

$$H(\mathbf{x},\mathbf{y})=\frac{1}{2}(p-\hat{p})(q-\hat{q}) \tag{28.43}$$

we obtain $\dfrac{\partial H}{\partial x_i}=p_i(q-\hat{q})$ and hence $\dot{x}_i=ax_i(f_i-\bar{f})$ with $f_i=\dfrac{\partial H}{\partial x_i}$.

Hence

$$\dot{H}=a(q-\hat{q})^2\mathrm{Var}\,\tilde{P}+b(p-\hat{p})^2\mathrm{Var}\,\tilde{Q} \tag{28.44}$$

so that H is a *Ljapunov function if a and b have the same sign.*

Exercise 1: Show that if a and b have the same sign, then (28.42) is a gradient with potential H. Relate this with Theorem 28.5.

Exercise 2: Analyse the nongeneric case $ab=0$.

If $N=M=2$ and $n=m=2$, the system admits a *constant of motion.* Let us consider this for the generic case where $P=p(11)-p(12)+p(22)-p(21)$ and $Q=q(11)-q(12)+q(22)-q(21)$ are nonzero. With $\hat{x}=P^{-1}(p(22)-p(12))$ and $\hat{y}=Q^{-1}(q(22)-q(12))$, (28.42) reduces to the equation

$$\begin{aligned}\dot{x}&=aP\,x(1-x)(x-\hat{x})(q-\hat{q})\\ \dot{y}&=bQ\,y(1-y)(y-\hat{y})(p-\hat{p})\end{aligned} \tag{28.45}$$

on the unit square. For the rational function

$$\varphi(x) = \frac{\hat{p} - p}{x(x-1)(x-\hat{x})} \tag{28.46}$$

one has the partial fraction expansion

$$\varphi(x) = \frac{\hat{\varphi}(0)}{x} + \frac{\hat{\varphi}(1)}{x-1} + \frac{\hat{\varphi}(\hat{x})}{x-\hat{x}}$$

where

$$\hat{\varphi}(s) = \lim_{x \to s} \varphi(x)(x-s)$$

for $s = 0{,}1$ and $\hat{x}$. Thus for $x \in (0{,}1)$

$$\sum_{s=0,1,\hat{x}} \log|x-s|^{\hat{\varphi}(s)}$$

is an antiderivative of φ. Clearly, its time-derivative is

$$\varphi(x)\dot{x} = aP(p-\hat{p})(q-\hat{q}) \quad . \tag{28.47}$$

Hence, with

$$\hat{\psi}(s) = \lim_{y \to s} (y-s) \frac{\hat{q} - q}{y(y-1)(y-\hat{y})} \tag{28.48}$$

we obtain as constant of motion

$$G(x,y) = \frac{1}{aP} \sum_{s=0,1,\hat{x}} \log|x-s|^{\hat{\varphi}(s)} - \frac{1}{bQ} \sum_{s=0,1,\hat{y}} \log|y-s|^{\hat{\psi}(s)} \; . \tag{28.49}$$

Exercise 3: Show that the points (x,y) where $p(x) = \hat{p}$ and $q(y) = \hat{q}$ are critical points of G.

Exercise 4: Compute the Jacobian of (28.45) at the genetic and the strategic equilibria.

Exercise 5: Analyse (28.45) for the Battle of the Sexes and the Prisoners Dilemma.

28.6 Notes

The paradigm of genetic and strategic models was formulated by Oster and Rochlin (1979). Critical remarks on the strategic modelling of evolution can be found in Auslander et al. (1978). The first papers that combine strategy and genetics are Stewart (1971), Hines (1980a), Maynard Smith (1981) and Hofbauer et al. (1982). The case of two strategies was studied by Eshel (1982) and by Lessard (1984), whose treatment is the basis for **28.3.** The compatibility with Shahshahani gradients (Theorem 28.2) is from Sigmund (1987b). Theorem 28.4 on the stability of ESS in genetic models is based on ideas of Hines and Bishop (1984a,b). Thomas (1985) attacks the problem on the genetic level (see Exercise 28.4.1). Semidominant patterns (Exercise 28.4.3) were studied by Cressman and Hines (1984). Exercise 28.4.4 is from Hofbauer et al. (1982). **28.5** follows Koth and Sigmund (1987) and Bomze et al. (1983).

29. Cycles and the Battle of the Sexes

29.1 Cross sections and Poincaré maps

Let $\mathbf{p}$ be a non-stationary point of the ODE $\dot{\mathbf{x}} = \mathbf{f}(\mathbf{x})$ and H a hyperplane through $\mathbf{p}$ which is not parallel to $\mathbf{f}(\mathbf{p})$. An open neighbourhood S of $\mathbf{p}$ in H is said to be a *cross section* if for every $\mathbf{x} \in \bar{S}$, $\mathbf{f}(\mathbf{x})$ is transverse to H.

If, in particular, $\mathbf{p}$ lies on some periodic orbit γ with period T, then the orbit of every $\mathbf{x}$ in some sufficiently small neighbourhood U of $\mathbf{p}$ will follow the orbit of $\mathbf{p}$ for some time and hence intersect the cross section S. Thus there exists for every $\mathbf{x} \in U$ a well defined $\tau_{\mathbf{x}} > 0$ such that $\mathbf{x}(\tau_{\mathbf{x}}) \in S$ and $\mathbf{x}(t) \notin S$ for $0 < t < \tau_{\mathbf{x}}$. This gives rise to a map $g : \mathbf{x} \to \mathbf{x}(\tau_{\mathbf{x}})$ from $S_0 = U \cap S$ to S, which is called the *Poincaré map* of the vectorfield near γ (see Fig. 29.1). Since the construction works for backward time as well, the map g is one to one and hence a diffeomorphism. It is easy to see that if we start from a different cross section S' at $\mathbf{p}' \in \gamma$ we obtain a Poincaré map g' which is equivalent in the sense that there exists a local diffeomorphism h from S to S' with $h(\mathbf{p}) = \mathbf{p}'$ and $g' \circ h = h \circ g$.

The local behaviour of the flow near the periodic orbit γ can be completely understood by analyzing the local behaviour of the Poincaré map g near its fixed point $\mathbf{p}$.

In particular, if all eigenvalues of $D_{\mathbf{p}}g$ are in the interior of the unit circle, then $\mathbf{p}$ is an asymptotically stable fixed point for g and the periodic orbit γ is an attractor, i.e. $d(\mathbf{x}(t), \gamma) \to 0$ as $t \to +\infty$ for all $\mathbf{x}$ near γ. One can show even more: every orbit $\mathbf{x}(t)$ near γ is *in phase* with γ in the sense that there exists a unique $\mathbf{z} \in \gamma$ such that $d(\mathbf{x}(t), \mathbf{z}(t)) \to 0$ for $t \to +\infty$.

Exercise: Show that

$$\dot{x} = x - y - x(x^2 + y^2)$$

$$\dot{y} = x + y - y(x^2 + y^2)$$

has a unique periodic orbit. Compute its Poincaré map (Hint: Use polar

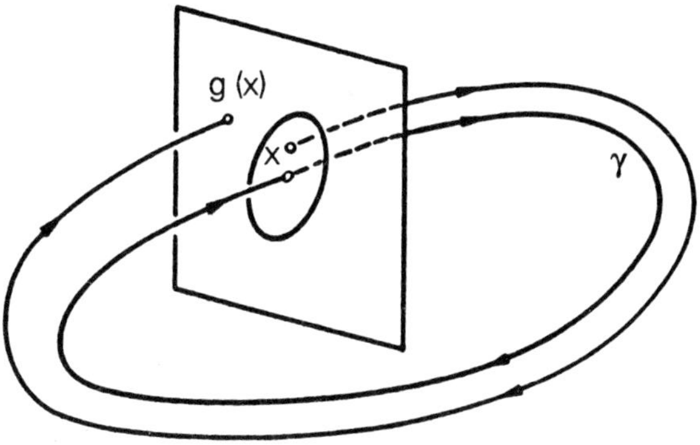

Figure 29.1

coordinates).

In most cases, it is no easy matter to compute a Poincaré map. In **29.4** we shall meet a special case where the system is a perturbation of an equation with a family of periodic orbits which are explicitly known through a constant of motion. But first, we shall study Poincaré maps for heteroclinic cycles, which can be handled more easily.

29.2 Poincaré maps for heteroclinic cycles

Let C be a heteroclinic cycle in the plane. For the sake of simplicity we shall assume that the connecting orbits are straight lines (as in the rock-scissors-paper game (19.8) where C is the boundary of S_3, or in the Battle of the Sexes (17.24) where C is the boundary of the unit square).

Let $\mathbf{F}_i$ be one of the saddle points. We can take local coordinates in such a way that $\dot{x} = \lambda_i x$ and $\dot{y} = -\mu_i y$ holds approximately near $\mathbf{F}_i$ (with $\mu_i, \lambda_i > 0$). We now consider two cross sections (see Fig. 29.2), namely $S_i = \{(x,y): y = 1\}$ "before" the saddle point $\mathbf{F}_i$ and $S_i' = \{(x,y): x = 1\}$ "after" $\mathbf{F}_i$. The units of length are chosen such that these sections are very close to $\mathbf{F}_i$. The transition map $\varphi_i : S_i \to S_i$ is given approximately by

$$(x,1) \to (xe^{\lambda_i t}, e^{-\mu_i t}) = (1,y)$$

such that

$$\varphi_i(x) \sim x^{\mu_i/\lambda_i} \quad . \tag{29.1}$$

The transition map ψ_i from S_i' to S_{i+1} , can be approximated (near the connecting line from $\mathbf{F}_i$ to $\mathbf{F}_{i+1}$) by a linear map

$$\psi_i(x) \sim a_i x \quad . \tag{29.2}$$

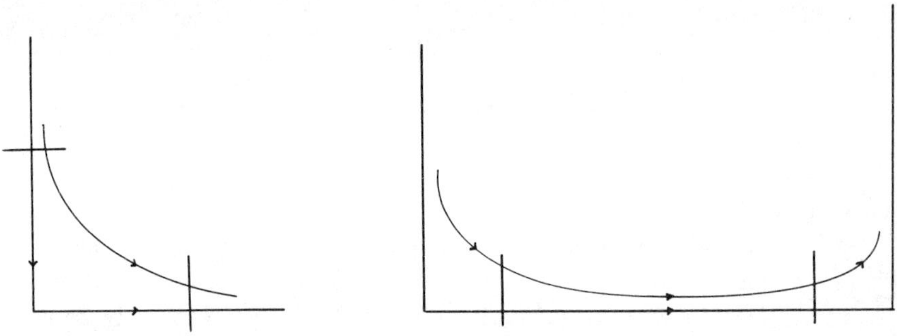

Figures 29.2, 29.3

Exercise 1: Show that by taking coordinates (x,y) with $0 \leq x \leq 1$, $y \geq 0$ as in Fig. 29.3, one obtains

$$a_i = \exp \int_0^1 c_i(x,0)\,dx \tag{29.3}$$

where $c_i(x,0) = \lim_{y \to 0} c_i(x,y)$ and

$$c_i(x,y) = \frac{\dot{y}}{y\dot{x}} + \frac{\mu_i}{\lambda_i}\frac{1}{x} - \frac{\lambda_{i+1}}{\mu_{i+1}}\frac{1}{1-x} \quad . \tag{29.4}$$

(Hint: Take cross sections at $x = \epsilon$ and $x = 1 - \delta$ and let $\epsilon,\delta \to 0$.)

By composing the $2n$ transition maps along the heteroclinic cycle, we obtain for the Poincaré map $g = \psi_n \circ \varphi_n \circ \cdots \circ \psi_1 \circ \varphi_1$ the approximate expression

$$g(x) \sim ax^{\rho} \tag{29.5}$$

where

$$\rho = \prod_{i=1}^{n} \frac{\mu_i}{\lambda_i} \tag{29.6}$$

and

$$a = a_n a_{n-1}^{\mu_n/\lambda_n} \cdots a_1^{\mu_n \cdots \mu_2/\lambda_n \cdots \lambda_2} \quad . \tag{29.7}$$

This Poincaré map g is defined on a small interval $[0,\alpha)$ and the fixed point $x = 0$ corresponds to the heteroclinic cycle C.

If $\rho > 1$ then 0 is a local (one sided) attractor for g and hence C is attracting (with a rate of convergence much larger than for the periodic orbit case described in **29.1**). If $\rho < 1$ then 0 and hence C are repelling. This confirms the stability result for heteroclinic cycles obtained earlier using average Ljapunov functions (cf. Ex. 22.1.6). The constant a in (29.5) is relevant only in the critical case $\rho = 1$.

The approximate form (29.5) of the Poincaré map contains enough information to treat an interesting bifurcation problem. Let us consider a one parameter family of vector fields $\dot{\mathbf{x}} = \mathbf{f}_{\mu}(\mathbf{x})$ in $\mathbf{R}^2$ which all have the same heteroclinic cycle (as in (19.8) or (17.24)). Suppose that $\rho > 1$ for $\mu < 0$ and $\rho < 1$ for $\mu > 0$. If $\frac{d\rho}{d\mu}(0) < 0$ we can rescale in such a way that

$$\rho(\mu) = 1 - \mu \quad . \tag{29.8}$$

The Poincaré map (29.5) then takes the form

$$g_{\mu}(x) = a(\mu)x^{1-\mu} \quad . \tag{29.9}$$

The map g_{μ} from the interval $[0,\alpha)$ to itself has two fixed points, namely $x = 0$ and

$$\bar{x}(\mu) = a(\mu)^{\frac{1}{\mu}} \quad . \tag{29.10}$$

By expanding $a(\mu)$ we obtain

$$a(\mu) = a + ab\mu + O(\mu^2) \tag{29.11}$$

where $a = a(0)$ and $O(\mu^n)$ denotes some function with the property that $\mu^{-n}O(\mu^n)$ is bounded for $\mu \to 0$. Hence

$$\log \bar{x}(\mu) = \frac{1}{\mu} \log a + b + O(\mu) \quad . \tag{29.12}$$

Thus $\bar{x}(\mu)$ is small and positive if either $a < 1$ and $\mu > 0$ or $a > 1$ and $\mu < 0$. The fixed point $\bar{x}(\mu)$ of the Poincaré map corresponds to a periodic orbit of the ODE bifurcating from the heteroclinic cycle. Generically (i.e. whenever $a \neq 1$) the bifurcation is either supercritical (C is attracting for $\mu \leq 0$, the periodic orbit exists and occurs for $\mu > 0$ and is stable) or subcritical (the periodic orbit exists and is unstable for $\mu < 0$ and C is unstable for $\mu \geq 0$).

This ***heteroclinic bifurcation*** is analogous to the familiar Hopf bifurcation described in **18.6.**

Exercise 2: Compute the Poincaré map of the heteroclinic orbit for the Battle of Sexes (17.24). Show that $a = \rho = 1$ in (29.5).

Exercise 3: Compute the Poincaré map for the rock-scissors-paper game. Show that $\rho < 1$ iff det $A > 0$. If $\rho = 1$ then $a = 1$ and the heteroclinic bifurcation is degenerate.

Exercise 4: Study the equation

$$\dot{x} = x(1-x)(a+bx+cy)$$
$$\dot{y} = y(1-y)(d+ex+fy)$$

on the unit square. For which choice of the coefficients is the boundary a heteroclinic cycle? Compute its Poincaré map. Show that both sub- and supercritical heteroclinic bifurcations occur and deduce the existence of limit cycles.

29.3 Heteroclinic cycles on the boundary of S_n

The generalization of the rock-scissors-paper-game to higher dimensions is a first order replicator equation on S_n with a heteroclinic cycle C connecting the vertices $\mathbf{e}_1 \to \mathbf{e}_2 \to \cdots \to \mathbf{e}_n \to \mathbf{e}_1$ along the edges. We shall assume that within bd S_n, this cycle is locally attracting. This means that at every vertex $\mathbf{e}_i$ there is one positive eigenvalue in direction $\mathbf{e}_{i+1}$, while the remaining $n-2$ eigenvalues are negative. These assumptions generate the "essentially hypercyclic" systems of **20.6.** The permanence criterion derived there can be reformulated as an explicit condition (in terms of M-matrices) determining whether the cycle C is attracting or repelling. In this section we describe an alternative method to derive this result and study the corresponding heteroclinic bifurcation.

For the sake of simplicity we exhibit the computations only in the case $n = 4$. We consider first a neighbourhood of a vertex in S_4 and transform it into the origin of $\mathbf{R}^3_+$. The linearized system then reads

$$\dot{x} = \lambda x \quad \dot{y} = -\mu y \quad \dot{z} = -\sigma z \tag{29.13}$$

with $\lambda,\mu,\sigma > 0$. We consider the two cross-sections $S = \{(x,y,z) : y = 1\}$ and $S' = \{(x,y,z) : x = 1\}$ (cf. Fig. 29.4). A point $(x,1,z) \in S$ is mapped by (29.13) to the point

$$(e^{\lambda t}x, e^{-\mu t}, e^{-\sigma t}z) = (1, x^{\mu/\lambda}, x^{\sigma/\lambda}z) \tag{29.14}$$

in S'. The map from S' to the cross-section "before" the next edge is again approximately given by a linear map

$$(y,z) \to (ay,bz) \quad .$$

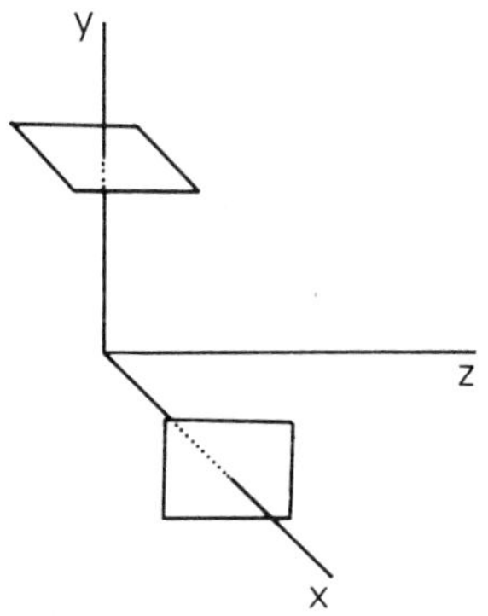

Figure 29.4

Let us apply this to a replicator equation with matrix

$$\begin{bmatrix} 0 & -b_2 & -c_3 & a_4 \\ a_1 & 0 & -b_3 & -c_4 \\ -c_1 & a_2 & 0 & -b_4 \\ -b_1 & -c_2 & a_3 & 0 \end{bmatrix} \quad . \tag{29.15}$$

This is an "essentially hypercyclic network" (cf. (20.23)). The eigenvalues at $\mathbf{e}_i$ along the cycle C are a_i and $-b_i$, while $-c_i$ is the "transversal" eigenvalue. We take now eight cross sections as above: S_i before $\mathbf{e}_i$

and S_i' after it. As coordinates we take $(x,y) = (x_{i+1},x_{i+2})$ at S_i and S_{i-1}'. Then $\varphi_i : S_i \to S_i'$ and $\psi_i : S_i' \to S_{i+1}$ are approximately given by

$$\begin{aligned}\varphi_i(x,y) &\sim (x^{c_i/a_i}y, x^{b_i/a_i})\\ \psi_i(x,y) &\sim (\alpha_i x, \beta_i y) \quad .\end{aligned} \tag{29.16}$$

The constants α_i and β_i can be determined just as in exercise 29.2.1. The Poincaré map g of the heteroclinic cycle C is the composite of these eight maps. Thus

$$g(x,y) \sim (Ax^\alpha y^\beta, Bx^\gamma y^\delta) \quad . \tag{29.17}$$

This is easier to analyze after a change of coordinates

$$z_1 = -\log x \qquad z_2 = -\log y \quad . \tag{29.18}$$

The heteroclinic cycle C corresponds to the limit $z_1, z_2 \to +\infty$. With $\mathbf{z} = (z_1,z_2)$ we obtain from g the return map

$$L : \mathbf{z} \to P\mathbf{z} + \mathbf{q} \quad . \tag{29.19}$$

Here P is the product of the four matrices

$$\begin{bmatrix} c_i/a_i & 1 \\ b_i/a_i & 0 \end{bmatrix} \qquad i = 4,3,2,1.$$

Exercise 1: Compute $\mathbf{q}$ in terms of $\alpha_i, \beta_i, a_i, b_i, c_i$.

On a higher dimensional simplex S_n, the same procedure works again: in (29.19), P is then an $(n-2) \times (n-2)$-matrix. The map (29.19) carries all the essential information about the flow near the heteroclinic cycle C.

Since P is a positive matrix, it has (by the Perron-Frobenius theorem, cf. **11.4**) a dominant eigenvalue $\rho > 0$ and positive left and right eigenvectors $\mathbf{v},\mathbf{u}$. If $\rho \neq 1$, we can forget about the translation term $\mathbf{q}$ in (29.19) by replacing $\mathbf{z}$ by $\mathbf{z} + \hat{\mathbf{z}}$, where $\hat{\mathbf{z}} = (1-P)^{-1}\mathbf{q}$ is the fixed point of (29.19).

If $\rho > 1$, then exercise 11.4 shows that

$$\rho^{-n}P^n\mathbf{z} \to \mathbf{u}$$

for every $\mathbf{z} > \mathbf{0}$. Hence $L^n\mathbf{z}$ grows to infinity in the direction of the right eigenvector $\mathbf{u} > \mathbf{0}$. This implies that the heteroclinic cycle is attracting.

If $\rho < 1$ then all orbits of $\mathbf{z} \to P\mathbf{z}$ converge to $\mathbf{0}$. Thus "infinity" is repelling for (29.19) and so is the heteroclinic cycle. Thus the stability criterion for heteroclinic cycles given in **29.2** generalizes to higher dimensions.

In order to decide which of these two cases holds, it is not necessary to compute ρ explicitly (an impossible task in general). Instead one considers the matrix

$$Q = P - I \tag{29.20}$$

which has nonnegative off-diagonal terms. Then **20.5** and Theorem **21.1** tell us that $\rho < 1$ iff Q is a stable matrix, which is the case iff $-Q$ is an M-matrix.

Exercise 2: Using this criterion, derive an explicit permanence criterion for (29.15) and compare it with the result of **20.5.**

Let us now study the heteroclinic bifurcation which occurs when C loses stability. We shall suppose again that the parameter μ of the vector field on S_n is related to the dominant eigenvalue $\rho = \rho(\mu)$ by

$$\rho(0) = 1 \qquad\qquad \frac{d\rho}{d\mu}\Big|_{\mu=0} < 0 \quad . \tag{29.21}$$

The fixed point of the map (29.19) is given (for $\mu \neq 0$) by

$$\hat{\mathbf{z}}(\mu) = (I - P(\mu))^{-1}\mathbf{q}(\mu) \quad . \tag{29.22}$$

Whenever $\hat{\mathbf{z}}(\mu)$ is positive and very large, it will correspond to a periodic orbit near the heteroclinic cycle. To see the asymptotic behaviour of $\hat{\mathbf{z}}(\mu)$ for $\mu \to 0$ we write $(I - P(\mu))^{-1}$ as the quotient of the adjoint matrix and its determinant. Since $\rho < 1$ for $\mu > 0$, the matrix $I - P(\mu)$ is positively stable for $\mu > 0$ and has exactly one negative eigenvalue for small $\mu < 0$. Thus $\det(I - P(\mu))$ has the sign of μ for μ near 0. The adjoint matrix of $I - P(\mu)$ is a small perturbation of $\operatorname{adj}(I - P(0))$. Since $I - P(0)$ has a zero eigenvalue of multiplicity 1, the adjoint matrix has rank one and is of the form cu_iv_j, where $\mathbf{u} > \mathbf{0}$ and $\mathbf{v} > \mathbf{0}$ are the right and left eigenvectors of $P(0)$ corresponding to the dominant eigenvalue $\rho(0) = 1$ (cf. exercise 21.2.10). The constant c is positive since $(I - P(\mu))^{-1}$ is a nonnegative matrix for $\mu > 0$ (by (M 10) in **21.1.**) Hence

$$\hat{\mathbf{z}}(\mu) \sim \frac{1}{\mu}\operatorname{adj}(I - P(0))\mathbf{q} = \frac{1}{\mu}(\mathbf{q}\cdot\mathbf{v})\mathbf{u} \quad . \tag{29.23}$$

Thus the fixed point $\hat{\mathbf{z}}(\mu)$ tends to infinity in the direction of $\mathbf{u} > \mathbf{0}$ if

$\mu \to 0$. If $\mathbf{q}\cdot\mathbf{v} > 0$ then the bifurcation is supercritical: the periodic orbits corresponding to $\hat{\mathbf{z}}(\mu)$ occur for $\mu > 0$ and are stable. If $\mathbf{q}\cdot\mathbf{v} < 0$ then the bifurcation is subcritical: the periodic orbits occur for $\mu < 0$ and are unstable.

Exercise 3: Compute the explicit conditions on a_i, b_i and c_i which make the heteroclinic bifurcation of (29.15) subcritical resp. supercritical.

Exercise 4: Compute the Poincaré map (29.19) for the hypercycle (12.7). Show that $\rho = 0$. (Hence C is "very repelling").

Exercise 5: Compute in a similar way the Poincaré map for the heteroclinic cycle in the 3-species system described in **22.1.** Choose coordinates $x \geq 0$ and $y \in \mathbf{R}$ such that x measures the distance from bd $\mathbf{R}^3_+$. Show that the Poincaré map takes the form

$$(x,y) \to (ax^\rho, x^\sigma f(x,y)) \tag{29.24}$$

where $a,\rho,\sigma > 0$ and f is affine linear in y. Derive a stability criterion for the heteroclinic cycle and study the heteroclinic bifurcation.

29.4 Averaging

We consider now a family of differential equations depending on a small parameter ϵ:

$$\begin{aligned} \dot{I} &= \epsilon\, f(I,\varphi,\epsilon) \\ \dot{\varphi} &= \omega(I) + \epsilon\, g(I,\varphi,\epsilon) \quad . \end{aligned} \tag{29.25}$$

This is a slight perturbation of

$$\begin{aligned} \dot{I} &= 0 \\ \dot{\varphi} &= \omega(I) \quad . \end{aligned} \tag{29.26}$$

We shall assume that (I,φ) behave like polar coordinates in $\mathbf{R}^2$: φ is a 2π-periodic "angular variable" and I a function from $\mathbf{R}^2$ into $\mathbf{R}$ with a strict maximum (or minimum) at 0. Then $\mathbf{0}$ is a fixed point for (29.26) surrounded by periodic orbits given by the level sets of I. The angular velocity is $\omega(I)$ and the period $T(I) = \dfrac{2\pi}{\omega(I)}$.

Let us compute the Poincaré map of the perturbed equation (29.25) along the cross section S defined by the positive x-axis. If I is monotone along S, then a point $\mathbf{p} \in S$ is uniquely determined by $I(\mathbf{p})$. The Poincaré map is given by

$$I(\mathbf{p}(T)) - I(\mathbf{p}) = \int_0^T \frac{dI}{dt} dt = \epsilon \int_0^{2\pi} f(I,\varphi,\epsilon) \frac{d\varphi}{\omega(I) + \epsilon g(I,\varphi,\epsilon)}$$

$$= \frac{\epsilon}{\omega(I)} \int_0^{2\pi} f(I,\varphi,0)\, d\varphi + O(\epsilon^2)$$

$$= \epsilon TF(I) + O(\epsilon^2) \tag{29.27}$$

where

$$F(I) = \frac{1}{2\pi} \int_0^{2\pi} f(I,\varphi,0)\, d\varphi \tag{29.28}$$

is a time average. Since the increment $I(T) - I(0) = \epsilon TF(I) + O(\epsilon^2)$ is small, we may relate it to the corresponding differential equation

$$\dot{J} = \epsilon F(J) \tag{29.29}$$

which is called the *averaged equation.*

It can be shown that in general the one dimensional equation (29.29) is a good approximation for the two dimensional equation (29.25) if ϵ is sufficiently small. In particular, if (29.29) has a hyperbolic fixed point at I_0, then (29.25) has a periodic orbit with the same stability properties. This is proven by finding a change of coordinates $J = I + \epsilon h(I,\varphi)$ which smoothes out the orbits of the perturbed equation (29.25) and transforms it into (29.29), up to terms of order ϵ^2.

As a typical example we consider the *van der Pol equation*

$$\dot{x} = y \qquad \dot{y} = -x + \epsilon(1 - x^2)y \tag{29.30}$$

which can be viewed as a perturbation of the linear equation

$$\dot{x} = y \qquad \dot{y} = -x \quad . \tag{29.31}$$

With the coordinates I,φ given by

$$2I = x^2 + y^2 \qquad \tan\varphi = \frac{y}{x} \quad , \tag{29.32}$$

(29.31) reads

$$\dot{I} = 0 \qquad \dot{\varphi} = -1$$

and (29.30) transforms into

$$\dot{I} = \epsilon(1 - x^2)y^2 = 2\epsilon I \sin^2\varphi (1 - 2I\cos^2\varphi) \quad .$$

The averaged equation can then be computed to

$$\dot{J} = \epsilon J(1 - \frac{J}{2}) \tag{29.33}$$

which is, for $\epsilon > 0$, a logistic equation with an asymptotically stable fixed point at $J_0 = 2$. The van der Pol equation has therefore, for small $\epsilon > 0$, a stable limit cycle near the circle $x^2 + y^2 = 4$.

29.5 A genetic model for the Battle of Sexes

In chapter 28, the gist of the results supports the view that the introduction of a genetic background does not fundamentally alter the outcome obtained by the asexual game dynamics. However, in the absence of ESS or gradients, this need not always be the case. The following example shows that the simplest genetic mechanism can give rise to a novel feature, which turns out to be both biologically interesting and mathematically nontrivial. In contrast to the diploid model in **28.4**, we use discrete dynamics and obtain stable oscillations.

We assume that male and female behaviour are regulated by genes at two different loci. The A-locus, with alleles A and a, regulates male behaviour and is not expressed in females. A corresponds to strategy E_1 (philanderer), and a to strategy E_2 (faithful). More precisely we assume that the homozygotes AA and aa play the pure strategies E_1 and E_2 respectively and that the heterozygote Aa plays the mixed strategy $\frac{1}{2}(E_1 + E_2)$. In a similar way the B-locus determines female behaviour: B corresponds to F_1 (coy) and b to F_2 (fast).

Let p_A, q_A be the frequencies of allele A in adult males and females respectively, and p_B, q_B be the corresponding frequencies of allele B. For simplicity we assume linkage equilibrium. Then the frequencies of AA, Aa, aa in the next generation are $p_A q_A$, $p_A(1 - q_A) + q_A(1 - p_A) = p_A + q_A - 2p_A q_A$ and $(1 - p_A)(1 - q_A)$ respectively. The frequency of gene A, which also measures the probability of strategy E_1 among males in the new generation, is then given by $p_A q_A + \frac{1}{2}(p_A + q_A - 2p_A q_A) = \frac{1}{2}(p_A + q_A)$. Similar results hold for the B-locus.

In order to simplify the mathematical analysis, we replace our original payoff matrix (17.23) by the following more symmetric one (this does not change the situation qualitatively):

$$A = \begin{bmatrix} 1-K & 1+K \\ 1+K & 1-K \end{bmatrix} \qquad B = \begin{bmatrix} 1+L & 1-L \\ 1-L & 1+L \end{bmatrix} \tag{29.34}$$

where the positive numbers K and L are small (certainly less than 1). This gives the following fitness values for males, depending on the frequencies of females:

$$w_{AA} = (1-K)\frac{p_B+q_B}{2} + (1+K)(1-\frac{p_B+q_B}{2}) = 1+K(1-p_B-q_B)$$

$$w_{aa} = (1+K)\frac{p_B+q_B}{2} + (1-K)(1-\frac{p_B+q_B}{2}) = 1-K(1-p_B-q_B) \quad (29.35)$$

$$w_{Aa} = \frac{1}{2}(w_{AA} + w_{aa}) = 1,$$

with similar results for females.

Hence, writing p_A' etc. for the allelic frequencies in the adults of the new generation, we obtain

$$p_A' = \frac{w_{AA}p_Aq_A+\frac{1}{2}w_{Aa}(p_A+q_A-2p_Aq_A)}{\bar{w}_A}$$

$$q_A' = \frac{p_Aq_A+\frac{1}{2}(p_A+q_A-2p_Aq_A)}{1} = \frac{p_A+q_A}{2} \quad ,$$

etc. The mean fitness of males at the A-locus, $\bar{w}_A$, is given by

$$\begin{aligned}\bar{w}_A &= 1 + K[\,p_Aq_A(1-p_B-q_B)-(1-p_A)(1-q_A)(1-p_B-q_B)] \\ &= 1-K(1-p_A-q_A)(1-p_B-q_B) \quad .\end{aligned}$$

This leads finally to the equations

$$\begin{aligned} p_A' &= \frac{\frac{1}{2}(p_A+p_A)+K(1-p_B-q_B)p_Aq_A}{1-K(1-p_A-q_A)(1-p_B-q_B)} \\ q_A' &= \frac{1}{2}(p_A+q_A) \\ p_B' &= \frac{1}{2}(p_B+q_B) \\ q_B' &= \frac{\frac{1}{2}(p_B+q_B)-L(1-p_A-q_A)p_Bq_B}{1+L(1-p_A-q_A)(1-p_B-q_B)} \quad . \end{aligned} \quad (29.36)$$

Since $0 \leq K,L < 1$ this is a well-defined transformation on the state space $[0,1]^4$.

Obviously $p_A = q_A = p_B = q_B = \frac{1}{2}$ is a fixed point of the map (29.36). It is easy to show that the fixed point is unstable, as for the haploid discrete time model (27.3).

Exercise 1: Compute the linearization of (29.36) at this fixed point. Show that there are two zero eigenvalues (biologically they imply a very quick convergence to the sex-ratio $\frac{1}{2}$) and a pair of complex eigenvalues with real part +1 (corresponding to (17.18)).

Exercise 2: Show that (29.36) is permanent. (Hint: Use $V = (p_A + q_A)\ (2 - p_A - q_A)(p_B + q_B)(2 - p_B - q_B)$ as average Ljapunov function. Show that every orbit on the boundary converges to a fixed point where $\frac{V'}{V} > 1$.)

Thus the attractor of (29.36) lies in the interior of $[0,1]^4$. Since the only fixed point in the interior is unstable, this attractor must be something else. We prove in the next section that it is a limit cycle.

29.6 The limit cycle

We begin with a linear change of variables which transforms the fixed point to the origin:

$$x{=}1{-}p_A{-}q_A,\ y{=}1{-}p_B{-}q_B,\ u{=}p_A{-}q_A,\ v{=}p_B{-}q_B \quad . \tag{29.37}$$

Then $(x,y,u,v) \in [-1,+1]^4$.

Furthermore we assume $K = L$ to make the analysis simpler and set $\epsilon{=}K/4$. Then (29.36) is transformed into

$$\begin{aligned} x' &= x - \epsilon y \frac{1-x^2-u^2}{1-4\epsilon xy} & y' &= y + \epsilon x \frac{1-y^2-v^2}{1+4\epsilon xy} \\ u' &= \epsilon y \frac{1-x^2-u^2}{1-4\epsilon xy} & v' &= -\epsilon x \frac{1-y^2-v^2}{1+4\epsilon xy} . \end{aligned} \tag{29.38}$$

Now $|u'|, |v'| \le \frac{\epsilon}{1+4\epsilon}$, and so u and v will be of order $O(\epsilon)$ for $\epsilon \to 0$, uniformly on the whole state-space, in all subsequent generations. (This corresponds to a rough equilibration of the sex-ratio.) Thus the four-dimensional system (29.38) reduces for small ϵ to a two-dimensional one:

$$\begin{aligned} x' &= x + \epsilon f(x,y) + \epsilon^2 \tilde{f}(x,y) + O(\epsilon^3) \\ y' &= y + \epsilon g(x,y) + \epsilon^2 \tilde{g}(x,y) + O(\epsilon^3) \end{aligned} \tag{29.39}$$

with

$$\begin{aligned} f(x,y) &= -y(1-x^2) & \tilde{f}(x,y) &= -4xy^2\,(1-x^2) \\ g(x,y) &= x(1-y^2) & \tilde{g}(x,y) &= -4x^2y(1-y^2). \end{aligned} \qquad (29.40)$$

For $\epsilon \to 0$ we obtain, as a first-order approximation, the differential equation

$$\begin{aligned} \dot{x} &= f(x,y) = -y(1-x^2) = -\frac{1}{2}\frac{\partial I}{\partial y} \\ \dot{y} &= g(x,y) = x(1-y^2) = \frac{1}{2}\frac{\partial I}{\partial x} \end{aligned} \qquad (29.41)$$

with

$$I = x^2 + y^2 - x^2y^2 = R^2 \quad . \qquad (29.42)$$

Therefore (29.41) is a Hamiltonian system with I as a constant of motion and all orbits of (29.41) in the square $[-1,+1]^2$ are periodic.

Exercise: Show that (29.41) corresponds to the haploid evolutionary dynamics (17.16) with special pay-off matrices (29.34). Show that $x_1x_2y_1y_2$ is the constant of motion corresponding to (29.42).

There exists then a canonical angle variable $\varphi = \varphi(x,y)$ which increases at a constant rate along the periodic orbits. In "action-angle" coordinates, this yields:

$$\dot{I} = 0\,, \qquad \dot{\varphi} = \omega(I) \quad . \qquad (29.43)$$

We will find an explicit expression for φ and the angle velocity $\omega(I)$ in terms of elliptic integrals in the next section.

Let us express our difference equation (29.39) in terms of the canonical variables $I = R^2$ and φ:

$$\begin{aligned} I' &= I(x',y') = I(x+\epsilon f+\epsilon^2\tilde{f}+\cdots,y+\epsilon g+\epsilon^2\tilde{g}+\cdots) = \\ &= I(x,y)+(\epsilon f+\epsilon^2\tilde{f})\frac{\partial I}{\partial x}+(\epsilon g+\epsilon^2\tilde{g})\frac{\partial I}{\partial y}+\frac{1}{2}[(\epsilon f+\cdots)^2\frac{\partial^2 I}{\partial x^2}+ \\ &+ 2(\epsilon f+\cdots)(\epsilon g+\cdots)\frac{\partial^2 I}{\partial x\partial y}+(\epsilon g+\cdots)^2\frac{\partial^2 I}{\partial y^2}]+O(\epsilon^3) \quad . \end{aligned}$$

Now the linear terms in ϵ vanish since (29.41) implies $\dot{I} = \frac{\partial I}{\partial x}f + \frac{\partial I}{\partial y}g = 0$. For the ϵ^2 -terms we obtain from (29.42)

$$\frac{\partial I}{\partial x} = 2x(1-y^2), \qquad \frac{\partial^2 I}{\partial x^2} = 2(1-y^2), \qquad \frac{\partial^2 I}{\partial x\partial y} = -4xy,\cdots$$

Therefore

$$I' = I+\epsilon^2 F(x,y)+O(\epsilon^3) \tag{29.44}$$

with

$$\begin{aligned} F(x,y) &= \tilde{f}\frac{\partial I}{\partial x}+\tilde{g}\frac{\partial I}{\partial y}+\frac{1}{2}(f^2\frac{\partial^2 I}{\partial x^2}+2fg\frac{\partial^2 I}{\partial y\partial x}+g^2\frac{\partial^2 I}{\partial y^2}) \\ &= -4xy^2(1-x^2)2x(1-y^2)-4x^2y(1-y^2)2y(1-x^2) \\ &\quad + y^2(1-x^2)^2(1-y^2)+xy(1-x^2)(1-y^2)4xy+x^2(1-y^2)^2(1-x^2) \\ &= (1-x^2)(1-y^2)[-16x^2y^2+y^2(1-x^2)+4x^2y^2+x^2(1-y^2)] \\ &= (1-R^2)(x^2+y^2-14x^2y^2) \\ &= \frac{1-R^2}{1-x^2}[R^2-14R^2x^2+13x^4] \quad . \end{aligned} \tag{29.45}$$

Similarly, using $\dot{\varphi} = \frac{\partial\varphi}{\partial x}f+\frac{\partial\varphi}{\partial y}g = \omega(I)$, we obtain

$$\begin{aligned} \varphi' &= \varphi(x',y') = \varphi(x+\epsilon f+\cdots,y+\epsilon g+\cdots) \\ &= \varphi(x,y)+\epsilon f\frac{\partial\varphi}{\partial x}+\epsilon g\frac{\partial\varphi}{\partial y}+O(\epsilon^2) \\ &= \varphi+\epsilon\omega(R^2)+O(\epsilon^2). \end{aligned} \tag{29.46}$$

So we end up with

$$\begin{aligned} I' &= I+\epsilon^2 F(I,\varphi)+O(\epsilon^3) \\ \varphi' &= \varphi+\epsilon\omega(I)+O(\epsilon^2) \quad . \end{aligned} \tag{29.47}$$

Imitating the averaging method, as described for differential equations in **29.4**, we eliminate now the φ-dependence of the ϵ^2-terms in (29.47). For this we try an Ansatz $J = I+\epsilon h(I,\varphi)$. Then

$$\begin{aligned} J' &= I'+\epsilon h(I',\varphi') \\ &= I+\epsilon^2 F(I,\varphi)+\epsilon h(I+\epsilon^2 F+\cdots,\varphi+\epsilon\omega(I)+\cdots)+\cdots \\ &= I+\epsilon h(I,\varphi)+\epsilon^2[F(I,\varphi)+\frac{\partial h}{\partial\varphi}\,\omega(I)]+O(\epsilon^3). \end{aligned}$$

Now we separate F into its mean value and its oscillatory part along the periodic orbits of (29.41):

$$F(I,\varphi) = G(I)+\tilde{F}(I,\varphi)$$

with

$$G(I) = \frac{1}{2\pi}\int_0^{2\pi} F(I,\varphi)\,d\varphi \quad . \tag{29.48}$$

Inserting this into (29.47) we may choose h in such a way that

$$\tilde{F}(I,\varphi) + \frac{\partial h}{\partial \varphi}\omega(I) = 0$$

or

$$\frac{\partial h}{\partial \varphi} = -\frac{\tilde{F}(I,\varphi)}{\omega(I)}. \tag{29.49}$$

(Since h is a 2π -periodic function in φ, the mean of its derivative, $\frac{1}{2\pi}\int_0^{2\pi}\frac{\partial h}{\partial \varphi}d\varphi$, equals zero. This will not in general be the case for the mean of $F(I,\varphi)$; hence we can eliminate only the oscillating part $\tilde{F}(I,\varphi)$ by averaging).

So with our new variable J instead of I we have simplified (29.47) to

$$\begin{aligned} J' &= J+\epsilon^2 G(J)+O(\epsilon^3) \\ \varphi' &= \varphi+\epsilon\omega(J)+O(\epsilon^2) \quad . \end{aligned} \tag{29.50}$$

The averaged function $G(J)$ will be computed in the next section in terms of elliptic integrals again. We shall show then that there exists a number $J_0 \in (0,1)$ such that $G(0) = G(1) = G(J_0) = 0$, $\frac{d}{dJ}G(J_0) < 0$, and

$$\begin{aligned} G(J) &> 0 \quad \text{for} \quad 0 < J < J_0 \\ G(J) &< 0 \quad \text{for} \quad J_0 < J < 1 \quad . \end{aligned} \tag{29.51}$$

If we neglect O-terms in (29.50), then $J = J_0$ is obviously the equation of an invariant globally attracting circle. Standard fixed point arguments can be used to prove that this circle persists under O-perturbations for small $\epsilon > 0$.

Thus we have shown that for small $\epsilon > 0$ (29.39) has an *attracting invariant curve* which is approximately given by

$$I = x^2 + y^2 - x^2y^2 = J_0 \quad . \tag{29.52}$$

Hence one particular periodic orbit of the Hamiltonian system (29.41), i.e. of the haploid model, subsists in the limiting case $\epsilon \to 0$ as limit cycle for the genetic model of the Battle of the Sexes.

29.7 Elliptic integrals in the Battle of the Sexes

We now fill in some of the technical details omitted in the last section. We first compute the angular variable φ and its angular velocity $\omega(I)$ in (29.43). Let us represent φ as a function of x and I. Then $\dot{\varphi} = \dfrac{\partial \varphi}{\partial x}\,\dot{x} + \dfrac{\partial \varphi}{\partial I}\,\dot{I}$ implies

$$\omega(I) = -\frac{\partial \varphi}{\partial x} y(1-x^2) \quad .$$

Eliminating y from (29.43) and writing $I = R^2$ henceforth, we obtain

$$\frac{\partial \varphi}{\partial x} = -\omega(R^2)(1-x^2)^{-\frac{1}{2}}(R^2-x^2)^{-\frac{1}{2}} \quad . \tag{29.53}$$

This leads to an elliptic integral. In particular we obtain

$$-\int_{-R}^{+R} \frac{\partial \varphi}{\partial x} dx = -\varphi\Big|_{-R}^{+R} = \pi = \omega(R^2) \int_{-R}^{+R} (1-x^2)^{-\frac{1}{2}}(R^2-x^2)^{-\frac{1}{2}} dx \; . \tag{29.54}$$

Now let us recall Legendre's formulas for complete elliptic integrals

$$\int_0^R (1-x^2)^{-\frac{1}{2}}(R^2-x^2)^{-\frac{1}{2}} dx = K(R) = \frac{\pi}{2}\sum_{n=0}^{\infty}\binom{1/2}{n}^2 R^{2n} \tag{29.55}$$

$$\int_0^R (1-x^2)^{\frac{1}{2}}(R^2-x^2)^{-\frac{1}{2}} dx = E(R) = \frac{\pi}{2}\sum_{n=0}^{\infty}\binom{-1/2}{n}\binom{1/2}{n} R^{2n} \tag{29.56}$$

$$\int_0^R (1-x^2)^{-3/2}(R^2-x^2)^{-\frac{1}{2}} dx = (1-R^2)^{-1}E(R) \; . \tag{29.57}$$

These series converge for $|R| < 1$. (29.54) and (29.55) imply the explicit formula for the angular velocity

$$\omega(R^2) = \frac{\pi}{2K(R)} = \left[\sum_{n=0}^{\infty}\binom{1/2}{n}^2 R^{2n}\right]^{-1} \quad . \tag{29.58}$$

The next problem is to evaluate the averaged function (29.48), namely

$$G(R^2) = \frac{1}{2\pi}\int_0^{2\pi} F(R^2,\varphi)\,d\varphi \quad . \tag{29.59}$$

Since we do not have an explicit expression for φ, we have to transform the integral. Using (29.53) and splitting the circle into 4 equal parts we obtain

$$G(R^2) = \frac{2\omega(R^2)}{\pi}\int_0^R \frac{F(R^2,x)}{y(1-x^2)}dx$$

$$= \frac{2}{\pi}\omega(R^2)\int_0^R F(R^2,x)(R^2-x^2)^{-\frac{1}{2}}(1-x^2)^{-\frac{1}{2}}dx\ . \tag{29.60}$$

So again we have to calculate an elliptic integral. From (29.45) we obtain

$$(1-R^2)^{-1}\int_0^R F(R^2,x)(1-x^2)^{-\frac{1}{2}}(R^2-x^2)^{-\frac{1}{2}}dx =$$

$$= \int_0^R (R^2-14R^2x^2 + 13x^4)(1-x^2)^{-3/2}(R^2-x^2)^{-\frac{1}{2}}dx =$$

$$= 13\int_0^R (1-x^2)^{\frac{1}{2}}(R^2-x^2)^{-\frac{1}{2}}dx + (14R^2-26)\int_0^R (1-x^2)^{-\frac{1}{2}}(R^2-x^2)^{\frac{1}{2}}dx +$$

$$+13(1-R^2)\int_0^R (1-x^2)^{-3/2}(R^2-x^2)^{-\frac{1}{2}}dx =$$

$$= 13E(R) + (14R^2-26)K(R) + 13E(R) =$$

$$= 26E(R) + (14R^2-26)K(R)\ . \tag{29.61}$$

Thus (29.60) together with (29.58) gives an "explicit" formula for G:

$$G(R^2) = (1-R^2)(14R^2-26 + 26\frac{E(R)}{K(R)})\ . \tag{29.62}$$

Inserting the series expansions (29.55), (29.56) into (29.62) we obtain after some calculation:

$$G(R^2) = \omega(R^2)R^2(1-R^2)g(R^2) \tag{29.63}$$

with

$$g(x) = \sum_{n=0}^{\infty}\frac{1}{n+1}(-52n^2 + 66n + 1)\binom{1/2}{n}^2 x^n \tag{29.64}$$

$$= 1 + \frac{15}{8}x - \frac{25}{64}x^2 - \ldots$$

Obviously all terms in $g(x)$ — up to the first two — have a negative coefficient. Hence $\frac{d^2 g(x)}{dx^2} < 0$ for $0 < x < 1$. Moreover, $G(R^2)(1-R^2)^{-1}|_{R=1} = -12 < 0$ from (29.62) implies $g(1) = -\infty$. Since $g(0) = 1$, the function $g(x)$ has a unique zero J_0 in the interval (0,1) and $G(R^2)$ is therefore of the desired form (29.51). The numerical value of J_0 can be found approximately from a table of elliptic functions: $J_0 \cong 0.46$.

29.8 Notes

For more background on cross sections and Poincaré maps see Hirsch and Smale (1974) and Guckenheimer and Holmes (1983). Robinson (1985) used Poincaré maps to prove stability of two-species systems. **29.2** and **29.3** are from Hofbauer (1987). The method of averaging is well described in Arnold (1978, 1983), Guckenheimer and Holmes (1983) and Sanders and Verhulst (1985). The genetic model for the Battle of the Sexes **29.5-6** is due to Maynard Smith and Hofbauer (1986). For the Legendre Formulas (29.55)-(29.57) see, e.g., Gradsteyn (1980).

Afterword

Arrived at the end of this book, this is our last opportunity to emphasize its limitations.

Evolution is an interplay of "chance and necessity", but we have almost totally neglected stochastic methods. In particular, two of our main notions, permanence and Shahshahani gradients, stress what happens near the boundary of the state space: this corresponds to situations where one or several populations are very small, and a deterministic description is likely to be misleading. As for the biological aspects, we have completely omitted quantitative genetics, the theory of modifier genes for genetic dominance or segregation distortion, the selection of mutation and recombination rates, the role of constraints, the evolution of sex, speciation, coevolution etc. ... But what can one do? That's life.

REFERENCES

Akin, E. (1979): The Geometry of Population Genetics. Lecture Notes in Biomathematics **31**. Berlin-Heidelberg-New York: Springer-Verlag.

Akin, E. (1980): Domination or equilibrium. Math. Biosciences **50**, 239-250.

Akin, E. (1982): Cycling in simple genetic systems. J. Math. Biology **13**, 305-324.

Akin, E. (1983): Hopf bifurcation in the two locus genetic model. Memoirs Amer. Math. Soc. No. **284**. Providence, R.I.

Akin, E. (1987): Cycling in simple genetic systems II: the symmetric case. In "Dynamical Systems", ed. A. Kurzhanski and K. Sigmund, Lecture Notes in Economics and Mathematical Systems **287**. Berlin - Heidelberg - New York: Springer.

Akin, E., and J. Hofbauer (1982): Recurrence of the unfit. Math. Biosciences **61**, 51-63.

Akin, E., and V. Losert (1984): Evolutionary dynamics of zero-sum games. J. Math. Biology **20**, 231-258.

Alberti, P.M., and Crell, B. (1984): Nonlinear evolution equations and H-theorems. J.Stat. Phys. **35**.

Amann, E. (1988): Permanence of catalytic networks (to appear).

Amann, E., and J. Hofbauer (1985): Permanence in Lotka-Volterra and replicator equations. In: Lotka-Volterra Approach to Cooperation and Competition in Dynamic Systems. Eds.: W. Ebeling and M. Peschel. Berlin: Akademie-Verlag.

Amann, H. (1983): Gewöhnliche Differentialgleichungen. Berlin - New York: Walter de Gruyter.

Anderson, W.W. (1971): Genetic equilibrium and population growth under density-regulated selection. Amer. Nat. **105**, 489-498.

Andronov, A., E. Leontovich, I. Gordon and A. Maier (1973): Qualitative Theory of Second-Order Dynamic Systems. New York: Halsted Press.

Armstrong, R.A., and R. McGehee (1976): Coexistence of two competitors on one resource. J. Theor. Biology **56**, 499-502.

Arneodo, A., P. Coullet and C. Tresser (1980): Occurrence of strange attractors in three dimensional Volterra equations. Physics Letters **79A**, 259-263.

Arnold, V.I. (1978): Mathematical Methods of Classical Mechanics. New York: Springer.

Arnold, V.I. (1983): Geometrical Methods in the Theory of Ordinary Differential Equations. Grundlehren math. Wissenschaften. Vol. **250**. New York: Springer.

Arrow, K.J. and F.H. Hahn (1971): General Competitive Analysis. San Francisco: Holden-Day.

Auslander, D., J. Guckenheimer, and G. Oster (1978): Random evolutionarily stable strategies. Theor. Population Biology **13**, 276-293.

Axelrod, R. (1984): The Evolution of Cooperation. New York: Basic Publ.

Baum, L.E. and Eagon, J.A. (1967): An inequality with applications to statistical estimation for probabilistic functions of Markov processes and to a model for ecology. Bull. Amer. Math. Soc. **73**, 360-363.

Bazykin, A.D. (1985): Mathematical Biophysics of Interacting Populations. Moscow: Nauka (in Russian).

Berman, A., and D. Hershkovitz (1983): Matrix diagonal stability and its applications. SIAM J. Alg. Discr. Math. **4**, 377-382.

Berman, A. and Plemmons, R.J. (1979): Nonnegative Matrices in the Applied Mathematical Sciences. New York: Academic Press.

Bishop, T., and C. Cannings (1978): A generalized war of attrition. J. Theor. Biology **70**, 85-124.

Block, L., J. Guckenheimer, M. Misiurewicz, and L.S. Young (1980): Periodic points and topological entropy of one-dimensional maps. Lecture Notes in Mathematics **819**, pp. 18-34. Berlin-Heidelberg-New York: Springer.

Bodmer, W. (1965): Differential fertility in population genetic models. Genetics **51**, 411-424.

Bomze, I. (1983): Lotka-Volterra equations and replicator dynamics: A two dimensional classification. Biol. Cybern. **48**, 201-211.

Bomze, I.M. (1986): Non-cooperative two-person games in biology: a classification. Int. J. Game Theory **15**, 31-57.

Bomze, I., P. Schuster, and K. Sigmund (1983): The role of Mendelian genetics in strategic models on animal behavior. J. Theor. Biology **101**, 19-38.

Bürger, R. (1983a): Nonlinear analysis of some models for the evolution of dominance. J. Math. Biology **16**, 269-280.

Bürger, R. (1983b): Dynamics of the classical genetic models for the evolution of dominance. Math. Biosciences **67**, 125-143.

Bürger, R. (1988): Mutation-selection balance and continuum-of-alleles models. Math. Biosci. To appear.

Butler, G.J., H.I. Freedman and P. Waltman (1982): Global dynamics of a selection model for the growth of a population with genotype fertility differences. J. Math. Biol. **14**, 25-35.

Butler, G., H.I. Freedman and P. Waltman (1986): Uniformly persistent systems. Proc. Amer. Math. Soc. **96**, 425-430.

Butler, G., and P. Waltman (1986): Persistence in dynamical systems. J. Diff. Equations **63**, 255-263.

Cech, T.R. (1986): RNA as an enzyme. Scientific American **255**, 76-84.

Charlesworth, B. (1978): The population genetics of anisogamy. J. Theor. Biology **73**, 347-357.

Chenciner, A. (1977): Comportement asymptotique de systèmes differentiels du type "compétition d'espèces". Comptes Rendus Ac. Sc. Paris **284**, 313-315.

Cheng, K.S. (1981): Uniqueness of a limit cycle for predator-prey systems. SIAM J. Math. Anal. **12**, 541-548.

Cheng, K.S., Hsu, S.B. and Lin, S.S. (1981): Some results on global stability of predator-prey system. J. Math. Biol. **12**, 115-126.

Clark, C.E. and T.G. Hallam (1982): The community matrix in three species community models. J. Math. Biology **16**, 25-31.

Comtet, L. (1974): Advanced Combinatorics. Dortrecht-Boston: Reidel.

Conway, E.D. and Smoller, J.A. (1986): Global analysis of a system of predator-prey equations. SIAM J. Appl. Math. **46**, 630-642.

Coppel, W. (1966): A survey of quadratic systems. J. Diff. Equations **2**, 293-304.

Coste J., J. Peyraud, and P. Coullet (1978): Does complexity favor the existence of persistent ecosystems? J. Theor. Biology **73**, 359-362.

Coste, J., J. Peyraud, and P. Coullet (1979): Asymptotic behavior in the dynamics of competing species. SIAM J. Appl. Math. **36**, 516-542.

Cottle, R.W. (1980): Completely Q-matrices. Math. Programming **19**, 347-351.

Cressman, R. Dash, A.T. and Akin, E. (1986): Evolutionary games and two species population dynamics. J. Math. Biol. **23**, 221-230.

Cressman, R. and Hines, W.G.S. (1984): Evolutionarily stable states of diploid populations with semidominant inheritance patterns. J. Appl. Prob. **21**, 1-9.

Cross, G.W. (1978): Three types of matrix stability. Linear Algebra and Appl. **20**, 253-263.

Crow, J.F., and M. Kimura (1970): An Introduction to Population Genetics Theory. New York: Harper and Row.

Darwin, Ch. (1859): On the Origin of Species by Means of Natural Selection. Cambridge-London. Reprinted in Harvard University Press (1964).

Dawkins, R. (1976): The Selfish Gene. Oxford University Press.

Dawkins, R. (1982): The Extended Phenotype. Oxford-San Francisco: Freeman.

Demetrius, L. Schuster, P. and Sigmund, K. (1985): Polynucleotide evolution and branching processes. Bull. Math. Biol. **47**, 239-262.

Devaney, R.L. (1986): An Introduction to Chaotic Dynamical Systems. Menlo Park: Benjamin/Cummings.

Dobzhanski, Th. (1937): Genetics and the Origin of Species. 1st edition. New York: Columbia University Press.

Domingo, E., R. Flavell, and Ch. Weissmann (1976): In vitro site-directed mutagenesis: Generation and properties of an infectious extracistronic mutant of bacteriophage Q_β. Gene **1**, 3-26.

Ebeling, W., and R. Feistel (1982): Physik der Selbstorganisation und Evolution. Berlin: Akademie-Verlag.

Eigen, M. (1971): Selforganization of matter and the evolution of biological macromolecules. Die Naturwissenschaften **58**, 465-523.

Eigen, M., W. Gardiner, P. Schuster, and R. Winkler-Oswatitsch (1981): The origin of genetic information. Scientific American **244**, 78-95.

Eigen, M., and P. Schuster (1979): The hypercycle: A principle of natural selforganization. Berlin-Heidelberg: Springer.

Eigen, M., and P. Schuster (1982): Stages of emerging life - five principles of early organization. J. Molecular Evolution **19**, 47-61.

Eigen, M., P. Schuster, K. Sigmund, and R. Wolff (1980): Elementary step dynamics of catalytic hypercycles. Biosystems **13**, 1-22.

Eshel, I. (1972): On the neighbourhood effect and the evolution of altruistic traits. Theor. Pop. Biol. **3** , 258-277.

Eshel, I. (1982): Evolutionarily stable strategies and viability selection in Mendelian populations. Theor. Pop. Biol. **22**, 204-217.

Eshel, I. and E. Akin (1983): Coevolutionary instability of mixed Nash solutions. J. Math. Biology **18**, 123-133.

Ewens, W.J. (1969): A generalized fundamental theorem of natural selection. Genetics **63**, 531-537.

Ewens, W.J. (1979): Mathematical Population Genetics. Berlin-Heidelberg-New York: Springer.

Feigenbaum, M.J. (1978): Quantitative universality for a class of nonlinear transformations. J. Stat. Phys. **19**, 25-52.

Fiedler, M. and V. Ptak (1962): On matrices with non-positive off-diagonal elements and positive minors. Czech. Math. J. **12**, 382-400.

Fisher, R.A. (1930): The Genetical Theory of Natural Selection. Oxford: Clarendon Press.

Ford, E.B. (1964): Ecological Genetics. 1st edition. London: Methuen.

Fox, S.W., and K. Dose (1972): Molecular evolution and the origin of life. San Francisco: Freeman.

Freedman, H. (1980): Deterministic Mathematical Models in Population Ecology. 2nd ed. Edmonton: HIFR Cons.

Freedman, H.I. and Waltman, P. (1977): Mathematical analysis of some three species food chain models. Math. Biosci. **33**, 257-273.

Freedman, H.I., and P. Waltman (1984): Persistence in models of three interacting predator-prey populations. Math. Biosci. **68**, 213-231.

Freedman, H.I., and P. Waltman (1985): Persistence in a model of three competitive populations. Math. Biosci. **73**, 89-101.

Gale, D., and H. Nikaido (1965): The Jacobian matrix and global univalence of mappings. Math. Annalen **159**, 81-93.

Gard, T., and T. Hallam (1979): Persistence in food webs I: Lotka-Volterra food chains. Bull. Math. Biology **41**, 877-891.

Gause, G.F. (1934): The Struggle for Existence. Baltimore: Williams and Wilkins.

Gilpin, M.E. (1979): Spiral chaos in a predator-prey model. Amer. Nat. **113**, 306-308.

Ginzburg, L.R. (1977): The equilibrium and stability for *n* alleles under the density-dependent selection. J. Theoret. Biol. **68**, 545-550.

Ginzburg, L.R. (1983): Theory of Natural Selection and Population Growth. Menlo Park: Benjamin-Cummings.

Goh, B.S. (1979): Stability in models of mutualism. Amer. Naturalist **113**, 261-275.

Gopalsamy, K. (1984): Global asymptotic stability in Volterra's population systems. J. Math. Biol. **19**, 157-168.

Gopalsamy, K. (1985): Persistence in periodic and almost periodic Lotka-Volterra systems. J. Math. Biol. **21**, 145-148.

Gradsteyn, I.S. (1980): Table of Integrals, Series and Products. New York: Academic Press.

Guckenheimer, J. (1977): On the bifurcation of maps of the interval. Inventiones Math. **39**, 165-178.

Guckenheimer, J. and P. Holmes (1983): Nonlinear Oscillations, Dynamical Systems and Bifurcations of Vector Fields. Applied Math. Sciences **42**. New York-Heidelberg-Berlin: Springer.

Hadeler, K.P. (1974): Mathematik für Biologen. Berlin-Heidelberg-New York: Springer.

Hadeler, K.P. (1981): Stable polymorphisms in a selection model with mutation. SIAM J. Appl. Math. **41**, 1-7.

Hadeler, K.P. (1983): On copositive matrices. Linear Algebra and Appl. **49**, 79-89.

Hadeler, K.P., and D. Glas (1983): Quasimonotone systems and convergence to equilibrium in a population genetic model. J. Math. Anal. Appl. **95**, 297-303.

Hadeler, K.P., and U. Liberman (1975): Selection models with fertility differences. J. Math. Biology **2**, 19-32.

Haigh, J. (1975): Game theory and evolution. Adv. Appl. Prob. **7**, 8-11.

Haldane, J.B.S. (1932): The Causes of Evolution. New York: Harper and Row.

Hallam, T., L. Svoboda and T. Gard (1979): Persistence and extinction in three species Lotka-Volterra competitive systems. Math. Biosciences **46**, 117-124.

Hamilton, W.D. (1964): The genetical evolution of social behavior. J. Theor. Biology **7**, 1-52.

Hammerstein, P. (1979): The role of asymmetries in animal contests. Animal Behaviour **29**, 193-205.

Hammerstein, P., and G. Parker (1982): The asymmetric war of attrition. J. Theor. Biol. **96**, 647-682.

Hardy, G. (1908): Mendelian proportions in a mixed population. Science **28**, 49-50.

Harrison, G.W. (1978): Global stability of food chains. Amer. Naturalist **114**, 455-457.

Hartman, Ph. (1964): Ordinary Differential Equations. New York: Wiley.

Hassard, B.D., N.D. Kazarinoff, and Y.-H. Wan (1981): Theory and Applications of Hopf Bifurcation. London Math. Society Lecture Note Series **41**. Cambridge University Press.

Hastings, A. (1978): Global stability in Lotka-Volterra systems with diffusion. J. Math. Biology **6**, 163-168.

Hastings, A. (1981a): Stable cycling in discrete-time genetic models. Proc. Nat. Acad. Sci. USA **11**, 7224-7225.

Hastings, A. (1981b): Simultaneous stability of $D = 0$ and $D \neq 0$ for multiplicative viabilities at two loci: An analytical study. J. Theor. Biology **89**, 69-81.

Hastings, A. (1985): Four simultaneously stable polymorphic equilibria in two-locus two-allele models. Genetics **109**, 255-261.

Hines, W.G.S. (1980a): An evolutionarily stable strategy model for randomly mating diploid populations. J. Theor. Biology **87**, 379-384.

Hines, W.G.S. (1980b): Three characterizations of population strategy stability. J. Appl. Prob. **17**, 333-340.

Hines, W.G.S. and Bishop, D.T. (1984a): Can and will a sexual population attain an evolutionarily stable state. J. Theor. Biol. **111**, 667-686.

Hines, W.G.S. and Bishop D.T. (1984b): On the local stability of an evolutionarily stable strategy in a diploid population. J. Appl. Prob. **21**, 215-224.

Hirsch, M.W. (1982): Systems of differential equations which are competitive or cooperative I: limit sets. SIAM J. Math. Anal. **13**, 167-179.

Hirsch, M., and S. Smale (1974): Differential Equations, Dynamical Systems and Linear Algebra. New York: Academic Press.

Hoekstra, R. (1982): On the asymmetry of sex: evolution of mating types in isogamous populations. J. theor. Biol. **98**, 427-451.

Hofbauer, J. (1981a): On the occurrence of limit cycles in the Volterra-Lotka equation. Nonlinear Analysis **5**, 1003-1007.

Hofbauer, J. (1981b): A general cooperation theorem for hypercycles. Monatsh. Math. **91**, 233-240.

Hofbauer, J. (1984): A difference equation model for the hypercycle. SIAM J. Appl. Math. **44**, 762-772.

Hofbauer, J. (1985a): Gradients versus cycling in genetic selection models. In: "Dynamics of Macrosystems". Eds. J.P. Aubin, D. Saari and K. Sigmund. Lecture Notes in Economics and Mathematical Systems **257**. Berlin-Heidelberg-New-York: Springer.

Hofbauer, J. (1985b): The selection mutation equation. J. Math. Biol. **23**, 41-53.

Hofbauer, J. (1987): Heteroclinic cycles on the simplex. Proc. Int. Conf. Nonlinear Oscillations. Janos Bolyai Math. Soc. Budapest.

Hofbauer, J. (1988a): B-matrices, Lotka-Volterra equations and the linear complementarity problem. To appear.

Hofbauer, J. (1988b): An index theorem for dissipative semiflows. To appear.

Hofbauer, J. (1988c): Saturated equilibria, permanence, and stability for ecological systems. In: Mathematical Ecology. Proc. Trieste 1986. Eds. L.J. Gross, T.G. Hallam, S.A. Levin. World Scientific.

Hofbauer, J. (1988d): Are there mixed ESS in asymmetric games? To appear.

Hofbauer, J., Hutson, V. and Jansen, W. (1987): Coexistence for systems governed by difference equations of Lotka-Volterra type. J. Math. Biol. **25**, 553-570.

Hofbauer, J., P. Schuster, and K. Sigmund (1979): A note on evolutionarily stable strategies and game dynamics. J. Theor. Biology **81**, 609-612.

Hofbauer, J., P. Schuster, and K. Sigmund (1981): Competition and cooperation in catalytic selfreplication. J. Math. Biol. **11**, 155-168.

Hofbauer, J., P. Schuster, and K. Sigmund (1982): Game dynamics for Mendelian populations. Biol. Cybern. **43**, 51-57.

Hofbauer, J., P. Schuster, K. Sigmund, and R. Wolff (1980): Dynamical systems under constant organization. Part 2: Homogeneous growth functions of degree 2. SIAM J. Appl. Math. **38**, 282-304.

Hofbauer, J. and Sigmund, K. (1987): Permanence for replicator equations. In: "Dynamical Systems", ed. A. Kurzhanski and K. Sigmund. Lecture Notes in Economics and Mathematical Systems **287**. Berlin-Heidelberg-New York: Springer.

Holling, C.S. (1973): The functional response of predators to prey density and its role in mimicry and population regulation. Mem. Ent. Soc. **45**, 3-60.

Hoppensteadt, F. (1975): Mathematical Theories of Populations. Demographics, Genetics and Epidemics. CBMS-NSF Regional Conf. Ser. Appl. Math. Philadelphia: SIAM.

Hsu, S., S. Hubbell, and P. Waltman (1978): Competing predators. SIAM J. Appl. Math. **35**, 617-625.

Hutson, V. (1984): A theorem on average Ljapunov functions. Monatsh. Math. **98**, 267-275.

Hutson, V. and Law, R. (1981): Evolution of recombination in populations experiencing frequency-dependent selection with time delay. Proc. R. Soc. London **B213**, 345-359.

Hutson, V. and R. Law (1985): Permanent coexistence in general models of three interacting species. J. Math. Biology **21**, 285-298.

Hutson, V. and Moran, W. (1982): Persistence of species obeying difference equations. J. Math. Biol. **15**, 203-213.

Hutson, V. and G.T. Vickers (1983): A criterion for permanent coexistence of species with an application to a two-prey one-predator system. Math. Biosci. **63**, 253-269.

Jansen, W. (1986): A permanence theorem for replicator and Lotka-Volterra systems. J. Math. Biology **25**, 411-422.

Johnson, C.R. (1974): Sufficient conditions for D-stability. J. Economic Theory **9**, 53-62.

Jones, B., R. Enns, and Ragnekar (1976): On the theory of selection of coupled macromolecular systems. Bull. Math. Biol. **38**, 15-28.

Karlin, S. (1966): A First Course in Stochastic Processes. New York: Academic Press.

Karlin, S. (1972): Some mathematical models of population genetics. Amer. Math. Monthly **79**, 699-739.

Karlin, S. (1975): General two-locus selection models: some objectives, results and interpretations. Theor. Pop. Biology **7**, 364-398.

Karlin, S. (1980): The number of stable equilibria for the classical one-locus multiallele selection model. J. Math. Biol. **9**, 189-198.

Karlin, S. (1984): Mathematical models, problems and controversies of evolutionary theory. Bull. Amer. Math. Soc. **10**, 221-273.

Karlin, S., and McGregor (1974): Towards a theory of the evolution of modifier genes. Theor. Pop. Biology **5**, 59-103.

Karlin, S. and S. Lessard (1986): Sex Ratio Evolution. Monographs in Population Biology **22**. Princeton, New Jersey: Princeton University Press.

Kimura, M. (1958): On the change of population fitness by natural selection. Heredity **12**, 145-167.

Kingman, J. (1961): A matrix inequality. Quarterly J. Math. **12**, 78-80.

Kingman, J.F.C. (1980): Mathematics of Genetic Diversity. CBMS NSF Regional Conf. Ser. in Appl. Math. 34. Philadelphia: SIAM.

Kirlinger, G. (1986): Permanence in Lotka-Volterra equations: linked prey-predator systems. Math. Biosciences **82**, 165-191.

Kolmogoroff, A. (1936): Sulla teoria di Volterra della lotta per l'esistenza. Giornale dell'Istituto Ital. Attuari **7**, 74-80.

Koth, M. and F. Kemler (1986): A one locus - two allele selection model admitting stable limit cycles. J. Theor. Biol. **122**, 263-267.

Koth, M. and Sigmund, K. (1987): Gradients for the evolution of asymmetric games. J. Math. Biol. **25**, 623-635.

Krebs, J.R. and Davies, N.B. (1984): Behavioral Ecology. 2nd edition. Sunderland: Sinauer.

Kun, L. and Lyubich, Y. (1979): Convergence to equilibrium under the action of additive selection in a multilocus multiallelic population. Soviet Math. Dokl. **20**, 1380-1385.

Küppers, B.O. (1983): Molecular Theory of Evolution. Berlin: Springer.

Lessard, S. (1984): Evolutionary dynamics in frequency dependent two-phenotype models. Theor. Pop. Biol. **25**, 210-234.

Levin, S. (1970): Community equilibria and stability, and an extension of the competitive exclusion principle. Amer. Naturalist **104**, 413-423.

Levin, S. (1978): Studies in Mathematical Biology. MAA Studies in Mathematics, Vol. **15** and **16**.

Lewontin, R. (1974): The Genetical Basis of Evolutionary Change. New York: Columbia Univ. Press.

Losert, V., and E. Akin (1983): Dynamics of games and genes: Discrete versus continuous time. J. Math. Biology **17**, 241-251.

Lotka, A.J. (1920): Undamped oscillations derived from the law of mass action. J. Am. Chem. Soc. **42**, 1595-1598.

Lyubich, Yu. I. (1986): On the dynamics of hypercycles. Soviet Math. Dokl. **33**, 849-851.

Lyubich, Yu. I., G.D. Maistrowskii and Yu. G. Ol'khovskii (1980): Selection-induced convergence to equilibrium in a single-locus autosomal population. Problems of Information Transmission **16**, 66-75.

MacArthur, R.H. (1962): Some generalized theorems of natural selection. Genetics **48**, 1893-1897.

MacArthur, R. (1970): Species packing and competitive equilibria for many species. Theor. Pop. Biol. **1**, 1-11.

MacArthur, R., and E.O. Wilson (1967): The theory of island biogeography. Princeton University Press.

Mallet-Paret, J. and H.L. Smith (1988): The Poincaré-Bendixson theorem for monotone cyclic feedback systems. To appear.

Mandel, S.P.H. (1959): The stability of a multiple allelic system. Heredity **13**, 289-302.

Margulis, L. (1981): Symbiosis in Cell Evolution. San Francisco: Freeman.

Marsden, J., and M. McCracken (1976): The Hopf Bifurcation and its Applications. Appl. Math. Sciences, Vol. **19** Berlin-Heidelberg-New York: Springer.

Masters, R.D. (1982): Evolutionary biology, political theory, and the state. J. Social Biol. Structures **5**, 439-450.

May, R.M. (1973): Stability and Complexity in Model Ecosystems. Princeton University Press.

May, R.M. (1976): Simple mathematical models with very complicated dynamics. Nature **261**, 459-467.

May, R.M., and W. Leonard (1975): Nonlinear aspects of competition between three species. SIAM J. Appl. Math. **29**, 243-252.

May, R.M., and Oster (1976): Bifurcation and dynamic complexity in simple ecological models. Amer. Naturalist **110**, 573-599.

Maynard Smith, J. (1964): Kin selection and group selection, Nature **201**, 1145-1147.

Maynard Smith, J. (1972): Game theory and the evolution of fighting. In: J. Maynard Smith, On Evolution. Edinburgh University Press.

Maynard Smith, J. (1974a): Models in Ecology. Cambridge University Press.

Maynard Smith, J. (1974b): The theory of games and the evolution of animal conflicts. J. Theor. Biol. **47**, 209-221.

Maynard Smith, J. (1975): The Theory of Evolution, 3rd edition. Harmondsworth: Penguin Books.

Maynard Smith, J. (1978): The Evolution of Sex. Cambridge University Press.

Maynard Smith, J. (1981): Will a sexual population evolve to an ESS? Amer. Naturalist **177**, 1015-1018.

Maynard Smith, J. (1982): Evolution and the Theory of Games. Cambridge University Press.

Maynard Smith, J. (1986): Problems in Biology. Oxford University Press.

Maynard Smith, J. and Hofbauer, J. (1987): The "battle of the sexes": a genetic model with limit cycle behavior. Theor. Pop. Biol. **32**, 1-14.

Maynard Smith, J., and G. Parker (1976): The logic of asymmetric contests. Animal Behavior **24**, 159-179.

Maynard Smith, J., and G. Price (1973): The logic of animal conflicts. Nature **246**, 15-18.

Mayr, E. (1963): Animal Species and Evolution. Cambridge, Mass.: Harvard Univ. Press.

Meisters, G.H. (1982): Jacobian problems in differential equations and algebraic geometry. Rocky Mountain J. Math. **12**, 679-705.

Mendel, G. (1866): Versuche über Pflanzenhybriden. Verhandlungen des Naturforscher-Vereins in Brünn.

Miller, S., and L. Orgel (1973): The Origins of Life on Earth. New Jersey: Prentice Hall.

Milnor, J.W. (1965): Topology from the Differentiable Viewpoint. Charlotteville: Univ. Press of Virginia.

Molchanov, A.M. (1961): Stability in the case of a neutral linear approximation. Doklady Akad. Nauk **141**, 24-27.

Monod, J. (1970): Le Hasard et la Necessité. Paris: Edition du Seuil.

Moon, J.W. and Moser, L. (1965): On cliques in graphs. Israel J. Math. **3**, 23-28.

Moran, P.A.P. (1976): Global stability of genetic systems governed by mutation and selection. Math. Proc. Cambridge Phil. Soc. **80**, 331-336.

Mulholland, H., and C. Smith (1959): An inequality arising in genetical theory. Amer. Math. Monthly **66**, 673-683.

Nagylaki, T. (1977): Selection in One- and Two-Locus Systems. Lecture Notes in Biomathematics **15**. Berlin-Heidelberg-New York: Springer.

Nagylaki, T., and J.F. Crow (1974): Continuous selective models. Theor. Pop. Biology **5**, 257-283.

Neumann, J.v., and O. Morgenstern (1953): Theory of Games and Economic Behavior. Princeton University Press.

Nikaido, H. (1968): Convex Structure and Economic Theory. New York: Academic Press.

Oster, G. and S. Rochlin (1979): Optimization models in evolutionary biology. AMS Lectures on Mathematics in the Life Sciences **11**, 21-88.

Owen, C. (1982): Game Theory. 2nd edition. New York: Academic Press.

Parker, G. (1978): Selection in non-random fusion of gametes during the evolution of anisogamy. J. Theor. Biol. **73**, 1-28.

Parker, G.A. and P. Hammerstein (1985): Game theory and animal behavior. In: Evolution. Essays in honour of John Maynard Smith. Eds. Greenwood, Harvey and Slatkin.

Parker, G., R. Baker, and V. Smith (1972): The origin and evolution of gamete dimorphism and the male-female-phenomenon. J. Theor. Biol. **36**, 529-553.

Peschel, M. and Mende, W. (1986): The Predator-Prey Model. Do We Live in a Volterra World? Wien-New York: Springer.

Pohley, H.-J. and B. Thomas (1983): Nonlinear ESS-models and frequency dependent selection. Biosystems **16**, 87-100.

Pollak, E. (1979): Some models of genetic selection. Biometrics **35**, 119-137.

Quirk, R. (1981): Qualitative stability of matrices and economic theory: a survey article. In: Computer-Assisted Analysis and Model Simplification (Proceedings of a symposium in Boulder, Colorado, 1980), H.J. Greenberg and J.S. Maybee, eds. New York: Academic Press.

Redheffer, R. (1985): Volterra Multipliers I., II. SIAM J. Alg. Discr. Meth. **6**, 592-623.

Riechert, S. and Hammerstein, P. (1983): Game theory in the ecological context. Ann. Rev. Ecol. Syst. **14**, 377-409.

Robinson, C. (1985): Phase plane analysis using the Poincaré map. Nonlinear Analysis **9**, 1159-1164.

Rosenzweig, M.L., and R.H. MacArthur (1963): Graphical representation and stability condition of predator-prey interaction. Am. Naturalist **97**, 209-223.

Roughgarden, J. (1971): Density-dependent natural selection. Ecology **52**, 453-468.

Roughgarden, J. (1979): Theory of Population Genetics and Evolutionary Ecology. New York: Mac Millan.

Roux, C. (1977): Fecundity differences between mating pairs for a single autosomal locus, sex differences in viabilities and nonoverlapping generations. Theor. Population Biology **12**, 1-9.

Sanders, J.A. and Verhulst, F. (1985): Averaging Methods in Nonlinear Dynamical Systems. Appl. Math. Sciences **59**, New York-Berlin-Heidelberg-Tokyo: Springer.

Scheuer, P., and S. Mandel (1959): An inequality in population genetics. Heredity **13**, 519-524.

Schuster, P. (1981): Prebiotic Evolution. In: Biochemical Evolution, Gutfreund, H. (Ed.): pp. 15-87. Cambridge University Press.

Schuster, P., and K. Sigmund (1980a): Self-organization of biological macromolecules and evolutionarily stable strategies. In: Dynamics of Synergetic Systems, ed. Haken, pp. 156-169. Berlin-Heidelberg-New York: Springer.

Schuster, P., and K. Sigmund (1980b): A mathematical model of the hypercycle. In: Dynamics of Synergetic Systems, ed. Haken, pp. 170-178. Berlin-Heidelberg-New York: Springer.

Schuster, P., and K. Sigmund (1981): Coyness, philandering and stable strategies. Anim. Behaviour **29**, 186-192.

Schuster, P., and K. Sigmund (1983): Replicator Dynamics. J. Theor. Biology **100**, 533-538.

Schuster, P., K. Sigmund, J. Hofbauer, and R. Wolff (1981a): Selfregulation of behavior in animal societies. Part I: Symmetric contests. Biol. Cybern. **40**, 1-8.

Schuster, P., K. Sigmund, J. Hofbauer, and R. Wolff (1981b): Selfregulation of behavior in animal societies II: Games between two populations with selfinteraction. Biol. Cybern. **40**, 9-15.

Schuster, P., K. Sigmund, J. Hofbauer, R. Gottlieb, and P. Merz (1981c): Selfregulation of behavior in animal societies. Part III: Games between two populations with selfinteraction. Biol. Cyb. **40** 17-25.

Schuster, P., K. Sigmund, and R. Wolff (1978): Dynamical systems under constant organization. Part 1: A model for catalytic hypercycles. Bull. Math. Biophysics **40**, 743-769.

Schuster, P., K. Sigmund, and R. Wolff (1979a): Dynamical systems under constant organization. Part 3: Cooperative and competitive behavior of hypercycles. J. Diff. Equations. **32**, 357-368.

Schuster, P., K. Sigmund, and R. Wolff (1979b): On ω-limits for competition between three species. SIAM J. Appl. Math. **37**, 49-54.

Schuster, P., K. Sigmund, and R. Wolff (1980): Mass action kinetics of selfreplication in flow reactors. J. Math. Anal. Appl. **78**, 88-112.

Schuster, P., and J. Swetina (1982): Selfreplication with errors: A model for polynucleotide replication. Biophys. Chem. **16**, 329-345.

Scudo, F., and J. Ziegler (1978): The Golden Age of Theoretical Ecology, 1923-1940. Lecture Notes in Biomathematics **22**. Berlin-Heidelberg-New York: Springer.

Selten, R. (1980): A note on evolutionarily stable strategies in asymmetrical animal conflicts. J. Theor. Biology **84**, 93-101.

Selten, R. (1983): Evolutionary stability in extensive two-person games. Math. Social Sciences **5**, 269-363.

Sharkovski, A. (1964): Coexistence of cycles of continuous mappings of the line into itself. Ukr. Mat. Z. **16**, 61-71.

Shahshahani, S. (1979): A new mathematical framework for the study of linkage and selection. Memoirs Amer. Math. Soc. **211**. Providence, R.I.

Shaw, R.F. and Mohler, J. (1953): The selective significance of the sex ratio. Am. Nat. **87**, 337-342.

Shigesada, N., K. Kawasaki and E. Teramoto (1984): The effects of interference competition on stability, structure and invasion of a multispecies system. J. Math. Biol. **21**, 97-113.

Sieveking (1983): Unpublished lectures on dynamical systems.

Sigmund, K. (1984): The maximum principle for replicator equations. In: "Lotka-Volterra Approach to Dynamical Systems", ed. W. Ebeling and M. Peschel. Berlin: Akademie Verlag.

Sigmund, K. (1987a): Game dynamics, mixed strategies and gradient systems. Theor. Pop. Biol. **32**, 114-126.

Sigmund, K. (1987b): A maximum principle for frequency dependent selection. Math. Biosci. To appear.

Sigmund, K. and Schuster, P. (1984): Permanence and uninvadability for deterministic population models, in "Stochastic Phenomana and Chaotic Behavior in Complex Systems", ed. P. Schuster. Synergetics **21**, New York: Springer.

Šiljak, D.D. (1978): Large-Scale Dynamic Systems: Stability and Structure. Amsterdam: North-Holland.

Simpson, G. (1953): The Major Features of Evolution. New York: Simon and Schuster.

Smale, S. (1976): On the differential equations of species in competition. J. Math. Biol. **3**, 5-7.

Smale, S., and R. Williams (1976): The qualitative analysis of a difference equation of population growth. J. Math. Biol. **3**, 1-4.

Smith, H. (1986a): On the asymptotic behavior of a class of deterministic models of cooperating species. SIAM J. Appl. Math. **46**, 368-375.

Smith, H.L. (1986b): Competing subcommunities of mutualists and a generalized Kamke theorem. SIAM J. Appl. Math. **46**, 856-874.

So, J. (1979): A note on global stability and bifurcation phenomenon of a Lotka-Volterra food chain. J. Theor. Biol. **80**, 185-187.

Spiegelman, S. (1971): An approach to the experimental analysis of precellular evolution. Quart. Rev. Biophysics **4**, 213-253.

Stewart, F. (1971): Evolution of dimorphism in a predator-prey model. Theor. Pop. Biol. **2**, 493-506.

Svirezhev, Y.M. (1986): Modern problems of mathematical ecology. In: Proc. Int. Cong. Math. 1983, pp. 1677-1693. Amsterdam: North-Holland.

Svirezhev, Y.M. and Logofet, D.O. (1983): Stability of Biological Communities. Moscow: Mir.

Takeuchi, Y. (1986): Global stability in generalized Lotka-Volterra diffusion systems. J. Math. Anal. Appl. **116**, 209-221.

Takeuchi, Y., N. Adachi and H. Tokumaru (1978): Global stability of ecosystems of the generalized Volterra type. Math. Biosci. **42**, 119-136.

Takeuchi, Y. and N. Adachi (1980): The existence of globally stable equilibria of ecosystems of the generalized Volterra type. J. Math. Biology **10**, 401-415.

Takeuchi, Y. and N. Adachi (1983): Existence and bifurcation of stable equilibrium in two-prey, one-predator communities. Bull. Math. Biology **45**, 877-900.

Takeuchi, Y. and N. Adachi (1984): Influence of predation on species coexistence in Volterra models. Math. Biosci. **70**, 65-90.

Taylor, P. (1979): Evolutionarily stable strategies with two types of players. J. Appl. Prob. **16**, 76-83.

Taylor, P., and L. Jonker (1978): Evolutionarily stable strategies and game dynamics: Math. Biosciences **40**, 145-156.

Thaxton, C., W. Bradley and R. Olsen (1984): The Mystery of Life's Origin. New York: Philosophical Library.

Thomas, B. (1985): Genetic ESS-models. II. Multi-strategy models and multiple alleles. Theor. Pop. Biol. **28**, 33-49.

Thompson, C., and J. Mc Bride (1974): On Eigen's theory of selforganization of matter and evolution of biological macromolecules. Math. Biosciences **21**, 127-145.

Trivers, R. (1985): Social Evolution. Menlo Park: Benjamin-Cummings.

Trivers, R., and H. Hare (1976): Haplodiploidity and the evolution of social insects. Science **191**, 249-263.

Turelli, M. (1984): Heritable genetic variation via mutation-selection balance: Lerch's zeta meets the abdominal bristle. Theor. Pop. Biol. **25** , 138-193.

van Valen, L. (1977): The red queen. Amer. Naturalist **111**, 809-10.

Verhulst, P.F. (1845): Recherches mathematiques sur les taux d'accroissement de la population. Mem. Acad. Roy. Belgique. **18**, 1-38.

Vickers, G.T. and Cannings, C. (1987): On the number of stable equilibria in a one-locus multi-allelic system. To appear.

Volterra, V. (1931): Lecons sur la theorie mathematique de la lutte pour la vie. Paris: Gauthier-Villars.

Wagner, G., and R. Bürger (1985): On the evolution of dominance modifiers II: a nonequilibrium approach to the evolution of genetic systems. J. Theor. Biology **113**, 475-500.

Waltman, P. (1983): Competition Models in Population Biology. CBMS-NSF Regional Conf. Ser. Appl. Math. **45**. Philadelphia: SIAM.

Watson, J.D. (1976): The Molecular Biology of the Genes. New York: Benjamin.

Weinberg, W. (1908): Über den Nachweis der Erblichkeit. Jb. Ver. Vaterl. Naturk. Württemberg **64**, 368-382.

Williams, G.C. (1966): Adaptation and Natural Selection. Princeton University Press.

Wilson, E. (1975): Sociobiology. Cambridge, Mass.: Belknap Press.

Wörz-Busekros, A. (1978): Global stability in ecological systems with continuous time delay. SIAM J. Appl. Math. **35**, 123-134.

Zeeman, E.C. (1980): Population dynamics from game theory. In: Global Theory of Dynamical Systems. Springer Lecture Notes in Mathematics **819**.

Zeeman, E.C. (1981): Dynamics of the evolution of animal conflicts. J. Theor. Biology **89**, 249-270.

Index